Physics of Manganites

FUNDAMENTAL MATERIALS RESEARCH

Series Editor: M. F. Thorpe, *Michigan State University*
East Lansing, Michigan

ACCESS IN NANOPOROUS MATERIALS
Edited by Thomas J. Pinnavaia and M. F. Thorpe

DYNAMICS OF CRYSTAL SURFACES AND INTERFACES
Edited by P. M. Duxbury and T. J. Pence

ELECTRONIC PROPERTIES OF SOLIDS USING CLUSTER METHODS
Edited by T. A. Kaplan and S. D. Mahanti

LOCAL STRUCTURE FROM DIFFRACTION
Edited by S. J. L. Billinge and M. F. Thorpe

PHYSICS OF MANGANITES
Edited by T. A. Kaplan and S. D. Mahanti

RIGIDITY THEORY AND APPLICATIONS
Edited by M. F. Thorpe and P. M. Duxbury

A Continuation Order Plan is available for this series. A continuation order will bring delivery of each new volume immediately upon publication. Volumes are billed only upon actual shipment. For further information please contact the publisher.

Physics of Manganites

Edited by

T. A. Kaplan and S. D. Mahanti
Michigan State University
East Lansing, Michigan

Kluwer Academic / Plenum Publishers
New York, Boston, Dordrecht, London, Moscow

Proceedings of the International Conference on Physics of Manganites,
held July 26–29, 1998, at Michigan State University, East Lansing, Michigan

ISBN 0-306-46132-3

©1999 Kluwer Academic / Plenum Publishers, New York
233 Spring Street, New York, N.Y. 10013

10 9 8 7 6 5 4 3 2 1

A C.I.P. record for this book is available from the Library of Congress

All rights reserved

No part of this book may be reproduced, stored in a retrieval system, or transmitted in any form or by any means, electronic, mechanical, photocopying, microfilming, recording, or otherwise, without written permission from the Publisher

Printed in the United States of America

SERIES PREFACE

This series of books, which is published at the rate of about one per year, addresses fundamental problems in materials science. The contents cover a broad range of topics from small clusters of atoms to engineering materials and involves *chemistry*, *physics*, *materials science* and *engineering*, with length scales ranging from Ångstroms up to millimeters. The emphasis is on basic science rather than on applications. Each book focuses on a single area of current interest and brings together leading experts to give an up-to-date discussion of their work and the work of others. Each article contains enough references that the interested reader can access the relevant literature. Thanks are given to the Center for Fundamental Materials Research at Michigan State University for supporting this series.

M.F. Thorpe, Series Editor
E-mail: thorpe@pa.msu.edu

PREFACE

This book records invited lectures given at the workshop on Physics of Manganites, held at Michigan State University, July 26-29, 1998. Doped manganites are an interesting class of compounds that show both metal-insulator and ferromagnetic to paramagnetic transitions at the same temperature. This was discovered in the early 1950s by Jonker and van Santen and basic theoretical ideas were developed by Zener (1951), Anderson and Hasegawa (1955), and deGennes (1960) to explain these transitions and related interesting observations. Realization that the *colossal* magnetoresistance later observed in doped manganites can be of practical importance has resulted in a flurry of experimental and theoretical activities in these materials since the early 1990s. Besides being of potential practical importance, these systems are extremely fascinating from a physics point of view because their physical properties depend on a strong interplay of charge, spin, orbital, and lattice degrees of freedom. Due to the intense activity in this field during the past three to four years, we felt a workshop on this subject to discuss the important physical issues was quite timely.

In this workshop there were twenty invited lectures equally distributed between experiment and theory and eleven poster presentations. Most of the invited lectures are represented in this book. Theoretical topics discussed were: phase separation in the double-exchange model (Dagotto), spin and charge excitations in the double-exchange model (Furukawa), role of electron-lattice interaction (Goodenough), role of orbital degeneracy on the optical conductivity (Horsch), exact calculation of spin waves (Kaplan), magnetism and electronic states of systems with strong Hund's coupling (Kubo), ferromagnetic states and first-order transitions driven by orbital ordering (Maekawa), effect of strain on the physical properties in bulk and thin film manganites (Millis), density functional studies of half-metallicity, lattice distortion, and transport (Pickett), ground and excited states of polarons (Trugman). Experimental topics covered were: new physics in two-dimensional manganites (Aeppli), polarons in manganites (Billinge), ferromagnetism vs. striped charge ordering (Cheong), spin injection in manganite-cuprate heterostructures (Goldman), role of polarons and double exchange in transport properties (Jaime), neutron scattering studies of spin dynamics (Lynn), optical properties (Noh), colossal magnetoresistance materials and mechanisms (Ramirez), metal-insulator phenomena relevant to charge/orbital ordering (Tomioka), evidence of polaronic charge carriers through oxygen isotope effect (Zhao).

Financial support for the Workshop was provided in part by the Center for Sensor Materials and the Center for Fundamental Materials Research at Michigan State University. We would like to thank the members of the advisory committee (G. Aeppli, K. Kubo, P. Horsch, Y. Tokura) for their thoughtful suggestions and encouragement. We are deeply indebted to Lorie Neuman for coordinating all the logistical aspects of the Workshop and to Janet King for editorial assistance and handling of the manuscripts for publication.

T. A. Kaplan
S. D. Mahanti

CONTENTS

Thermodynamics of the Double Exchange Systems 1
   Nobuo Furukawa

Phase Separation in Models for Manganites: Theoretical
   Aspects and Comparison with Experiments 39
   E. Dagotto, S. Yunoki, and A. Moreo

Two Ferromagnetic States in Magnetoresistive Manganites:
   First Order Transition Driven by Orbitals 57
   S. Maekawa, S. Ishihara, and S. Okamoto

Magnetism and Electronic States of Systems
   with Strong Hund Coupling 71
   K. Kubo, D.M. Edwards, A.C.M. Green, T. Momoi, and H. Sakamoto

Density Functional Studies of Magnetic Ordering,
   Lattice Distortion, and Transport in Manganites 87
   W.E. Pickett, D.J. Singh, and D.A. Papaconstantopoulos

Optical Conductivity of Doped Manganites: Comparison of Ferromagnetic
   Kondo Lattice Models with and without Orbital Degeneracy 103
   Frank Mack and Peter Horsch

Electron-Lattice Interactions in Manganese-Oxide Perovskites 127
   J.B. Goodenough

Spin Waves in Doped Manganites 135
   T.A. Kaplan and S.D. Mahanti

Spin Dynamics of Magnetoresistive Oxides 149
   J.W. Lynn

Metal-Insulator Phenomena Relevant to Charge/Orbital
   Ordering in Perovskite Manganites 155
   Y. Tomioka, A. Asamitsu, H. Kuwahara, and Y. Tokura

Optical Properties of Colossal Magnetoresistance Manganites 177
    T.W. Noh, J.H. Jung, and K.H. Kim

Polarons in Manganites; Now You See Them Now You Don't 201
    Simon J.L. Billinge

Oxygen Isotope Effects in Manganites: Evidence
    for (Bi)polaronic Charge Carriers 221
    Guo-meng Zhao, H. Keller, R.L. Greene, and K.A. Müller

Electronic Transport in La-Ca Manganites 243
    Marcelo Jaime and Myron B. Salamon

Spin Injection in Manganite-Cuprate Heterostructures 269
    V.A. Vas'ko, K.R. Nikolaev, V.A. Larkin, P.A. Kraus, and A.M. Goldman

List of Participants 287

Index 293

Physics of Manganites

THERMODYNAMICS OF THE DOUBLE EXCHANGE SYSTEMS

Nobuo Furukawa

Dept. of Physics, Aoyama Gakuin University,
Setagaya, Tokyo 157-8572, Japan

I. INTRODUCTION

The concept of the double-exchange (DE) interaction was introduced by Zener[1] in order to explain the ferromagnetism of the perovskite manganites AMnO$_3$. He considered a Kondo-lattice type model

$$\mathcal{H} = -\sum_{ij,\sigma} t_{ij} \left(c_{i\sigma}^\dagger c_{j\sigma} + h.c. \right) - J_{\rm H} \sum_i \vec{\sigma}_i \cdot \vec{S}_i, \tag{1}$$

where t and $J_{\rm H} > 0$ are $e_{\rm g}$ electron's hopping and (ferromagnetic) Hund's coupling between $e_{\rm g}$ and $t_{2\rm g}$ electrons, respectively. An effective Hamiltonian in the limit $J_{\rm H} \to \infty$ was introduced by Anderson and Hasegawa[2] in the form

$$\mathcal{H} = -\sum_{ij} t(\vec{S}_i, \vec{S}_j) \left(\tilde{c}_i^\dagger \tilde{c}_j + h.c. \right). \tag{2}$$

Mean-field type arguments of these models, including those by de Gennes,[3] helped us to discuss the magnetism of manganite compounds.

However, recent reinvestigations of these models which are motivated by the observation of colossal magnetoresistance (CMR) phenomena[4,5] in manganites revealed that such simplified treatments are insufficient to discuss quantitative nature of the models. Let us show an example. Curie temperature $T_{\rm c}$ of the model has been estimated by Millis et al.[6] using the mean-field type discussion. Using some appropriate values for t and $J_{\rm H}$, they found that the mean-field $T_{\rm c}$ of the model is much larger than those for perovskite manganites (La,Sr)MnO$_3$, and concluded that double-exchange alone is insufficient to explain the thermodynamics of these manganites. However, as we will show in this article, accurate treatments for the model give suppression of $T_{\rm c}$ from the mean-field value, and show consistency with (La,Sr)MnO$_3$ data.

Another important point which will be discussed here is the systematic understanding of experiments. For example, it is known that CMR manganites exhibit metal-insulator transition at Curie temperature. This statement is, however, inaccurate. By varying compositions, varieties of phases with different properties have been known.[7] A typical reference compound $La_{2/3}Sr_{1/3}MnO_3$ shows no metal-insulator transition. The resistivity always increases by the increase of temperature, i.e. $d\rho(T)/dT > 0$ even above T_c. Its absolute value in the paramagnetic phase is around the Mott's limit value, and may be explained by the DE model. On the other hand, $La_{2/3}Ca_{1/3}MnO_3$ which is often reffered to as an "optimal" CMR material shows insulating behavior above T_c with much larger $\rho(T)$. Thus we have to be mindful to make distinctions between various compositions of manganites in a systematic manner.

The purpose of this article is to solve such misunderstandings and confusions in both theoretical and experimental studies. We review the investigation of the DE model by the author.[8-16] The first part of this article shows the finite temperature behaviors in the DE systems. In §2 we introduce the model and the method (dynamical mean-field approach and the Monte Carlo calculation). In §3 the results for infinitely large spin (classical spin limit) is presented, and the $1/S$ correction is shown in §4. The second part is devoted for discussions with respect to comparisons of model behavior with experimental data. Comparison with experiments. are shown in §5. Section 6 is devoted for concluding remarks.

II. MODEL AND METHODS

2.1 Double Exchange Model in the Large Spin Limit

Model. The compound $LaMnO_3$ has four 3d electrons per atom in the $(t_{2g})^3(e_g)$ configuration. Due to Hund's coupling, these electrons have the high spin state, i.e. spin parallel configuration. By substitution of the La site with alkaline-earth divalent ions, holes are doped as carriers which is considered to enter the e_g orbitals.

As briefly mentioned in the previous section, Zener[1] introduced a Kondo-lattice type Hamiltonian (1) with ferromagnetic spin exchange J_H between localized spins and itinerant electrons. In manganites $(R,A)MnO_3$, Hund's coupling J_H is estimated to be a few eV while electron hopping t be in the order of 0.1eV. Therefore, we have to deal with a strong coupling region $J_H \gg t$. In order to discuss the ferromagnetism of the model in the metallic region, Zener introduced the notion of the "double exchange" interaction.

In some cases, the limit $J_H \to \infty$ first studied by Anderson and Hasegawa[2] is called the DE model. However, in order to avoid complications, we call the model (1) in the strong coupling region as DE model. In the weak coupling limit $J_H/t \ll 1$, the model is often referred to as the s-d model and studied as a model hamiltonian for the magnetic semiconductors.

The classical rotator limit, or equivalently large spin limit $S = \infty$, has been introduced by Anderson and Hasegawa.[2] In manganites, we consider the case where the localized spin is in a high-spin state ($S = 3/2$) with the ferromagnetic coupling, the effect of quantum exchange seems to be less relevant compared to thermal fluctuations, at least in low energy physics. However, the role of quantum exchanges might give non-trivial effects in the system in the region with less thermal fluctuations. Such issues are left for future studies.

In this article, we introduce a model with finite J_H and infinite S,

$$\mathcal{H}_{S=\infty} = -\sum_{ij,\sigma} t_{ij} \left(c_{i\sigma}^\dagger c_{j\sigma} + h.c. \right) - J_H \sum_i \vec{\sigma}_i \cdot \vec{m}_i. \tag{3}$$

Hereafter we express the localized (classical) spin by $\vec{m}_i = (m_i^x, m_i^y, m_i^z)$ with the normalization $|\vec{m}|^2 = 1$.

Previous investigations. Anderson-Hasegawa[2] as well as de Gennes[3] studied the magnetism of the model with localized spins treated as static, *i.e.* neglecting spin fluctuations, they made a search for energetically favored spin configurations as well as a mean-field calculation for finite temperatures. However, if we consider the system at finite temperature, especially around $T \sim T_c$, spin fluctuations δS_i become important. Especially, in our case of $J_H \gg t$, such spin fluctuations give large effects to electronic structures since $J_H \delta S \gg t$.

The point of interest for us in relationship with CMR phenomena is the change of resistivity ρ at around T_c. Mean-field type treatments are not justified for this purpose. An alternative approach has been made by Kubo and Ohata.[17] They phenomenologically assumed the electron self-energy Σ in the form

$$\tau^{-1} = \mathcal{I}m\,\Sigma \propto 1 - M^2, \tag{4}$$

where τ is the quasiparticle lifetime, and calculated the resistivity using the Drude's formula $\rho \propto \tau^{-1}$. Within this phenomenological treatment, the result qualitatively helps us to understand the magnetoresistance (MR) of the DE model via spin disorder scattering mechanism. However, in a quantitative way, it fails to reproduce the MR of (La,Sr)MnO$_3$ as we will show in §5.1.

Beyond the previous theories. Thus, in order to understand the behaviour of the DE model and to make direct comparisons with experimental data, it is very important to treat the model in an accurate way. The methods have to be an unbiased non-perturbative approaches since $J_H \gg t$, and must take into account the effect of spin fluctuations to calculate finite temperature behaviors.

Here we introduce two methods: The dynamical mean-field (DFM) theory and the Monte Carlo calculation. Both of these methods give us the electronic states at finite temperature including $T \sim T_c$. They are unbiased and become exact in the limit of large coordination number and large system size, respectively. One of the advantages of these methods, in the viewpoint of comparison with experiments, is that it is easy to obtain dynamical quantities such as density of states (DOS) and optical conductivity $\sigma(\omega)$.

2.2 Dynamical mean-field theory

We introduce the DMF method for the double-exchange model. For a general review of the field, see the review articles in refs. 18 and 19. Within the general scheme of the DMF, we treat a lattice system by considering a single-site coupled with "electron bath", or the time-dependent mean-field $\tilde{G}_0$. This method becomes exact in the large coordination number limit or equivalently large spatial dimension limit.[20, 21]

Generic part of the DMF treatment is as follows. Solving a model-specific single-site problem, we obtain the self-energy $\tilde{\Sigma}(i\omega_n)$ from $\tilde{G}_0$. Lattice Green's function is approximated by

$$G(k, i\omega_n) = [i\omega_n - (\varepsilon - \mu) - \tilde{\Sigma}(i\omega_n)]^{-1}. \tag{5}$$

The local Green's function is defined by

$$G_{\text{loc}}(i\omega_n) \equiv \frac{1}{N} \sum_k G(k, i\omega_n) \tag{6}$$

Since $\tilde{\Sigma}$ is k-independent, k dependence of $G(k, i\omega_n)$ comes through the energy dispersion and we have

$$G_{\text{loc}}(i\omega_n) = \int d\varepsilon\, N_0(\varepsilon)[i\omega_n - (\varepsilon - \mu) - \tilde{\Sigma}(i\omega_n)]^{-1}. \tag{7}$$

$N_0(\varepsilon)$ is the DOS for the noninteracting lattice system. The information of the lattice geometry is included through the noninteracting DOS in eq. (7). The method is applicable to finite size systems by taking the sum over k points in eq. (6) in the discrete k-space. We self-consistently obtain the time-dependent mean-field $\tilde{G}_0$ as

$$\tilde{G}_0(i\omega_n) = \left(G_{\text{loc}}^{-1}(i\omega_n) + \tilde{\Sigma}(i\omega_n)\right)^{-1}. \tag{8}$$

Let us now discuss the model-specific part.[8] For the present system (3), the action of the effective single-site model is described as

$$\begin{aligned}\tilde{S} &= -\int_0^\beta d\tau_1 \int_0^\beta d\tau_2\ \Psi^*(\tau_1)\tilde{G}_0^{-1}(\tau_1-\tau_2)\Psi(\tau_2) \\ &\quad -J_{\text{H}} \int_0^\beta d\tau\ \vec{m}\cdot\Psi^*(\tau)\vec{\sigma}\Psi(\tau).\end{aligned} \tag{9}$$

Here $\Psi^* = (c_\uparrow^*, c_\downarrow^*)$ is the Grassmann variables in the spinor notation. Green's function in the imaginary time is calculated as

$$\begin{aligned}\tilde{G}(i\omega_n) &= \left\langle \left(\tilde{G}_0^{-1}(i\omega_n) + J_{\text{H}}\vec{m}\cdot\vec{\sigma}\right)^{-1}\right\rangle \\ &= \frac{1}{\tilde{Z}} \int d\Omega P(\vec{m}) \left(\tilde{G}_0^{-1}(i\omega_n) + J_{\text{H}}\vec{m}\cdot\vec{\sigma}\right)^{-1}.\end{aligned} \tag{10}$$

$P(\vec{m})$ is the Boltzmann factor for the spin

$$P(\vec{m}) = \frac{1}{\tilde{Z}} \exp[-\tilde{S}_{\text{eff}}(\vec{m})], \tag{11}$$

where $\tilde{S}_{\text{eff}}$ is the effective action for the spin

$$\begin{aligned}\tilde{S}_{\text{eff}}(\vec{m}) &= -\log \text{Tr}_F \exp(-\tilde{S}) \\ &= -\sum_n \log \det \left[\frac{1}{i\omega_n}(\tilde{G}_0^{-1}(i\omega_n) + J_{\text{H}}\vec{m}\vec{\sigma})\right] e^{i\omega_n 0+}.\end{aligned} \tag{12}$$

$\tilde{Z}$ is the partition function

$$\begin{aligned}\tilde{Z} &= \int d\Omega_{\vec{m}} \int \mathcal{D}\Psi^* \mathcal{D}\Psi \exp(-\tilde{S}) \\ &= \int d\Omega_{\vec{m}} \exp[-\tilde{S}_{\text{eff}}(\vec{m})].\end{aligned} \tag{13}$$

The self energy for the single-site system $\tilde{\Sigma}$ is obtained from $\tilde{\Sigma}(i\omega_n) = \tilde{G}_0^{-1}(i\omega_n) - \tilde{G}^{-1}(i\omega_n)$.

Magnetization of the local spin is obtained by

$$\langle \vec{m} \rangle = \int d\Omega_{\vec{m}} P(\vec{m})\vec{m}. \tag{14}$$

Hereafter we take the axis of the magnetization in z direction and the order parameter is expressed as $M = \langle m_z \rangle$. Transport properties are obtained through the Kubo formula. Conductivity in $D=\infty$ is calculated as[22-24]

$$\sigma(\omega) = \sigma_0 \sum_\sigma \int d\omega'\ I_\sigma(\omega', \omega'+\omega) \frac{f(\omega') - f(\omega'+\omega)}{\omega}, \tag{15}$$

where
$$I_\sigma(\omega_1, \omega_2) = \int N_0(\epsilon) d\epsilon \, W^2 A_\sigma(\epsilon, \omega_1) A_\sigma(\epsilon, \omega_2). \tag{16}$$

Here, the spectral weight function is defined by
$$A_\sigma(\epsilon, \omega) = -\frac{1}{\pi} \mathrm{Im} G_\sigma(\epsilon, \omega + i\eta), \tag{17}$$

while f is the Fermi distribution function. The constant σ_0 gives the unit of conductivity. In this formula, we used that the vertex correction cancels in the conductivity calculation at infinite-dimensional limit.[25] Thermopower is also calculated: The Seebeck coefficient is obtained as[26, 18]
$$S = \frac{1}{eT} \frac{L_2}{L_1}, \tag{18}$$

where L_k ($k = 1, 2$) is defined by
$$L_k = \sum_\sigma \int d\omega \left(-\frac{\partial f(\omega)}{\partial \omega} \right) I_\sigma(\omega, \omega)(\beta\omega)^{k-1}. \tag{19}$$

The method is easily expanded to a Bethe lattice with two-sublattice symmetry. In this case, magnetic phases with ferromagnetic and antiferromagnetic order parameters can be considered simultaneously. We introduce $\alpha = A, B$ sublattice indices for the Weiss fields $\tilde{G}_{0\alpha}(i\omega_n)$, and solve coupled self-consistency equations. The formula to calculate the Green's function is now given by
$$\tilde{G}_\alpha(i\omega_n) = \int d\Omega_\alpha P_\alpha(\vec{m}) \left(\tilde{G}_{0\alpha}^{-1}(i\omega_n) + J\vec{m} \cdot \vec{\sigma} \right)^{-1}. \tag{20}$$

Here, the Boltzmann weight for the configuration of local spin $P_\alpha(\vec{m})$ is calculated from the effective action $\tilde{S}_\alpha$,
$$\tilde{S}_\alpha(\vec{m}) = -\log \mathrm{Tr}_F \exp[-\tilde{S}(\tilde{G}_{0\alpha}, \vec{m})], \tag{21}$$
$$P_\alpha(\vec{m}) = \exp[-\tilde{S}_\alpha(\vec{m})]/\tilde{Z}_\alpha, \tag{22}$$
$$\tilde{Z}_\alpha = \int d\Omega_\alpha \exp[-\tilde{S}_\alpha(\vec{m})]. \tag{23}$$

Integration over DOS as in eq. (8) gives the self-consistent mapping relation[27]
$$\begin{aligned}\tilde{G}_{0A}^{-1}(i\omega_n) &= i\omega_n + \mu - \tilde{G}_B(i\omega_n) \, W^2/4, \\ \tilde{G}_{0B}^{-1}(i\omega_n) &= i\omega_n + \mu - \tilde{G}_A(i\omega_n) \, W^2/4.\end{aligned} \tag{24}$$

Now the self-consistency equations (20)-(24) form a closed set. Within this approach, we can study the instability and the formation of magnetic ordering with ferromagnetic, antiferromagnetic, and canted antiferromagnetic symmetries.

2.3 Analytical solutions of the dynamical mean-field theory

In some limiting cases, analytical solutions are available for the DMF calculations. Analytical expressions are in general quite useful to obtain intuitions.

Paramagnetic phase. For the paramagnetic solution, Green's function is a scalar function with respect to spin rotation, so we have $\tilde{G}_0(i\omega_n) = \tilde{g}_0(i\omega_n)I$ where I is the 2×2 eigenmatrix. Then, we have

$$\tilde{G}(i\omega_n) = \left\langle \frac{\tilde{g}_0(i\omega_n)^{-1}I - J_H \vec{m}\vec{\sigma}}{\tilde{g}_0(i\omega_n)^{-2} - J_H^2 |\vec{m}|^2} \right\rangle. \tag{25}$$

Since $\langle \vec{m} \rangle = 0$ and $\langle |\vec{m}|^2 \rangle = 1$, we have

$$\tilde{G}(i\omega_n) = \frac{\tilde{g}_0(i\omega_n)^{-1}}{\tilde{g}_0(i\omega_n)^{-2} - J_H^2}I = \frac{1}{2}\left(\frac{1}{\tilde{g}_0(i\omega_n)^{-1} + J_H} + \frac{1}{\tilde{g}_0(i\omega_n)^{-1} + J_H}\right)I. \tag{26}$$

The self-energy is then given by

$$\Sigma(i\omega_n) = \tilde{G}_0^{-1}(i\omega_n) - \tilde{G}^{-1}(i\omega_n) = J_H^2 \tilde{G}_0(i\omega_n). \tag{27}$$

From the derivation, we see that the Green's function (26) is the same as that of the system with the Ising substrate spin $\vec{m} = (0, 0, \pm 1)$. Furthermore, the Green's function (26) shares the same analytical structure as that of the infinite-dimensional Falicov-Kimball model (FKM) (or, simplified Hubbard model)

$$\mathcal{H}_{\text{FKM}} = -\sum_{ij} t_{ij}(c_i^\dagger c_j + \text{h.c.}) + U\sum_i c_i^\dagger c_i f_i^\dagger f_i \tag{28}$$

at $\langle n_f \rangle = 1/2$. Green's function of the FKM in infinite dimension is described as[28, 29]

$$\tilde{G}_{\text{FKM}}(i\omega_n) = \frac{1 - \langle n_f \rangle}{\tilde{G}_0^{-1}(i\omega_n)} + \frac{\langle n_f \rangle}{\tilde{G}_0^{-1}(i\omega_n) - U}. \tag{29}$$

Then, the Green's functions (26) and (29) share the same analytical structure at $\langle n_f \rangle = 1/2$. In the FKM, the c-electrons are scattered by the charge fluctuations of the localized f-electrons, which corresponds to the scattering process of the itinerant electrons by the localized spins in the DE model.

Hence, thermodynamical properties of the DE model in $D = \infty$ and $S = \infty$ can be understood from the nature of the FKM in $D = \infty$ which has been studied intensively. For example, the imaginary part of the self-energy at the fermi level is finite $Im\,\Sigma(0) \neq 0$ in the paramagnetic phase.[22, 29]

In Fig. 1(a) we schematically illustrate the spectral function $A(k, \omega)$ in the paramagnetic phase. As in the case for FKM, the spectral weight is split into two parts at $\omega \sim \pm J_H$ for sufficiently large Hund's coupling $J_H \gg W$. For the semi-circular density of states with the bandwidth W, we have the metal-insulator transition[30] at $J_H^c = 0.5W$, which has the Hubbard-III like nature. Kondo resonance peak, which is seen in the Hubbard model[31-33] is missing in this model, since the quantum exchange process is absent. As the magnetic moment is induced, the imaginary part of the self-energy decreases because the thermal fluctuation of spins decreases. At the ground state, spins are magnetically ordered. For the ferromagnetic ground state, there exists two free electron bands which are energetically split by J_H exchange interactions.

Infinite J_H limit –Green's function–. In the case of Lorentzian DOS, Green's function is easily obtained in the limit $J_H \to \infty$.[10] We consider the hole doped region where $\mu \sim -J_H$. In the case of the Lorentzian DOS, the self-consistency equation gives

$$G_0(\omega + i\eta) = (\Omega - J_H + iW)^{-1}. \tag{30}$$

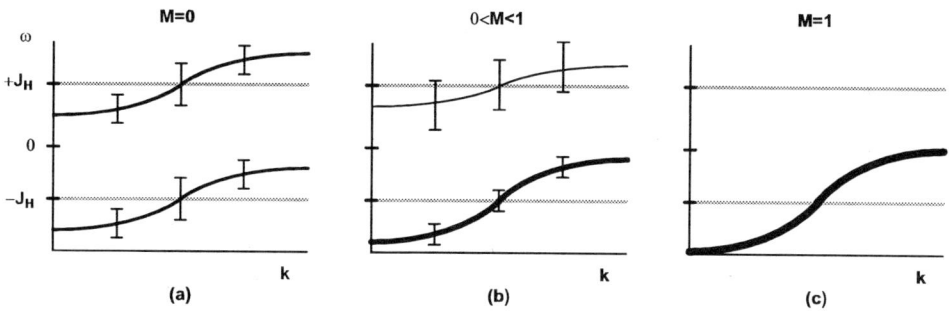

Figure 1: Schematic behavior in the spectral function for up-spin electrons, at (a) paramagnetic state $M = 0$, (b) ferromagnetic state at $0 < T < T_c$, and (c) at the ground state $T = 0$ and $M = 1$. Solid curves illustrate peak positions of $A(k,\omega)$. Width of the curves represent height of the peak (quasiparticle weight), while error bars represent the linewidth (inverse of lifetime). Grey lines are guides to eyes. Spectral functions for down-spin electrons are obtained by exchanging upper part ($\omega \sim J_H$) and lower part ($\omega \sim -J_H$). Also see Fig. 4 in §3.1 for actual data.

Here, chemical potential is $\mu = -J_H + \delta\mu$ where $\delta\mu = O(W)$, and $\Omega \equiv \omega + \delta\mu = O(W)$ is the energy which is measured from the center of the lower sub-band $-J_H$.

Magnetic field in the z direction is applied to the localized spins in the paramagnetic phase, and the induced magnetization is expressed as $M = \langle m_z \rangle$. Since C_0 in eq. (30) is spin independent even in the spin polarized cases, eqs. (25) and (30) gives

$$G_\sigma(\omega + i\eta) = \frac{(\Omega - J_H + iW) - J_H M\sigma}{(\Omega - J_H + iW)^2 - J_H^2}$$
$$= \frac{1 + M\sigma}{2} \frac{1}{\Omega + iW} + O(1/J_H). \quad (31)$$

At $J_H/W \to \infty$, the spectral weight is calculated as

$$A_\sigma(\omega) = -\frac{1}{\pi} \mathcal{I}m\, G_\sigma(\omega + i\eta) = \frac{1 + M\sigma}{2} \cdot \frac{1}{\pi} \frac{W}{\Omega^2 + W^2}. \quad (32)$$

We see that the center of the spectral weight is indeed shifted to $-J_H$. The amplitude of A_σ is proportional to the population of the local spins parallel to σ, which indicates that the electronic states that are anti-parallel to the local spin are projected out. The self-energy is calculated from eqs. (30) and (31) as

$$\Sigma_\sigma(\omega + i\eta) = -J_H - \frac{1 - M\sigma}{1 + M\sigma}(\Omega + iW). \quad (33)$$

Eq. (33) gives $\mathcal{R}e\,\Sigma \sim -J_H$, so the shift in μ is self-consistently justified again.

Similarly, Green's function at $\omega' \sim 2J_H$, namely at the upper subband, is described as follows. Using $\omega' = \omega + 2J_H$, where $\omega = O(W)$, Green's function is given by

$$G_\sigma(\omega + 2J_H + i\eta) = \frac{1 - M\sigma}{2} \frac{1}{\Omega + iW} + O(1/J_H), \quad (34)$$

and the spectral weight is calculated as

$$A_\sigma(\omega + 2J_H) = \frac{1 - M\sigma}{2} \cdot \frac{1}{\pi} \frac{W}{\Omega^2 + W^2}. \quad (35)$$

From eqs. (32) and (35), we see the transfer of the spectral weight by magnetization.
From above equations, Green's function is given in the form

$$G_\sigma(k,\omega;M) = \frac{z_\sigma^{(l)}(M)}{\omega + J_H + \mu - \zeta_\sigma^{(l)}(k;M) + i\Gamma_\sigma^{(l)}(M)} + \frac{z_\sigma^{(u)}(M)}{\omega - J_H + \mu - \zeta_\sigma^{(u)}(k;M) + i\Gamma_\sigma^{(u)}(M)}, \quad (36)$$

asymptotically at $J_H/W \to \infty$. In this limit, Green's function is a sum of contributions from lower subbandrs (l) and upper subband (u). Quasiparticle residue is given by

$$z_\sigma^{(l)}(M) = P_\sigma^+(M), \qquad z_\sigma^{(u)}(M) = P_\sigma^-(M), \quad (37)$$

quasiparticle dispersion relation is described as

$$\zeta_\sigma^{(l)}(k;M) = P_\sigma^+(M)\varepsilon_k, \qquad \zeta_\sigma^{(u)}(k;M) = P_\sigma^-(M)\varepsilon_k, \quad (38)$$

and the quasiparticle linewidth is in the form

$$\Gamma_\sigma^{(l)}(M) = P_\sigma^-(M)W, \qquad \Gamma_\sigma^{(u)}(M) = P_\sigma^+(M)W. \quad (39)$$

Here, P are the function of spin polarization

$$P_\sigma^+(M) \equiv \frac{1+M\sigma}{2}, \qquad P_\sigma^-(M) \equiv \frac{1-M\sigma}{2}. \quad (40)$$

In Fig. 1 we schematically show the spectral function for the up-spin electron $A_\uparrow(k,\omega) = -\mathcal{I}m\, G_\uparrow(k,\omega)/\pi$ calculated from eq. (36). For the down-spin electrons, $A_\downarrow(k,\omega)$ is obtained by replacing upper and lower subbands, e.g. $z_\downarrow^{(l)} = z_\uparrow^{(u)}$, etc. In the paramagnetic phase (Fig. 1(a)), quasiparticle dispersion are split into lower and upper subbands at $\omega \sim \pm J_H$ with quasiparticle weight $z_\uparrow^{(l)} = z_\uparrow^{(u)} = 1/2$. As magnetization is increased below T_c, the lower subband gains quasiparticle weight $z_\uparrow^{(l)} = (1+M)/2$ and that for upper subband decreases as $z_\uparrow^{(u)} = (1-M)/2$. At the ground state with perfect spin polarization $M = 1$, the electronic Hamiltonian describes a free electron system under Zeeman splitting field. Hence there exists only lower subband for the up spin electron. This limit is described as $z_\uparrow^{(l)} = 1$ and $z_\uparrow^{(u)} = 0$.

In the paramagnetic phase $M = 0$ we see $\mathcal{I}m\,\Sigma = -W$, which means that the quasiparticle excitation is very incoherent; the lifetime of a quasi-particle is comparable with the time scale that an electron transfers from site to site. This result justifies us to take the $D = \infty$ limit which is essentially a single-site treatment.

In the following section (§3.1), we will discuss the behavior of the spectral function for general cases.

Infinite J_H limit –Curie temperature–. In the semicircular DOS case, we consider the ferromagnetic state under doping at $J_H \gg W$. We set

$$G_0^{-1}(i\omega_n) = (i\omega_n + R_n)\hat{I} + Q_n\hat{\sigma}_z \quad (41)$$

and $\mu = -J_H + \delta\mu$. In order to keep the carrier concentration finite, we take the limit $J_H \to \infty$ with keeping $\delta\mu = O(W)$. In this limit, we have

$$R_n = -\frac{W^2}{8}\left\langle \frac{1}{z_n + R_n + m_z Q_n}\right\rangle, \qquad Q_n = -\frac{W^2}{8}\left\langle \frac{m_z}{z_n + R_n + m_z Q_n}\right\rangle, \quad (42)$$

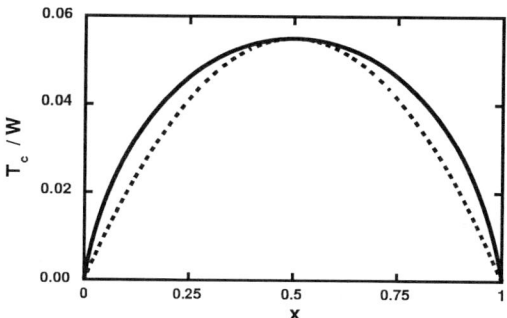

Figure 2: Curie temperature for the semicircular DOS at $J_H = \infty$ (solid curve). Dotted curve is the form $T_c \propto x(1-x)$.

where $z_n = i\omega_n + \delta\mu$. The Boltzmann weight is calculated from eq. (12) as

$$P(\vec{m}) \propto \exp\left[\sum_n \log\left(1 + \frac{R_n + Q_n m_z}{z_n}\right)\right]. \qquad (43)$$

This equation tells us that the model is not simply mapped to the Heisenberg model. The Boltzmann weight of the Heisenberg model in infinite dimension is expressed as $P_{\text{Heis}}(\vec{m}) \sim \exp(-\beta h_{\text{eff}} m_z)$, which contradicts with that of the DE model, i.e. $P(\vec{m}) \neq P_{\text{Heis}}$. This reflects the fact that the itinerant ferromagnet has smaller T_c due to the spin fluctuation affecting the itinerant electrons, compared to the insulating ferromagnet with the same spin stiffness.

Let us calculate T_c. At $T \sim T_c$, we have $M = \langle m \rangle \ll 1$ and

$$R_n = -\frac{W^2}{8}\frac{1}{z_n + R_n} + O(M), \qquad (44)$$

$$Q_n = -\frac{W^2}{8}\left\langle \frac{m_z}{z_n + R_n} - \frac{m_z^2}{(z_n + R_n)^2}Q_n \right\rangle + O(M^2). \qquad (45)$$

By solving eqs. (44) and (45), we have

$$R_n = \sqrt{z_n^2 - W^2/4} - z_n, \qquad \frac{Q_n}{M} = \frac{R_n}{1 - 8\langle m_z^2\rangle R_n^2/W^2}. \qquad (46)$$

Here, $\langle m_z^2 \rangle \equiv \int d\Omega\, m_z^2 = 1/3$ at $S = \infty$. The Boltzmann weight is given by

$$P(\vec{m}) \propto \exp\left[\sum_n \frac{Q_n}{z_n + R_n} m_z\right] = \exp(-\beta J_{\text{eff}} M m_z) \qquad (47)$$

where

$$J_{\text{eff}}(\beta) = \frac{1}{\beta}\sum_n \frac{8R_n^2/W^2}{1 - 8R_n^2/(3W^2)}. \qquad (48)$$

Then, the partition function is identical to that of the Heisenberg model with exchange coupling J_{eff}, and T_c is obtained from

$$T_c = \frac{1}{\beta}\sum_n \frac{8R_n^2}{3W^2 - 8R_n^2}\bigg|_{\beta=1/T_c}. \qquad (49)$$

Solving eqs. (46) and (49) self-consistently, we obtain T_c as a function of μ, while the carrier number is calculated directly from Green's function $G(i\omega_n)$. In Fig. 2 we plot T_c as a function of doping x. We see that T_c has a maximum at $x = 0.5$ with $T_c \sim 0.05W$. Then, using $\beta W \ll 1$ we may approximate the summation in eq. (49) by integration,

$$T_c = \frac{1}{2\pi} \int_{-\infty}^{\infty} \mathrm{d}x \frac{8R^2(\mathrm{i}x + \delta\mu)}{3W^2 - 8R^2(\mathrm{i}x + \delta\mu)} \tag{50}$$

where $R(z) = \sqrt{z^2 - W^2/4} - z$.

The result is particle-hole symmetric, namely $T_c(x) = T_c(1-x)$. At $x \to 0$ and $x \to 1$, T_c diminishes because the ferromagnetism of this system is due to the kinetic energy of the conduction electron. In Fig. 2, we also depict a curve $T_c \propto x(1-x)$ proposed in ref. 34 for comparison. This simple form roughly reproduces the doping dependence of T_c. In the following section, we discuss more precisely about the Curie temperature at finite J_H/W.

2.4 Monte Carlo method for finite size clusters

On a finite-size clusters, it is possible to investigate the double-exchange system by numerical methods. The result is unbiased and exact within the numerical errors.

The partition function of the present model with localized spins treated as classical rotators is defined by

$$Z = \mathrm{Tr}_S \mathrm{Tr}_F \exp\left(-\beta[\mathcal{H}(\{\vec{m}_i\}) - \mu\hat{N}]\right), \tag{51}$$

where Tr_S and Tr_F represent traces over spin and fermion degrees of freedom, respectively. In the finite size system, Z is obtained by taking the trace over fermion degrees of freedom first and spin degrees of freedom afterwards. Fermion trace is directly calculated from the diagonalization of $2N \times 2N$ Hamiltonian matrix, where N is the number of sites. Trace over spin degrees of freedom is replaced by the Monte Carlo summation over spin configurations $\{\vec{m}_i\}$.

For a fixed configuration of classical spins $\{\vec{m}_i\}$, the Hamiltonian is numerically diagonalized and we obtain

$$\mathrm{Tr}_F \exp(-\beta[\mathcal{H}(\{\vec{m}_i\}) - \mu N]) = \prod_{\nu=1}^{2N} [1 + \exp(-\beta(E_\nu(\{\vec{m}_i\}) - \mu))]. \tag{52}$$

E_ν ($\nu = 1, \ldots, 2N$) are eigenvalues of the Hamiltonian matrix for a given configuration $\{\vec{m}_i\}$. We have the effective action for the classical spin system

$$S_{\mathrm{eff}}(\{\vec{m}_i\}) = -\sum_\nu \log\left(1 + e^{-\beta(E_\nu - \mu)}\right), \tag{53}$$

which gives

$$Z = \mathrm{Tr}_S \exp(-S_{\mathrm{eff}}). \tag{54}$$

Monte Carlo update of spin configurations is performed using the Boltzmann weight of the state $\{\vec{m}_i\}$,

$$P(\{\vec{m}_i\}) \propto \exp(-S_{\mathrm{eff}}(\{\vec{m}_i\})). \tag{55}$$

In a Monte Carlo unit step, orientations of each spins are updated using the Metropolis algorithm. Since the spins are classical, spin updates can be performed ergodically.

Thermodynamic quantities are stochastically calculated. Quantities which are associated with localized spins are obtained directly from the thermal average of spin configurations. Electronic quantities are calculated from the eigenvalues and eigenfunctions of $\mathcal{H}(\{\vec{m}_i\})$.

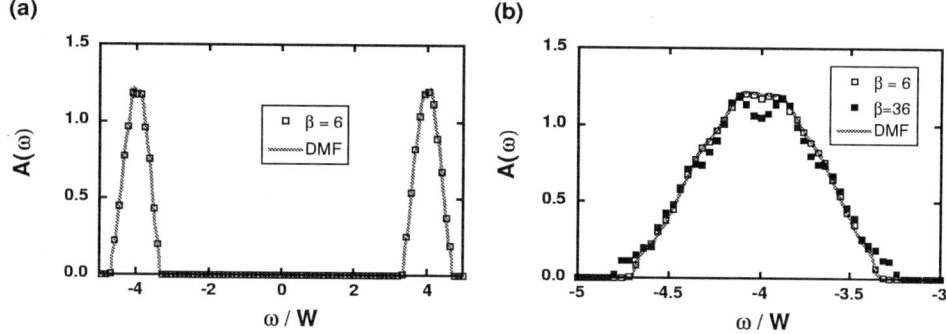

Figure 3: Monte Carlo results for DOS on $6 \times 4 \times 4$ cubic lattice at $J_H/W = 4$, at $\beta \equiv W/T = 6$ and 36. Error bars are within the symbol size. Curves in the figure show the DMF results. (left) Two peak structure at $\omega \sim \pm J_H$ is seen for both DMF and Monte Carlo results. (right) Lower subband at $\omega \sim -J_H$.

One of the advantages for taking the classical spin limit is that there exists no "negative sign problem", which is present in quantum spin models. In the classical spin limit, the spin degrees of freedom is completely decoupled from those of fermions, and the fermionic trace in eq. (52) is obtained by solving the noninteracting lattice fermion system with random static potential. Another advantage is that the real frequency dynamics of electronic properties are directly obtained, since eigenvalues of the Hamiltonian are calculated through the Monte Carlo procedures. There is no difficulties of analytical continuations form imaginary frequencies, as is present in some quantum Monte Carlo methods. A disadvantage of this method is that the auxiliary field is static *i.e.* non-local in imaginary time, so it is not possible to make a local spin flip using the imaginary time Green's function as in the case for the Hubbard model.[35, 36]

Let us now compare the dynamical mean-field theory and the cluster Monte Carlo method. Here we calculate the electron DOS

$$A(\omega) = -\frac{1}{N} \sum_{k\sigma} \mathcal{I}m\, G_\sigma(k, \omega + i\eta)/\pi \qquad (56)$$

on a finite size cluster system in Fig. 3. We treat $N = 6 \times 4 \times 4$ cubic lattice at $J_H/W = 4$ and $\mu = -J_H$, where the dynamical mean-field approach gives $T_c = 0.028W$. In order to avoid delta-function singularities in finite size systems, we use an adiabatic factor $\eta = 10^{-2}$ to smooth the spectra. In the Monte Carlo calculation, we calculate at $\beta \equiv W/T = 6$ and 36, and compare with the dynamical mean-field result in the paramagnetic phase $T > T_c$. The result shows that the DMF calculation is very accurate.

III. RESULTS

In this section, we show the results for the classical spin limit $S = \infty$ of the DE system using the DMF approach as well as the Monte Carlo calculation. Hereafter, the electronic bandwidth is taken to be $W \equiv 1$ as a unit of energy. For the carrier electron number, we express by $x = 1 - \langle n \rangle$. We define $M = \langle m_z \rangle$, where $0 \leq M \leq 1$. Also, we describe the total moment, or sum of moments of localized spin and electron spin, as $M_{\text{tot}} = \langle \frac{3}{2} m_z + \frac{1}{2} \sigma_z \rangle$. In a normalized form we describe $M^* = M_{\text{tot}}/M_{\text{sat}}$, where M_{sat}

is the saturation value of M_{tot} at the ground state. We make distinctions between M and M^*, since analytical calculation is better understood by M while the comparison with experiments should be done by M^*. Nevertheless, in the strong coupling region $J_{\text{H}} \gg W$ we have $M \simeq M^*$ so effectively there exists no major differences.

3.1 Electronic structures

Spectral function and the density of states. In Fig. 4 we show the spectral function for the up-spin electron $A_\uparrow(k,\omega)$ on a cubic lattice, where $k/\pi = (\zeta,\zeta,\zeta)$. From the particle-hole symmetry and the spin symmetry, down-spin part $A_\downarrow(k,\omega)$ is reproduced by the relation

$$A_\downarrow(k,\omega) = A_\uparrow(Q - k, -\omega), \qquad (57)$$

where $Q = (\pi,\pi,\pi)$.

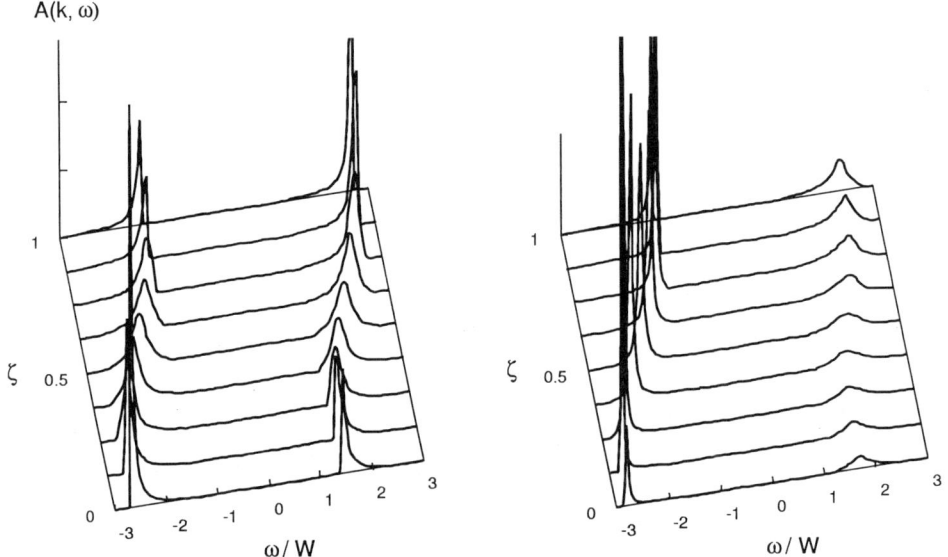

Figure 4: Spectral function $A_\uparrow(k,\omega)$ on a cubic lattice at $J_{\text{H}}/W = 2$ and x=0.3, (left) in the paramagnetic phase $T = 1.05T_{\text{c}}$, and (right) in the ferromagnetic phase $T = 0.5T_{\text{c}}$. Here, $k/\pi = (\zeta,\zeta,\zeta)$.

In Fig. 4 we show the quasiparticle excitation structure and its temperature dependence. There exists two-peak structure at around $\omega \sim \pm J_{\text{H}}$. Above T_{c}, peaks at upper and lower bands are symmetric and equally weighted. Below T_{c}, the structure remains split but becomes asymmetric. For the up-spin electron, the integrated weight is transferred from upper band to lower band. We also see the change of the quasiparticle linewidth Γ. The lower band peak becomes sharper, which means the reduction of Γ or the enlonged quasiparticle lifetime. On the other hand, Γ for the upper band peak increases.

To investigate the change of the spectral weight in further detail, we calculate the DOS $A_\sigma(\omega)$ by k-integrating the spectral weight. In Fig. 5 we show the DOS as a function of temperature in the paramagnetic and ferromagnetic phases. Two subband structure at $\omega \sim \pm J_{\text{H}}$ reflects the quasiparticle structure as described above.

At the ground state, lower subband is composed of up-spin only, and down-spin band exists only at the high-energy region. The bandwidth of the DOS becomes narrower as temperature becomes higher. The band center is fixed at $\omega \sim \pm J_H$ which is the energy level of the atomic limit $t = 0$.

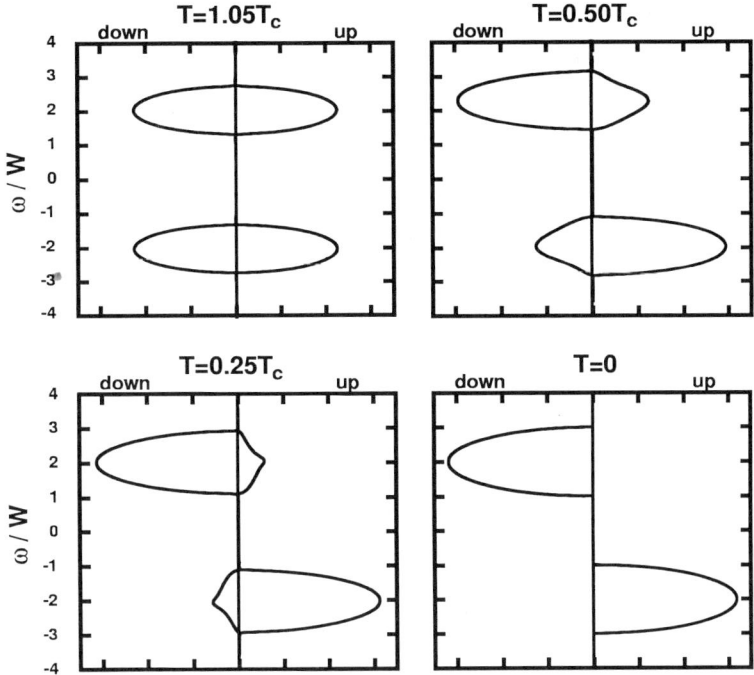

Figure 5: Temperature dependence of DOS for $J_H/W = 2$ and $x = 0.3$, where $T_c = 0.019W$. Peak structures are observed at $\omega \sim \pm J_H$.

These results are in agreement with the exact result in a limiting case (Lorentzian DOS with $J_H \to \infty$) discussed in §2.3. Let us focus on the lower subband. The DOS is nearly proportional to the spin polarization of the localized spin, namely

$$A_\uparrow(\omega) \sim z_\uparrow^{(l)}(M) = (1 + M)/2,$$
$$A_\downarrow(\omega) \sim z_\downarrow^{(l)}(M) = (1 - M)/2, \qquad (58)$$

where z is the quasiparticle weight discussed in §2.3. We see that z is determined by the magnetization through $(1 \pm M)/2$, which is the probability that the local spin is parallel to the itinerant electron with up (down) spin. This is easily understood from the nature of the double-exchange interaction which projects out the antiparallel component of the spins. This spin-dependent projection may be viewed as the evolution of the majority-minority band structure. The up (down) spin band becomes a majority (minority) band below T_c or under magnetic field. Shift of the spectral weight from the minority band to the majority band occurs as temperature is lowered. At the ground state ($M = 1$), minority band completely loses its weight.

Change of the electron bandwidth is understood qualitatively through Anderson-Hasegawa's picture.[2] Electron hopping amplitude is proportional to $\cos(\theta/2)$ where θ is the relative angle of the localized spins. At high temperature, θ deviates from zero due to spin fluctuation, and the amplitude of electron hopping matrix element and hence the bandwidth decreases. This is also shown by the virtual crystal approximation.[37]

In order to account for the width of $A(k,\omega)$, or the quasiparticle lifetime, we have to go beyond a mean-field picture like in Anderson-Hasegawa approach. The origin of the linewidth Γ is the thermal fluctuation of the spins. In the strong coupling region $J_H \gg W$, spin scattering phenomena in the paramagnetic phase is so large that quasi-particles lose their coherence due to the inelastic scattering by thermally fluctuating spins. In the ferromagnetic phase, the spin fluctuation decreases as temperature is decreased. For the majority band, this decreases Γ, and in the limit $T \to 0$ the majority band becomes a free electron band. However, for the minority band, the spin projection causes further loss of the coherence, which leads to the increase of Γ as well as the decrease of the quasiparticle weight z. Asymptotically in the limit $T \to 0$, Γ approaches to a constant $\sim W$, and at the same time $z \to 0$. Then, at $T = 0$, the minority band with finite Γ is projected out.

Half metal. Metal with a DOS structure shown in Fig. 5 where only one of the spin species have the Fermi surface is called a half metal.[38, 39] Namely, because of the 'Zeeman splitting' due to Strong Hund's coupling $J_H \gg W$, ferromagnetic ground state of the DE model shows a perfect spin polarization and thus is a half-metal. Experimentally, spin-resolved photoemission investigation[40, 41] shows that the conduction band of the doped manganites is a half-metal. Artificial trilayer junction of manganites[42] also shows a large tunneling magnetoresistance phenomena, and the spin polarization is estimated to be more than 80%. Such a DOS structure creates a phenomena called tunneling magnetoresistance (TMR), which will be discussed in §5.1.

Shift of the chemical potential. A direct consequence of the change of the bandwidth controlled by the magnetization will be observed in the shift of the chemical potential. In this case, the change of the DOS structure is in a way such that the band center is pinned by the Hund's coupling energy $\pm J_H$ and the band edge shifts away from the center as magnetization is increased. Then, for a hole doped case the position of the chemical potential increases by increasing the magnetization.

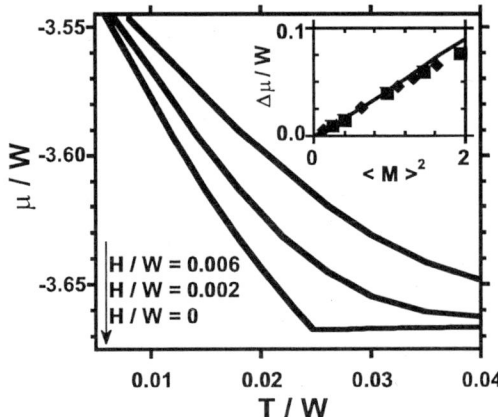

Figure 6: Temperature dependence of μ at $J_H/W = 4$ and $x = 0.20$ under various magnetic field. Inset: $\Delta\mu/W$ as a function of M^2. Lines show the result at $T < T_c$ for $H = 0$. Squares and diamonds are data at $T = 1.01T_c$ and $1.2T_c$ by applying H, respectively.

In Fig. 6 we show the temperature dependence of the chemical potential at $x = 0.2$ under various magnetic field.[13] At $H = 0$, chemical potential μ is nearly temperature independent above T_c. Below T_c, μ shifts as a function of temperature. We also calculate

μ and M, (i) at $H = 0$ by changing temperature in the region $T \leq T_c$, and (ii) at fixed temperature above T_c by changing H. In the inset of Fig. 6 we plot μ as a function of magnetic moment M^2 for both cases. As a result, we see the scaling relation

$$\Delta\mu/W \propto M^2, \qquad (59)$$

where $\Delta\mu \equiv \mu(T, H) - \mu(T = T_c, H = 0)$. We see that $\Delta\mu$ can be as large as $0.1W$.

Thus, for a fixed band filling, the total change of the DOS width in the entire energy range causes the shift of μ. The change in such a large energy scale controlled by magnetization produce the characteristic feature of the shift of μ in DE systems; namely, that the shift of μ is as large as a few tenth of W and that the scaling relation (59) is satisfied up to such a large energy scale.

Such a large shift of μ might possibly be applied to electronic devices which controls the MOS gate voltage by the magnetic field.

3.2 Magnetic structure and transport properties

Magnetic transition temperature. In the limit $J_H \to \infty$, Curie temperature T_c of the DE model is determined by the electron kinetic energy. Indeed, DMF calculation shows that T_c is scaled by electron hopping, i.e. $T_c \propto W$ for $J_H = \infty$ limit (see §2.3). Here we show the case for finite J_H/W.

In Fig. 7 we show the Curie temperature T_c as a function of doping x for various values of J_H/W. At finite J_H/W, T_c is reduced from $J_H = \infty$ values. We also see that T_c systematically increase as x is increased and have maximum at around $x \sim 0.5$, which is due to the increase of the kinetic energy.

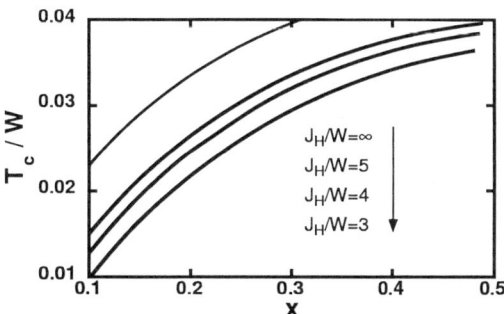

Figure 7: Curie temperature T_c as a function of J_H/W and x.

At half-filling $x = 0$, there exists antiferromagnetic order. In Fig. 8 we show the phase diagram at $x = 0$ for the $D = \infty$ Bethe lattice. In the weak coupling region $J_H \ll W$, the Nèel temperature T_N is equivalent to the results from the SDW mean-field type equation with J_H/W dependence in an essential singular function. At $J_H \gg W$, T_N is determined from the Heisenberg model with the exchange coupling $J_{AF} \sim t^2/J_H \propto W^2/J_H$.

Thus we see the antiferromagnetic order at $x = 0$ and ferromagnetic ground state at sufficiently doped case. In the underdoped region $x \ll 1$, de Gennes discussed the existence of the canted state[3]. However, in §3.4 we show that the canted state is unstable against phase separation, i.e. mixed phase of $x = 0$ antiferromagnetic region and $x > 0$ ferromagnetic region.[16]

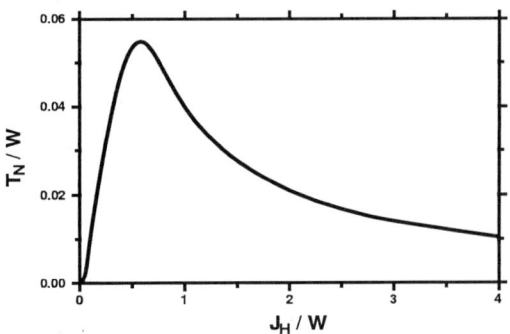

Figure 8: Nèel temperature at $x = 0$.

Resistivity as a function of magnetization. Resistivity $\rho(T)$ as well as total magnetization M_{tot} as a function of temperature is given in Fig. 9. Here, ρ_0 is a constant of resistivity which corresponds to the Mott's limit value (inverse of the Mott's minimum conductivity) in three dimension. M_{sat} is the saturated magnetization at $T = 0$.

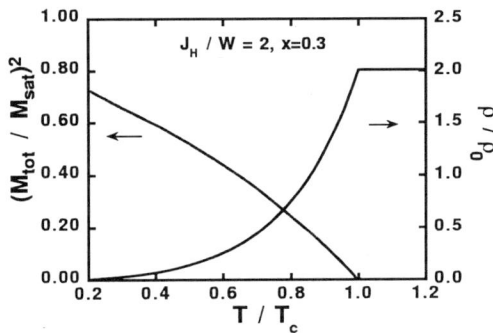

Figure 9: Resistivity ρ and magnetization M_{tot} as a function of temperature.

Above T_c, the value of the resistivity is in the order of the Mott limit $\rho(T) \sim \rho_0$, with small T-dependence. Below T_c, resistivity drops quickly as magnetization increases. From the DMF calculation,[8,11] it has been made clear that the resistivity behaves as

$$\rho(M)/\rho(M=0) = 1 - CM^2, \tag{60}$$

where C is a temperature/field independent constant. Namely, temperature and magnetic field dependences come from the magnetization $M = M(H,T)$. As a function of magnetization M, all the T and H dependent values $\rho(T,H)$ converge on a universal curve in (60). In other words, the origin of the resistivity is due to spin fluctuation, or more precisely spin disorder scattering, discussed by Kasuya[43] and later by Fisher and Langer.[44]

In the Born approximation (weak coupling limit), we see $C = 1$.[43] A phenomenological treatment by Kubo and Ohata which estimates the resistivity from the spin fluctuation,

$$\rho \propto (\delta S)^2 \propto 1 - M^2, \tag{61}$$

also gives $C = 1$. However, in the DMF at $J_H \gg W$, we have $C > 1$ which indicates the strong coupling behavior. In the next section, the relation with experimental MR is discussed in more detail.

Above T_c, small T-dependence in $\rho(T)$ is observed within the DMF treatment. This result might be an artifact of the approximation since the local spin fluctuation is saturated above T_c.

3.3 Charge and spin dynamics

Optical conductivity. Temperature dependence of the optical conductivity is shown in Fig. 10. In the paramagnetic phase, the spectrum splits into two peaks due to the 2-subband structure of the DOS. Namely, intraband particle-hole channel creates a Drude-like peak at $\omega \sim 0$, while interband channel creates a peak at around $\omega \sim 2J_H$. In the inset of Fig. 10, we show the weight of the interband process as a function of $1 - M^{*2}$. The integrated weight of the interband optical process at $\omega \sim 2J_H$ defined by

$$S = \int_{\omega_c}^{\infty} d\omega \, \sigma(\omega), \qquad (62)$$

where cutoff frequency is taken as $\omega_c = J_H$. We see a scaling relation

$$S \propto 1 - M^{*2}. \qquad (63)$$

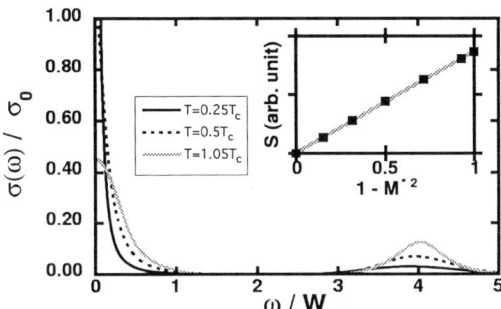

Figure 10: Temperature dependence of the optical conductivity for $J_H/W = 2$ and $x = 0.3$. Inset: Integrated weight S at high energy part $\omega \sim 2J_H$.

This is explained by the temperature dependence of DOS. Below Curie temperature, the DOS changes as in Fig. 5. Interband optical process at $\omega \sim 2J_H$ is constructed from a process of making a pair of lowerband hole and upperband electron. Then, the transfer of spectral weight by magnetization creates the change in the optical spectra as follows. For the up spin electrons, the spectral weight of lower and upper subbands are proportional to $(1 + M^*)/2$ and $(1 - M^*)/2$, respectively. Since this optical process conserves quasiparticle spin, the total weight of the optical spectrum is proportional to the product of the initial-state weight at lower subband and the final-state weight at the upper subband, and hence scaled as $1 - M^{*2}$. Contribution from the down-spin band is also the same, and we have $S \propto 1 - M^{*2}$.

Stoner excitation. Stoner susceptibility is calculated by

$$\chi(q, z) = \frac{1}{\beta N} \sum G(k + q, i\omega_n + z) G(k, i\omega_n). \qquad (64)$$

Here, correlation effects are taken into account through the self-energy correction in G. In Fig. 11 we show q-dependence of $\text{Im}\chi(q, \omega)$ for various temperatures, at $J_H/W = 2$

and $x = 0.3$. We see two-peak structure at $\omega \sim 0$ and $\omega \sim 2J_H$ which is explained from the J_H-split DOS structure. The Stoner absorption is produced from a particle-hole pair excitations with spin flip, which produces a peak at low energy from intra-band processes and another peak at $\omega \sim 2J_H$ from interband processes.

We see a weak q dependence in the low frequecy part, especially at $T \ll T_c$. This part of the Stoner process is dominated by the combination of the majority-minority quasiparticles. Since the quasiparticles in the minority band is incoherent, $Im\chi(q,\omega)$ is weakly q dependent. On the other hand, high energy part of $Im\chi(q,\omega)$ at $T \ll T_c$ have larger q dependence. This part is dominated by the majority-majority quasiparticle channel. Thus q dependence of $Im\chi(q,\omega)$ reflects the band structure of the quasiparticles.

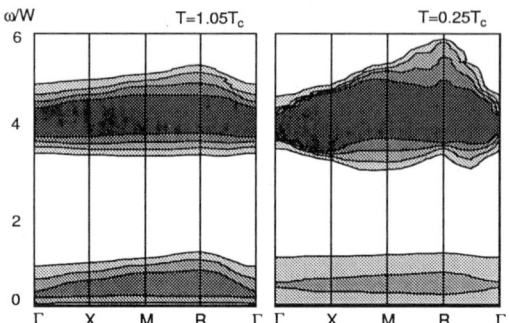

Figure 11: Contour plot of the Stoner absorption $\chi(q,\omega)$ on a cubic lattice at $J_H/W = 2$ and $x = 0.3$, in the paramagnetic phase $T = 1.05T_c$ (left) and in the ferromagnetic phase $T = 0.25T_c$ (right). At $T = 0$, the low energy part at $\omega \sim 0$ disappears.

Let us see the ω dependence at the low frequency region. In Fig. 12 we show $\chi_{ZB}(\omega) = \chi(Q,\omega)$ where $Q = (\pi,\pi,\pi)$. We see that at small ω we have ω-linear relation, i.e. $Im\chi \propto \omega$ at $\omega \ll W$. Coefficients for ω-linear part decrease by decreasing the temperature, and we find[14]

$$Im\chi(Q,\omega) \propto (1 - M^{*2})\omega \tag{65}$$

for small values of ω. The relation (65) is observed at all values of q with weak q dependence.

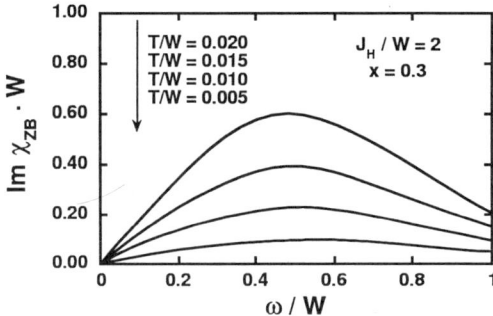

Figure 12: Stoner absorption $\chi(Q,\omega)$ at the zone boundary $Q = (\pi,\pi,\pi)$. At $J_H/W = 2$ and $x = 0.3$, transition is at $T_c/W = 0.019$.

Low energy Stoner absorption is constructed from a minority particle and majority hole channel. Since the majority and the minority bands have the spectral weight proportional to $(1 + M^*)/2$ and $(1 - M^*)/2$, respectively, the low energy part of the Stoner absorption is proportional to their product, $1 - M^{*2}$. The ω-linear behavior comes from the Fermi distribution function. The incoherence of the minority band gives the weak q dependence. Thus we have the scaling relation $\mathcal{I}m\,\chi \propto (1 - M^{*2})\omega$. The weak q dependence is in large contrast with the conventional weak ferromagnet where minority band is also coherent, which gives strong q dependence through its band structure.

3.4 Phase separation*

Magnetic phase diagram of the weakly doped DE model has been studied by de Gennes.[3] Assuming the homogeneity of the doped carriers, he concluded that the spin canted phase is the most energetically favorable state. However, we have recently shown that there exists an instability toward phase separation,[16] and the assumption of uniformly doped charges by de Gennes is not valid.

One of the ways to discuss the phase separation is to make a grand canonical calculation of the particle number x as a function of the chemical potential μ. If there exists a jump of $x(\mu)$ at the critical value $\mu = \mu_c$ in the thermodynamic limit, it implies that two phases with different doping x coexist at $\mu = \mu_c$. We have shown[16] the jump of $x(\mu)$ in the DE model for sufficiently large J_H/W. The jump occurs from $x = 0$ to a state with finite x. The calculation has been performed within the DMF approach ($D = \infty$), and cross checked by the Monte Carlo method for the $D = 1$ and 2 clusters in the extrapolated limit of $T \to 0$.

Phase boundary is simply determined from the jump of $x(\mu)$. In Fig. 13 we show the x-T phase diagram. At the low temperature region, we see mixed phases of $x = 0$ AF state and doped ($x > 0$) state with either paramagnetic or ferromagnetic state, depending on temperature.

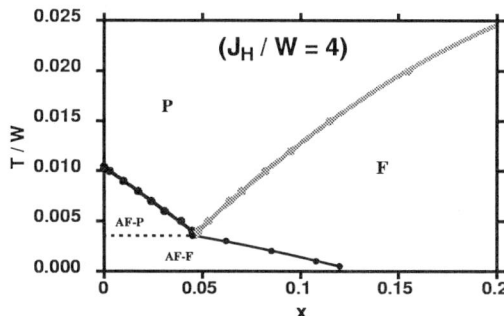

Figure 13: Phase diagram at $J_H/W = 4$. P (F) represents paramagnetic (ferromagnetic) region. At $x = 0$ we have antiferromagnetic (AF) phase. Regions labeled by AF-P (AF-F) are phase separated region with mixed phases of AF and P (AF and F) phases.

The mechanism of the phase separation may be understood as follows. In Fig. 14 we show the change of the density of states for different doping concentration. We see that the bandwidth substantially differs for different magnetic states, while the band center

---

* Works shown here concerning the issue of phase separation has been made in collaboration with S. Yunoki, A. Moreo and E. Dagotto.

remains at $\omega \sim \pm J_\text{H}$. As discussed in §3.1, this typical DOS structure is a consequence of the DE half-metallic system. Let us consider the zero temperature limit. At the $x = 0$ AF state, we have $\mu = 0$. In order to hole dope this AF state, we need to decrease μ to a AF gap-value, $\mu = -\Delta_\text{AF}$. However, as we see in Fig. 14, it is also possible to make a F state at $\mu = -\Delta_\text{AF}$ with rich hole density x. AF state gains the exchange energy $\sim t^2/J_\text{H}$, while the F state gains the kinetic energy $\sim tx$. Thus, in the limit $J_\text{H} \gg t$, the doped F state becomes energetically favorable. In such a case, there exists the chemical potential μ_c such that $-\Delta_\text{AF} < \mu_\text{c} < 0$, where the energies of AF state with $x = 0$ and F state with $x > 0$ are equal. Level crossing from AF to F states occurs at this critical point $\mu = \mu_\text{c}$. Doped F state is realized before doping the AF state. Thus we have a jump in the hole density from $x = 0$ to finite x.

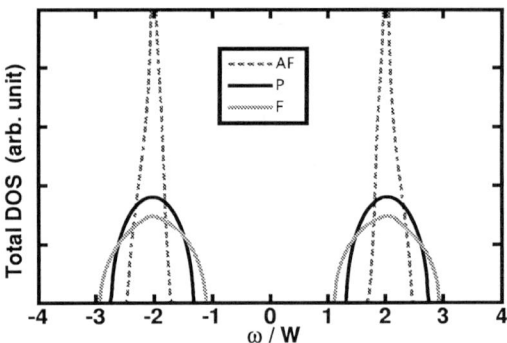

Figure 14: Density of states for ferromagnetic (F), paramagnetic (P) and antiferromagnetic (AF) states at $x = 0.2$, 0.1 and 0, respectively, for $J_\text{H}/W = 2$ and $T/W = 0.005$.

The discussion based on the DOS structure is quite generic. In strongly correlated systems with bistable phases, macroscopic change of the order parameters creates a large change in the electronic DOS. This gives a macroscopic jump in $\langle n \rangle$ from one phase to another when μ is fixed. Then, phase separation is associated with the density-driven phase transition.

Let us discuss the issue from a different viewpoint. In Fig. 15 we show the equal time spin correlation

$$S(q) = \frac{1}{N} \sum_{ij} \langle \vec{S}_i \cdot \vec{S}_j \rangle e^{iq(i-j)} \tag{66}$$

on a one dimensional system at $L = 40$, $J_\text{H}/W = 4$ and $\beta = 150 W^{-1}$. We clearly see the crossover from the antiferromagnetic state at $n = 1$ with a peak of $S(q)$ at $q = \pi$, and the ferromagnetic state at $n \sim 0.7$ with the peak at $q = 0$. Around $n \sim 0.9$, where large change of $n(\mu)$ is observed, we see two peak structure at $q = \pi$ and $q = 0$. This result is consistent with mixed phase of ferromagnetic and antiferromagnetic states. On the other hand, magnetic states with incommensurate momentum, typically at $q = 2k_\text{F}$, have been observed for the weak coupling region $J_\text{H} \lesssim W$. Thus the tendency toward phase separation is prominent at the strong coupling region.

Recently, the issue of phase separation is also investigated by various methods.[45-47] The model in the weak coupling region $J_\text{H} \ll W$ has also been studied.[48] However, the mechanism of the phase separation might be different from the strong coupling region where the half-metallic DOS plays an important role. In manganites, several experiments claim the existence of phase separation.[49,50] We will later discuss this issue in comparison with experiments in detail.

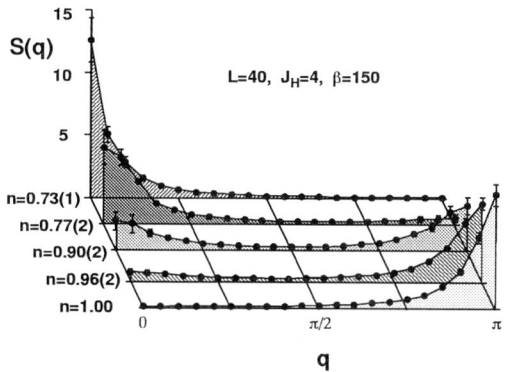

Figure 15: Equal time spin correlation for various electron concentration. Shaded area are guides to eyes. Error bars of the concentrations in the last digits are given in parenthesis.

IV. 1/S CORRECTIONS

Spin wave expansion. We introduce a linear spin wave theory. The spin wave operators are introduced from

$$S_i^+ \simeq \sqrt{2S} a_i, \qquad S_i^- \simeq \sqrt{2S} a_i^\dagger, \qquad S_i^z = S - a_i^\dagger a_i. \tag{67}$$

Hereafter we restrict ourselves to the lowest order terms of the $1/S$ expansion at $T = 0$, where the localized spins are perfectly polarized so that $\langle a_i^\dagger a_i \rangle = 0$. We consider the half-metallic ground state, i.e. $f_{k\downarrow} = 0$ where $f_{k\sigma}$ is the Fermi distribution function.

The spin wave self-energy in the lowest order of $1/S$ expansion is obtained diagrammatically as[12]

$$\begin{aligned}\Pi(q,\omega) &= -\frac{2J_H^2}{S} \frac{1}{N\beta} \sum_{k,n} G_\uparrow(k, i\omega_n) G_\downarrow(k+q, i\omega_n + i\nu) \\ &\quad - \frac{J_H}{S} \frac{1}{N\beta} \sum_{k,n} (G_\uparrow(k,i\omega_n) - G_\downarrow(k,i\omega_n)) e^{i\omega_n 0^+} \\ &= \frac{1}{SN} \sum_k f_{k\uparrow} \left(J_H - \frac{2J_H^2}{2J_H - (\omega + \varepsilon_k - \varepsilon_{k+q})} \right).\end{aligned} \tag{68}$$

Here, $G_\sigma(k, i\omega_n) = (i\omega_n - \varepsilon_k + \sigma J_H)^{-1}$ is the fermion Green's function, and $\beta = 1/T$.

The spin wave dispersion relation ω_q is obtained self-consistently from $\omega_q = \Pi(q, \omega_q)$. We have

$$\omega_q = \frac{1}{SN} \sum_k f_{k\uparrow} \left(J_H - \frac{2J_H^2}{2J_H - (\varepsilon_k - \varepsilon_{k+q})} \right) + O(1/S^2), \tag{69}$$

where $f_{k\uparrow}$ is the fermi distribution function of the majority band.

Let us consider the strong Hund coupling limit $J_H \gg t$. If we assume a simple cubic lattice with nearest-neighbor electron hopping,

$$\varepsilon_k = -2t \left(\cos k_x + \cos k_y + \cos k_z \right), \tag{70}$$

we have

$$\omega_q \simeq E_{sw} \frac{3 - \cos q_x - \cos q_y - \cos q_z}{6} \tag{71}$$

where $E_{\text{sw}} \equiv \omega_{q=Q} - \omega_{q=0}$ is the spin wave bandwidth, given by

$$E_{\text{sw}} = \frac{6t}{SN} \sum_k f_{k\uparrow} \cos k_x. \tag{72}$$

We see that in the strong coupling region the spin wave bandwidth is determined only by the electron transfer energy.

In the isotropic case, the spin stiffness is defined via $\omega_q = Dq^2$ in the long wavelength limit $q \to 0$. From eq. (69) we have

$$D = \frac{1}{2S} \frac{1}{N} \sum_k f_{k\uparrow} \left[\frac{1}{2} \frac{\partial^2 \varepsilon_k}{\partial k^2} - \frac{1}{2J_{\text{H}}} \left(\frac{\partial \varepsilon_k}{\partial k} \right)^2 \right]. \tag{73}$$

We emphasize that the expansion given here is not with respect to $J_{\text{H}}/(tS)$ but to $1/S$, due to energy denominator $2J_{\text{H}}$ between up- and down- spin electrons. This is understood from the fact that J_{H} plays an role of a projection and does not enter the energy scale by itself in the limit $J_{\text{H}} \to \infty$. Therefore, the calculation is valid even in the large Hund's coupling limit $J_{\text{H}} \gg t$ as long as we restrict ourselves to $T = 0$. Indeed, the result for $J_{\text{H}} \to \infty$ obtained by the present approach is equal to those in the projection limit $J_{\text{H}} = \infty$ shown by Kubo and Ohata.[17]

Discussion. In the DE model with $J_{\text{H}} \gg W$ we observe a short-range spin interaction, in contradiction with the case of $J_{\text{H}} \ll W$ (s-d model) where the well-known RKKY interaction is long ranged with a power-law decay. The difference comes from the electronic structure.

In the case $J_{\text{H}} \ll W$ where electronic DOS is the ordinary one (not half-metal), the gapless quasi-particle excitation of the electronic part creates a particle-hole spin excitation channel with $2k_{\text{F}}$ singularities. Interactions among localized spins are mediated by this gapless channel and thus long ranged.

On the other hand, for the DE model with $J_{\text{H}} \gg W$ the half-metallic DOS creates a gap due to J_{H} splitting, and unlike the former case the particle-hole spin channel is massive and short ranged. The qualitative difference comes from the half-metallic structure. In the real-space picture, we consider a perfectly spin polarized state at $T = 0$ and twist a spin at site i_0. Spin polarization of itinerant electrons are along the total polarization axis except for the i_0-th site where it orients toward the local spin direction $\vec{S}_{i_0}$. In the strong coupling limit $J_{\text{H}} \gg W$, the electron at site i_0 is localized because it has different spin orientation from the spins in neighboring sites. Since the effective interaction between localized spins are mediated by the motion of electrons, the effective spin-spin interaction is short ranged. As J_{H}/W increases, electrons become more localized so the range of effective interaction becomes shorter. In the extreme limit $J_{\text{H}} \to \infty$, the interaction is nonzero only for the nearest neighbors which gives a cosine-band dispersion.

Let us discuss the higher orders of $1/S$ expansion terms. In an ordinary (insulating) Heisenberg ferromagnet, the first $1/S$ term is the relevant term with respect to the one-magnon dispersion ω_q. Higher orders of $1/S$ expansion only give magnon-magnon interaction terms, and thus irrelevant within the one magnon Hilbert space. However, for the DE model, higher order terms also give one-magnon kinetic terms. We need to take into account the asymptotic $1/S$ expansions even for the one-magnon dispersion relations. Such higher order terms may be considered as vertex corrections to the self-energy term $\Pi(q, \omega)$.

Numerically, Kaplan et al.[51] studied the $S = 1/2$ case at $J_H \to \infty$. They observed cosine-band type behaviors in the well-doped cases, and the deviation from them in the limit $n \to 0$ and $n \to 1$. The result might be understood from the Migdal's discussion. Let us consider the electron kinetic energy E_{kin} and the energy scale of magnons $\langle \omega_q \rangle$ of the DE model. The deviation from a cosine-band comes from the vertex correction which is relevant if $E_{kin} \lesssim \langle \omega_q \rangle$, which occurs in the lightly hole/electron doped region $n \to 1$ or $n \to 0$. On the other hand, in the well-doped region we have $E_{kin} \gtrsim \langle \omega_q \rangle$ and the vertex corrections are small. In the region where doped manganites show ferromagnetism, vertex corrections do not seem to be important. This explains the consistency between experiments and $1/S$ results.

V. COMPARISON WITH EXPERIMENTS

Abbreviations: Hereafter we use these abbreviations:
$La_{1-x}Sr_xMnO_3$ (LSMO), $La_{1-x}Pb_xMnO_3$ (LPMO), $La_{1-x}Ca_xMnO_3$ (LCMO), $Pr_{1-x}Sr_xMnO_3$ (PSMO), $Nd_{1-x}Sr_xMnO_3$ (NSMO), $Pr_{1-x}Ca_xMnO_3$ (PCMO).

5.1 Experimental –varieties of properties in "CMR manganites"–

Let us briefly mention the varieties of phenomena in CMR manganites. For a review of recent experiments, readers are referred to ref. 7. It is emphasized that systematic studies of A-site substitution is quite important to understand the complex behaviors in manganites. Extrinsic effects due to grain/domain boundaries are also discussed.

A-site substitution. Recent improvements in precise control of the A-site cations substitutions in $AMnO_3$ revealed a complex phase diagram as a function of substitution, temperature and magnetic field. They exhibit various phases with magnetic, charge, orbital and lattice orderings. For example, phase diagram for doping (x) vs. temperature (T) is well known for $La_{1-x}Ca_xMnO_3$ (LCMO),[52] as well as $La_{1-x}Sr_xMnO_3$ (LSMO),[53] $Nd_{1-x}Sr_xMnO_3$ (NSMO) and $Pr_{1-x}Ca_xMnO_3$ (PCMO).[54] Effects of A-site substitution is also studied for a fixed doping.[55-57]

Major effects of A-site substitution are the bandwidth control and the carrier number control. It is well understood that the ratio of rare-earth (3+) ions and alkaline-earth (2+) ions determines the nominal values of the carrier number x. At the same time, change of the average radius of the A-site ions $\langle r_A \rangle$ by chemical substitutions gives the "bandwidth control" through the chemical pressure. Such kind of chemical control creates a large change in the nature of the compounds. In general, compounds with larger $\langle r_A \rangle$ have higher T_c.[55-57] It is considered to be due to wider effective bandwidth for e_g electrons in larger $\langle r_A \rangle$ compounds, since it gives less Mn-O octahedra tilting.

However, we should note that it is still controversial whether the phase diagram is controlled mostly by the bandwidth alone. For example, ionic size variation $\sigma(r_A)$ also plays some role to change T_c,[58] as well as the fact that decrease of T_c for small $\langle r_A \rangle$ is substantially larger than the estimated value from the change of bond angles.

Let us concentrate on the region $x \sim 1/3$ where it is far from antiferromagnetic insulating phase at $x \sim 0$ and the region with charge and orbital ordering at $x \sim 0.5$. The compounds are roughly classified as follows:

- High T_c compounds: e.g. LSMO.
  A canonical example for the high T_c compounds is (La,Sr)MnO$_3$ (LSMO) with $T_c \sim 380K$. Resistivity shows a small value at lowest temperature ($\rho_0 \sim 10^2 \mu\Omega$cm).

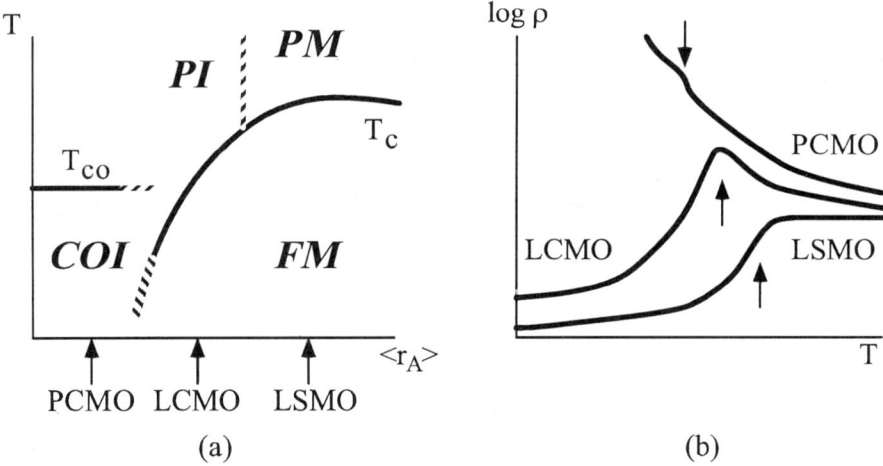

Figure 16: (a) Schematic phase diagram at $x \sim 1/3$ by ionic radius of A-site $\langle r_A \rangle$. Abbreviations are: paramagnetic metal (PM), paramagnetic insulator (PI), ferromagnetic metal (FM), charge ordered insulator (COI), as well as Curie temperature (T_c) and charge ordering temperature (T_{CO}). (b) Qualitative behaviors in $\rho(T)$ for $(La,Sr)MnO_3$ (LSMO), $(La,Ca)MnO_3$ (LCMO) and $(Pr,Ca)MnO_3$ (PCMO) at $x \sim 1/3$. Uparrows ($\uparrow$) in the figure show T_c, while downarrows ($\downarrow$) indicates T_{CO}.

At $T \sim T_c$, $\rho(T)$ takes much larger value but still in the order of Mott's limit $\rho = 2 \sim 4 m\Omega cm$. Above T_c, $\rho(T)$ shows a metallic behavior, i.e. $d\rho(T)/dT > 0$. Namely, this compound is a good metal below T_c and become an incoherent metal above T_c with the absolute value for $\rho(T)$ being near Mott's limit.[53,59]

- Low T_c compounds: e.g. LCMO, PSMO.
  LCMO is the most well-investigated compound. Ca substitution creates smaller $\langle r_A \rangle$ and larger $\sigma(r_A)$. It has lower $T_c \sim 280K$ compared to LSMO, and shows metal to insulator transition at around T_c.[52,4,5,60]

- Compounds with charge ordering instability: e.g. PCMO.
  As $\langle r_A \rangle$ is further decreased, compounds show a charge ordering at $T \sim 200K$.[61,54,62,7] In the zero-field cooling process, ferromagnetic metal phase does not appear.

At $x \sim 1/2$, the phase diagram becomes more complex. At the lowest temperature, the tendency of the competition between ferromagnetism and charge ordering[63,64] driven by $\langle r_A \rangle$ control remains the same. It has been recently discovered that in the intermediate region, a new phase of A-type antiferromagnetic metal region exists in the narrow vicinity of $x = 1/2$.[65,66] Temperature dependence is also complex. Behaviors of the resistivity above T_c are roughly the same with the case of $x \sim 1/3$. For example, metallic behavior above T_c is observed for NSMO at $x = 1/2$ as well as compounds with larger $\langle r_A \rangle$. Substitution of Nd by Sm reduces $\langle r_A \rangle$ and makes the paramagnetic phase insulating.[67]

Extrinsic effects in polycrystal samples. In samples with multiple grain structures, it has been shown that there exist so-called tunneling magnetoresistance (TMR) phenomena through the spin valve mechanism.[60,68-70] Magnetoresistance of the material with artificially controlled grain/interface boundaries are also studied to realize a low-field MR device through TMR.[42,71,72]

In manganites, the half-metallic behavior due to DE interaction is considered to cause a large amplitude in MR. Spin polarizations in grains are schematically depicted in Fig. 17. In the zero field case, each grain have different spin orientations. Namely, the spins of the half-metallic electrons are different from one grain to another. Due to the nearly perfect polarization nature of itinerant electrons, inter-grain hopping amplitude is suppressed by such random spin polarizations. Under the magnetic field, spin polarization of grains become parallel to the external field. Inter-grain electron hopping becomes larger in this case. For multi-grain system with half-metallic states such spin valve phenomena becomes prominent and gives MR effect in a low field range.

Resistivity in polycrystal samples seems to be dominated by such extrinsic effects. One should be careful about discussing the experimental data from a microscopic point of view. Such TMR behavior in these perovskite manganites should be discussed in relation with other half-metallic materials such as CrO_2 or $Tl_2Mn_2O_7$.[39,73]

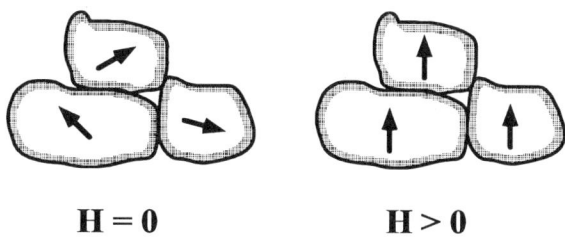

Figure 17: Grains of the polycrystal samples. (a) at zero magnetic field $H = 0$, (b) under magnetic field $H > 0$. Arrows indicate the orientation of the magnetization for each grains. For the relation with low field TMR, see the text.

5.2 Comparison with high Curie temperature compounds

Here, we will show the comparison of theoretical results with experimental data of high T_c compounds such as LSMO and LPMO. We discuss that the DE Hamiltonian alone explains most of the thermodynamics of these manganites, including Curie temperature and resistivity.

Curie temperature. In Fig. 18 we plot the x dependence of the Curie temperature for LSMO, together with the fitting curves obtained by DMF.[9] From the fitting, we see that with parameters $W \sim 1eV$ and $J_H/W \sim 4$ the value of T_c as well as its x-dependence is reproduced. The bandwidth of $W \sim 1eV$ is a typical value for 3d transition metal oxides, and is consistent with the band calculation estimate for manganites.[74-76] It is also consistent with the value obtained from the spin wave dispersion fit,[12] shown later in this section.

Thus the value of T_c in LSMO is reproduced by the DE model alone. This result is consistent with other methods, e.g. high-temperature expansion by Röder et al.[78] as well as by the Monte Carlo method.[16]

Millis et al.[6] discussed that the magnitude of T_c as well as its x dependence for LSMO cannot be accounted for by the DE model alone, and discussed the importance of the dynamic Jahn-Teller effect. However, this part of their discussion is due to an

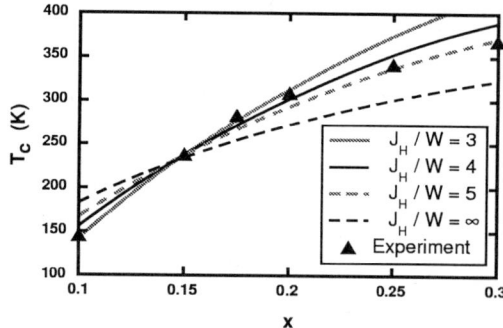

Figure 18: Curie temperature T_c of the DE model in comparison with those for La$_{1-x}$Sr$_x$MnO$_3$. Experimental data are from ref. 77.

inappropriate estimate of T_c which is based on a calculation of effective spin coupling at $T = 0$. In the itinerant systems, T_c is reduced by spin fluctuation, in general. Although T_c scales with W, it is quite small compared to W due to the prefacter, whose typical value is $T_c \lesssim 0.05W$ (See Fig. 2). Thus bandwidth of 1eV creates T_c of the order of room temperature. Decrease of T_c as x decreases is explained by the reduction of the kinetic energy.[34]

Resistivity and magnetoresistance. Resistivity of the high-T_c compounds in single crystals at sufficiently large doping $x \sim 1/3$ are different from those of low-T_c compounds or polycrystal samples. For LSMO at $x \sim 1/3$,[53] residual resistivity ρ_0 is in the order of a few $10\mu\Omega$cm, and the temperature dependence $\rho(T)$ shows a monotonously increase, i.e. $d\rho/dT > 0$ even above T_c. (For LPMO, see ref. 79. Recent experiment by Cheong et al.[59] shows that the resistivity of LSMO continuously increase up to 1000K, without saturation or metal-semiconductor transition.) The value of resistivity at T_c is typically $\rho(T_c) = 2 \sim 4$mΩcm, which is in the order of the Mott limit. In short, LSMO is a good metal at $T \ll T_c$, and an incoherent metal at $T \gtrsim T_c$. The DE model reproduces these data (See Fig. 9). Similar temperature dependences of resistivity are observed in a wide class of materials of half-metals such as CrO$_2$ and Heusler alloys.[39]

It has been discussed that the DE model cannot explain the resistivity of LSMO in its absolute value as well as the temperature dependence.[6] However, it is now clear that if one compares data for a high quality single crystal of LSMO (not polycrystal, or other compounds with lower T_c), DE alone does account for the resistivity.

In Fig. 19 we show magnetoresistance of LSMO at $x = 0.175$. Universal behavior of the magnetoresistance in the form

$$-\Delta\rho/\rho_0 = CM^{*2} \tag{74}$$

is observed, where $C \sim 4$ is temperature/field independent constant.[53,77] Namely, the resistivity is directly related to the magnetization. In Fig. 19 we also show the DMF results. The curve reproduces the experimental data.[11] The relation $\Delta\rho/\rho(0) = -CM^2$ is obtained theoretically in eq. (60). The value $C > 1$ shows that the system is in the strong coupling region.

Note that in LCMO polycrystal samples, a different scaling relation

$$\rho(M) = \rho(0)\exp(-\alpha M^2/T) \tag{75}$$

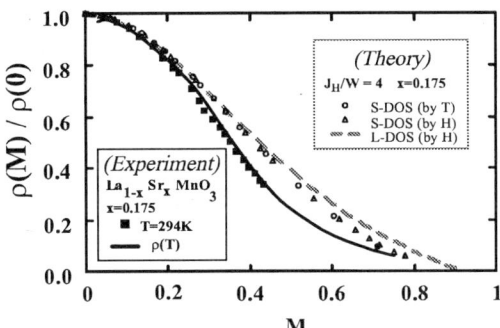

Figure 19: Magnetoresistance, plotted by magnetization vs. resistivity. Solid curves and filled symbols are the data for $La_{0.875}Sr_{0.175}MnO_3$ from ref. 77. Open symbols and grey curve show DMF results for $J_H/W = 4$ and $x = 0.175$, for semicurcular (S-DOS) and Lorentzian (L-DOS).

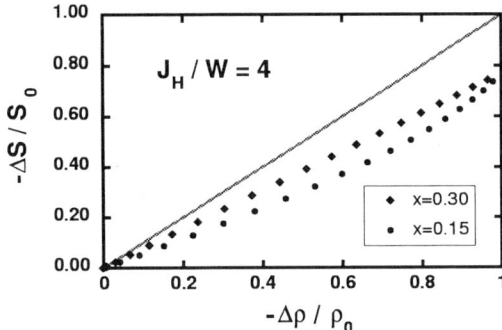

Figure 20: Seebeck coefficient under magnetic field, scaled by resistivity. The line is a guide to eyes.

has been reported.[80] From such an activation-type temperature dependence, polaronic origin of the resistivity has also been discussed. The difference between the scaling relation in LSMO (74) and LCMO (75) indicates the qualitative difference for the origin of CMR.

Thermoelectric power. Seebeck coefficient S for LSMO has been reported.[81] They show non-universal behaviors, including the change of the sign. However, in the vicinity of T_c, a scaling behavior in the form $-\Delta S/S(0) \simeq -\Delta\rho/\rho(0)$ irrespective of doping is reported, where ΔS and $\Delta\rho$ are the change of Seebeck coefficient and resistivity under magnetic field, and $S(0)$ and $\rho(0)$ are their zero-field values, respectively.[81] In Fig. 20 we plot the DMF result for the Seebeck coefficient S on a Lorentzian DOS at $J_H/W = 4$. The data is plotted in the form $-\Delta S/S(0)$ vs. $-\Delta\rho/\rho(0)$. We see $-\Delta S/S(0) \simeq -\Delta\rho/\rho(0)$ for different values of x.

Spin excitation. From the neutron inelastic scattering experiment, spin wave dispersion relation of $La_{0.7}Pb_{0.3}MnO_3$ (LPMO at $x = 0.3$) is investigated.[82] LSMO in the ferromagnetic metal region has also been studied.[14,83-85]

Perring et al.[82] found that the experimental data of spin wave dispersion relation for LPMO fits the cosine band form. They phenomenologically argued that a ferromagnetic

Heisenberg model with nearest-neighbor spin exchange couplings $2J_{\text{eff}}S \sim 8.8\text{meV}$ is a good candidate for the effective spin Hamiltonian, although the material is a metal.

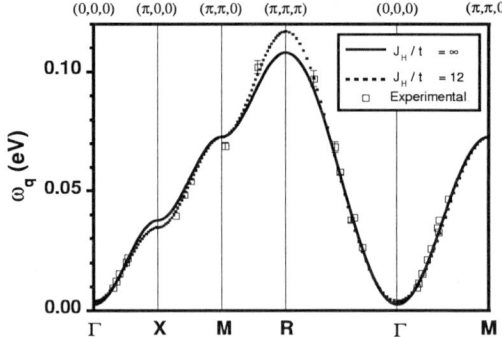

Figure 21: Spin wave dispersion at $x = 0.3$. The experimental data for $\text{La}_{0.7}\text{Pb}_{0.3}\text{MnO}_3$ is from ref. 82.

Now we discuss the DE model results.[12,51,86] In Fig. 21, we plot the theoretical results together with the neutron inelastic scattering experiment data. As we have discussed previously, cosine-band dispersion is obtained in the limit $J_H \to \infty$. This gives the identical fit with the analysis by Perring *et al.*, and the fitting parameter determines electron hopping as $t \sim 0.3\text{eV}$. Here, finite gap at $q = 0$ is artificially introduced as done in ref. 82, due to some experimental inaccuracies.

At the zone boundary $q \sim (\pi, \pi, \pi)$ we see the deviation from the cosine band, or softening of the dispersion. The data is well accounted for by introducing some finite value for J_H/t. In Fig. 21 we show the result for $J_H/t = 12$. From fitting we obtain $t \sim 0.26\text{eV}$ and $J_H \sim 3.1\text{eV}$. Although it is not easy to estimate the systematic errors of the fitting, it is plausible to say that the DE model with $t = 0.2 \sim 0.3\text{eV}$ and $J_H \gtrsim 3\text{eV}$ explains the spin wave dispersion relation of LPMO. These values are the effective hopping energy of the double-exchange model with reduced degrees of freedom. Nevertheless, these values of t are consistent with those estimated from the band calculation.[74-76]

Anomalous damping of zone boundary magnons are also reported.[82] Within the DE model, spin wave excitation at finite temperature interacts with the Stoner continuum as shown in Fig. 12. From eq. (65), we see that

$$\Gamma(q,T) \propto (1 - M^{*2})\omega_q, \tag{76}$$

where $\Gamma(q,T)$ is the linewidth (inverse lifetime) of the magnon. This explains that the magnons are damped at finite temperature, especially at the zone boundary. Indeed such a scaling relation (76) is observed in LSMO.[14]

Optical conductivity. Optical conductivity for $\text{La}_{0.6}\text{Sr}_{0.4}\text{MnO}_3$ by Moritomo *et al.*[87] is shown in Fig. 22(a). Here, temperature independent part which is discussed to be due to d-p charge-transfer type excitation is subtracted. There exists a peak at $\omega \sim 3\text{eV}$, and its temperature dependence is in a way that it vanishes at $T \to 0$.

By choosing $W = 1\text{eV}$, $J_H = 1.5\text{eV}$ and $x = 0.4$, it is possible to reproduce the experimental data by the DE model. In Fig. 22(b), we show the DMF results for the semicircular DOS. The peak structure as well as its temperature dependence is in agreement.

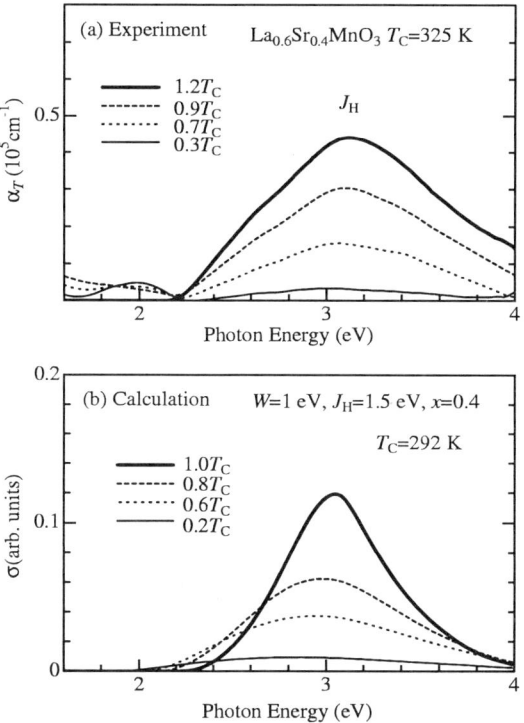

Figure 22: Optical conductivity. (a) Experimental data for $La_{0.6}Sr_{0.4}MnO_3$ taken from ref. 87, where temperature independent part is subtracted as a background. (b) DMF result for $W = 1eV$, $J_H = 1.5eV$ and $x = 0.4$.

Let us discuss more details about the scaling relation.[15,87] Experimentally, the integrated spectral weight

$$S_{\text{exp}} \equiv \int_{2.2\text{eV}}^{4.0\text{eV}} d\omega \sigma(\omega) \tag{77}$$

shows a scaling relation

$$S_{\text{exp}} \propto 1 - M^{*2}. \tag{78}$$

This is explained by the transfer of DOS by magnetic fluctuation in the DE model, as shown in eq. (63). (See also Fig. 10 in §3.3.)

Thus, optical spectra show experimental evidence for the shift of DOS which is a typical phenomena in DE model, and give the estimate of the parameter $J_H \sim 1.5$eV.

Summary: LSMO as a canonical DE system. As long as the high-T_c compounds are concerned, e.g. LSMO at $x = 0.2 \sim 0.4$, the DE Hamiltonian accounts for various experimental data concerning the ferromagnetism and the transport

- Curie temperature.[9,16,78]

- Resistivity (absolute value and magnetoresistance).[8,11]

- Spin and charge excitation spectra.[11,12]

- Scaling behaviors in charge and spin dynamics.[14,15]

The mechanism of MR in the single-crystal of these high T_c compounds are understood from double-exchange alone, namely due to spin disorder scattering.[8,17,43,44,88]

There still exist controversies with some experiments and the DE model. An example is the loss of spectral weight near the Fermi level observed by photoemission experiments,[40,89] the other is the optical conductivity spectrum[90] at $\omega \lesssim 1\,eV$ which shows substantial loss of the Drude weight (integrated oscilator strength in the low frequency region) in contradiction with the specific heat behavior which does not show such a large mass enhancement. It should be noted that these experiments are highly sensitive to surface states, and there always remains an open question whether these experiments really measure the bulk states or are merely observing the surface states. Recently, Takenaka et al.[91] reported that the optical spectrum of the "clean" surface prepared by cleaving shows larger Drude part compared to those prepared by surface filing, and is consistently large with respect to the specific heat estimate. Although the issue is still far from conclusive results, we should be very careful about such experimental details.

Let us discuss the value of J_H in manganites estimated from these calculations. From experiments concerning spin and charge dynamics, data fitting gives $J_H = 1.5 \sim 2eV$. On the other hand, static and low frequency experiments such as T_c and $\rho(T)$ measurements are well reproduced by $J_H = 3 \sim 5eV$. Such a change in the value of J_H may be understood by the renormalization of interaction. While high energy experiments (such as $\sigma(\omega)$) directly measure the bare value of J_H, low frequency measurement is affected by the dressed quasiparticle involving the vertex renormalization. In this case, particularly, the effect of Coulomb interaction should be important. Since our DE Hamiltonian does not involve on-site Coulomb repulsion terms, elimination of the double occupancy of itinerant electrons are underestimated. This is compensated by increasing the value of J_H which also energetically prohibits the on-site double occupancy. Thus the vertex renormalization due to Coulomb repulsion may increase the value of J_H in the low-frequency scale.

Discussion: Absence of polaronic behaviors in LSMO. Millis et al.[6] discussed the discrepancies in DE model and experiments of LSMO with respect to (i) Curie temperature (T_c), (ii) temperature dependence of resistivity ($\rho(T)$), and (iii) absolute value of resistivity ($\rho(T \sim T_c)$). They concluded that DE alone does not explain the MR of LSMO.

What we have shown here is, however, that the accurate calculation of the DE model and the measurements of high quality crystal samples indeed give consistent results. Polaron effects, if any, should be small enough to be irrelevant. For example, Röder et al.[92] discussed the decrease of T_c due to the reduction of electron hopping caused by the polaronic effect. Thus, if polaronic interaction is included to the DE model hamiltonian, T_c of the model becomes inconsistently small compared to the experimental value.

Another point to be discussed is the experimental difference between LSMO and LCMO (see also §5.1). Since LCMO shows different behaviors in resistivity, e.g. temperature dependence of $\rho(T)$ as well as the MR scaling relations, different physics must be taking place. If one assumes the polaronic arguments, crossover from metallic (LSMO) to insulating (LCMO) behavior with respect to the resistivity above T_c could be understood in a following way. Insulating behavior might be due to small-polaron formation in systems with large electron-lattice coupling, while systems with smaller electron-lattice coupling might show large-polaron behaviors.

However, this argument is in contradiction with experimental data. Pair distribution function (PDF) measurement of atomic displacements observed by inelastic neutron scattering[93] shows that there exist large Jahn-Teller distortions in a dynamic way even in the metallic phase of LSMO above T_c. As long as high T_c compound LSMO is concerned, lattice distortion does not seem to affect transport properties as well as magnetic behaviors (e.g., T_c). Since large lattice distortion is observed both in LSMO and LCMO,[94,95] polaronic discussion alone is not sufficient to explain the qualitative difference in $\rho(T)$.

5.3 CMR effects in low Curie temperature compounds

What is the origin of the difference in resistivity behavior of LCMO compared to LSMO? As shown in §5.2, the DE model is sufficient to explain LSMO, and should be a good reference model to discuss LCMO. We discuss the half-metallic behavior in ferromagnetic states and the competition between ferromagnetic metals and charge ordered insulators.

Theories of CMR based on the change in transfer integral. Several theories based on microscopic models have been proposed to explain the CMR phenomena in LCMO. Polaron effects of Jahn-Teller distortions are introduced to explain the metal-insulator transition from the point of view of large polaron to small polaron crossover by magnetism.[92,96,97] The idea of Anderson localization due to spin disorder as well as diagonal charge disorder has also been discussed.[34,88,98-102]

Many of these proposals relate magnetism and transport through the change of the hopping matrix element

$$t_{\text{eff}}(T, H) \propto \cos(\theta/2), \tag{79}$$

where θ is the angle between nearest neighbor spins, as discussed by Anderson and Hasegawa.[2] Namely, in general, these scenarios discuss the existence of the critical value of hopping t_c. Change of the mean magnetic structure controls the value of t_{eff}, and it is considered that some kind of transition occurs at $t_{\text{eff}} = t_c$.

In a small polaron scenario, there exists a competition between the electron kinetic energy $E_{\text{kin}} \propto t_{\text{eff}}$ and the polaron energy $E_{\text{pol}}(\propto \lambda)$. It is characterized by the critical value for hopping t_c where transition from small polaron state (small t_{eff} region) and large polaron state (large t_{eff} region) occurs. Such a change in the behavior of carriers affect the conductivity through the self trapping mobility or the percolation of localized carriers. In the context of Anderson localization, critical value t_c is determined by the relationship between the mobility edge and the Fermi level. In either cases, electron hopping t_{eff} is considered to be controlled by the magnetic structure through the Anderson-Hasegawa's DE mechanism. In the presence of the competition between extended and localized states, metallic states are realized at the large hopping region $t_{\text{eff}}(T, H) > t_c$, and the system becomes insulating at small hopping region $t_{\text{eff}}(T, H) < t_c$.

However, in order to explain the experiments for resistivity, these scenario need additional explanations to justify themselves. Let us point out two issues here:

- Does the transition point t_c really exist at the well-doped system $x = 1/3 \sim 1/2$?

- Why does the "metal-insulator" transition occur only in the vicinity of T_c, irrespective of carrier concentration and bandwidth?

Localization phenomena in low carrier concentration is explained by the suppressed overlaps of wavefunctions. For example, in a low carrier small polaron system, a self-trapped polaron has a short confinement length scale compared to mean polaron-polaron distances. On the other hand, if we consider the region $x \sim 1/3$, every two sites out of total six neighbors in a cubic lattice is occupied by other polarons. In such a high density system, quantum mechanical overlaps of polaron wavefunctions have to be quite large. In other words, in order to assume a localized state, the gain of self-trapping potential energy has to be unrealistically large to compensate the loss of kinetic energy by such a localization with small length scale. For example, dynamic Jahn-Teller scenario requires that the polaron binding energy be larger than the electron kinetic energy to make insulating behaviors for the carrier doped case.[97] Thus, it is difficult to understand whether the localized state really exists in a realistic parametrization of the Hamiltonian.

Another point which is hard to explain is the fact that the metal-insulator transition always occurs in the vicinity of T_c for various A-site substitutions in both ionic radius and average valence changes. In order to explain the metal-insulator transition scenarios require the "pinning" of the critical point at $T \sim T_c$, i.e., irrespective of carrier concentration as well as A-site bandwidth and randomness, $t_c \sim t_{\text{eff}}(T = T_c)$ always has to be satisfied. Note that t_{eff} is determined by the short range spin correlation $\langle S_i \cdot S_j \rangle$ and is a smooth and continuous function of temperature without an anomaly at T_c. It is also required to explain the complete absence of the metal-insulator transition in high-T_c compounds such as LSMO at $x \sim 1/3$ within some realistic parametrizations.

Magnetic inhomogeneities and the nanodomain TMR mechanism. In manganites, especially in the low-T_c regions, magnetic inhomogeneities are experimentally observed. Here we discuss the importance of such inhomogeneities.

Unconventional feature in the low-T_c compounds is the presence of the central peak well below T_c,[103] which indicates the presence of the magnetic cluster and its diffusive dynamics. From the spin diffusion constant, the correlation length of the spin clusters are estimated to be $\xi \sim 10$Å. Neutron elastic scattering measurements also observed the ferromagnetic cluster with correlation length $\xi \sim 20$Å.[104,105]

Magnetic inhomogeneities also show up at the linewidth of the spin wave $\Gamma_q \sim q^2$,[84] which systematically increases as T_c is suppressed by A-site substitution. It is speculated to be due to inhomogeneity effect through spin stiffness distribution.[14,15] For lower T_c compounds, there exist much prominent broadening of the spin wave dispersion at the zone boundary.[106]

Optical conductivity measurements[56,107-110] show that the A-site substitution causes changes in the spectrum at the peak structure around ~ 1.5eV and the infrared quasi-Drude (incoherent) structures. Formation of lattice polaron at lower T_c compounds which causes $\omega \sim 1.5$eV peak is discussed. Inhomogeneities in charge and magnetic structures observed by μSR and X-ray measurements[111-113] suggest that such polaronic cluster remain even at at low temperatures. Possible micrograin formation due to charge segregation as well as phase separation between ferromagnetic and antiferromagnetic domains has also been discussed.[49,50,114]

Thus in most compounds inhomogeneous behaviors are observed experimentally. The tendency is that inhomogeneity is more prominent in the compounds with lower T_c due to smaller $\langle r_A \rangle$ (or larger $\sigma(r_A)$ as well).

Let us now focus on the case $x \sim 1/3$. As discussed in §5.1, decrease of T_c is much larger than the estimate from the reduction of the bandwidth by $\langle r_A \rangle$. Inhomogeneity may be playing an important role to this behavior. Since wide bandwidth compound LSMO shows a ferromagnetic metal phase while narrow band compound PCMO is a charge-ordered insulator, there exist a competition as well as a bistability of these phases in the intermediate bandwidth region. Inhomogeneous behaviors in LCMO suggests that coexistence of microscopic domains with ferromagnetic and charge-ordering correlations might happen. In the paramagnetic phase, neither correlations are long-ranged in a macroscopic sense. They should be short ranged and/or dynamic.

In such nanodomain structures with microscopic phase separation, magnetic phase transition occurs when inter-domain correlations become long ranged. Although intra-domain correlations may begin at higher temperature determined by the DE mechanism, true long range order is controlled by the domain-domain interactions mediated by the junction structures. Intermediate region with charge ordering reduces the magnetic coupling between ferromagnetic nanodomains. Then T_c should be substantially reduced. Such nanodomain structure creates two different energy scales for magnetisms, *i.e.* intra-domain and inter-domain interactions.

It also shows up in two correlation lengths. One is the intra-domain correlation length, which becomes the domain size at low temperature regions. The other is the inter-domain correlation length which is the length scale to determine the nature of the magnetic phase transition. It has been reported that in the low T_c compounds the magnetic correlation length does not diverge at T_c but stays constant, *e.g.* ~ 20Å for NSMO.[105] This unconventional behavior is understood if the typical length scale of the nanodomain structure is ~ 20Å. Due to the resolution of the triple-axis experiment, it seems that the macroscopic correlation length which should diverge at T_c was not detected.

Now we discuss the relationship with CMR phenomena. In the presence of nanodomain structures with ferromagnetic metals and (charge ordered) insulators, several mechanisms create magnetoresistance. A possible mechanism is the percolation of metallic nanodomains. If the external magnetic field is applied, the system gains energy by increasing the volume fraction of the ferromagnetic nanodomains. At the percolation threshold, there exists a metal-insulator transition. This scenario is, however, unlikely in the sense that metal-insulator transition occurs only at the percolation threshold, while the CMR phenomena is widely observed for various composition range away from some critical point.

Another possibility to be discussed here is the nanodomain TMR phenomena, as is commonly the case for polycrystals.[68] At $T \sim T_c$, the intra-domain spin correlation is well developed so each nanodomain can behave as a half-metallic domains. Then, application of magnetic field controls the spin valve transports between nanodomains. This scenario is most likely in the sense that it naturally explains generic MR behavior at $T \sim T_c$ and does not assume any critical points. The conductivity is controlled by magnetism through the spin valve channels. Huge sensitivity to the external field is due to the fact that each metallic nanodomain already forms a ferromagnetic cluster. The idea is consistent with the phenomenological explanation of resistivity in LCMO by the Two-fluid model proposed by Jaime and Salomon.[115] They discuss the coexistence of a metallic conductivity path and an activation-type polaronic conductivity.

Let us finally mention the origin of such nanodomain structures. Instability of electronic phase separation[16] is a candidate for the initial driving force for such phenomena, which is stabilized to form droplet structures due to long range Coulomb interactions. Another possibility is the effect of static potential disorder due to A-site cation R^{3+}-A^{2+} distributions which causes charge inhomogeneities,[116] as well as self-trapping effect of lattice polarons.

VI. SUMMARY AND CONCLUDING REMARKS

In this article we discussed the thermodynamics of the DE model. We treated the model at finite temperature non-perturbatively with respect to the spin fluctuations. Various thermodynamic quantities including magnetic and transport properties are calculated. Modifications to the mean-field type treatment by Anderson-Hasegawa and de Gennes are made.

These new results are found to be very important in comparing the experimental data for manganites with the DE model. As long as the high T_c compounds (*e.g.* LSMO) in the metallic phase are concerned, the DE model accounts for various experimental properties including magnetic transition and resistivity. Comparison with respect to low T_c compounds (*e.g.* LCMO) are also discussed.

There still remains many open questions. One is the roles of other interactions such as lattice distortion[6] or orbital fluctuations.[73,117] It is experimentally clear that these are quite important and indeed make some long range orderings in the insulating phases. However, as we have shown in this article, such interactions does not show up in thermodynamic properties in the high T_c metallic phases. It is quite interesting how and why such "screening" of interactions occur in the metallic phase.

In other words, if we start from a realistic model for manganites with orbital and lattice degrees of freedom as well as Coulomb interactions etc., we somehow end up with the double-exchange model in the metallic phase as a consequence of neglecting high-energy excitations. Therefore, we should always regard the model as the renormalized model with parameters $t = t_{\text{eff}}$ and $J = J_{\text{eff}}$ in eq. (3). Or, in a strict sense, we should consider the action of the double-exchange model in eq. (9) with renormalized Green's function G_{eff} and the coupling strength J_{eff}. Such a detailed renormalization studies will help us to understand other intermediate phases.

Another point of interest is to understand the nature of the inhomogeneous systems with nanodomain structures. Such study with respect to both microscopic and mesoscopic length scale might help us to understand the generic features of the physics in strongly correlated oxides.

REFERENCES

[1] C. Zener, Phys. Rev. **82**, 403 (1951).

[2] P. W. Anderson and H. Hasegawa, Phys. Rev. **100**, 675 (1955).

[3] P. G. de Gennes, Phys. Rev. **118**, 141 (1960).

[4] K. Chahara, T. Ohno, M. Kasai, and Y. Kozono, Appl. Phys. Lett. **63**, 1990 (1993).

[5] S. Jin *et al.*, Science **264**, 413 (1994).

[6] A. J. Millis, P. B. Littlewood, and B. I. Shrainman, Phys. Rev. Lett. **74**, 5144 (1995).

[7] A. P. Ramirez, J. Phys. **CM9**, 8171 (1997).

[8] N. Furukawa, J. Phys. Soc. Jpn. **63**, 3214 (1994).

[9] N. Furukawa, J. Phys. Soc. Jpn. **64**, 2754 (1995).

[10] N. Furukawa, J. Phys. Soc. Jpn. **64**, 2734 (1995).

[11] N. Furukawa, J. Phys. Soc. Jpn. **64**, 3164 (1995).

[12] N. Furukawa, J. Phys. Soc. Jpn. **65**, 1174 (1996).

[13] N. Furukawa, J. Phys. Soc. Jpn. **66**, 2523 (1997).

[14] N. Furukawa and K. Hirota, Physica B **241-243**, 780 (1998).

[15] N. Furukawa, Y. Moritomo, K. Hirota, and Y. Endoh, cond-mat/9808076.

[16] S. Yunoki *et al.*, Phys. Rev. Lett. **80**, 845 (1998).

[17] K. Kubo and N. Ohata, J. Phys. Soc. Jpn. **33**, 21 (1972).

[18] Th. Pruschke, M. Jarrell, and J. Freericks, Advs. in Phys. **44**, 187 (1995).

[19] A. Georges, G. Kotliar, W. Krauth, and M. J. Rozenberg, Rev. Mod. Phys. **68**, 13 (1996).

[20] W. Metzner and D. Vollhardt, Phys. Rev. Lett. **62**, 324 (1989).

[21] E. Müller-Hartmann, Z. Phys. **B74**, 507 (1989).

[22] G. Möller, A. E. Ruckenstein, and S. Schmitt-Rink, Phys. Rev. **B46**, 7427 (1992).

[23] Th. Pruschke, D. L. Cox, and M. Jarrel, Europhys. Lett. **21**, 593 (1993).

[24] Th. Pruschke, D. L. Cox, and M. Jarrel, Phys. Rev. **B47**, 3553 (1993).

[25] A. Khurana, Phys. Rev. Lett. **64**, 1990 (1990).

[26] H. Schweitzer and G. Czycholl, Phys. Rev. Lett. **67**, 3724 (1991).

[27] M. J. Rozenberg, G. Kotliar, and X. Y. Zhang, Phys. Rev. **B49**, 10181 (1994).

[28] U. Brandt and C. Mielsch, Z. Phys. B **75**, 365 (1989).

[29] Q. Si, G. Kotliar, and A. Georges, Phys. Rev. **B46**, 1261 (1992).

[30] P. G. J. van Dongen, Phys. Rev. **B45**, 2267 (1992).

[31] A. Georges and G. Kotliar, Phys. Rev. **B45**, 6479 (1992).

[32] M. Jarrel and Th. Pruschke, Z. Phys. **B90**, 187 (1993).

[33] O. Sakai and Y. Kuramoto, Solid State Commun. **89**, 307 (1994).

[34] C. M. Varma, Phys. Rev. **B54**, 7328 (1996).

[35] J. E. Hirsch, D. J. Scalapino, R. L. Sugar, and R. Blankenbecler, Phys. Rev. Lett. **47**, 1628 (1981).

[36] J. E. Hirsch, R. L. Sugar, D. J. Scalapino, and R. Blankenbecler, Phys. Rev. **B26**, 5033 (1982).

[37] K. Kubo, J. Phys. Soc. Jpn. **33**, 929 (1972).

[38] R. de Groot, Phys. Rev. Lett. **50**, 2024 (1983).

[39] V. Yu Irkhin and M. I. Katsnel'son, Physics Uspekhi **37**, 659 (1994).

[40] J.-H. Park *et al.*, Nature **392**, 794 (1998).

[41] J.-H. Park *et al.*, Phys. Rev. Lett. **81**, 1353 (1998).

[42] J. Sun *et al.*, Appl. Phys. Lett. **69**, 3266 (1996).

[43] T. Kasuya, Prog. Theo. Phys. **16**, 58 (1956).

[44] M. E. Fisher and J. Langer, Phys. Rev. Lett. **20**, 665 (1968).

[45] J. Riera, K. Hallberg, and E. Dagotto, Phys. Rev. Lett. **79**, 713 (1997).

[46] M. Kagan, D. Khomskii, and M. V. Mostovoy, cond-mat / 9804213.

[47] D. P. Arovas, G. Gomez-Santos, and F. Guinea, cond-mat/9805399.

[48] E. L. Nagaev, Physica B **230-232**, 816 (1997).

[49] G. Allodi *et al.*, Phys. Rev. **B56**, 6036 (1997).

[50] M. Hennion *et al.*, Phys. Rev. Lett. **81**, 1957 (1998).

[51] T. A. Kaplan and S. D. Mahanti, J. Phys. Cond. Mat. **9**, L291 (1997).

[52] P. Schiffer, A. Ramirez, W. Bao, and S.-W. Cheong, Phys. Rev. Lett. **75**, 3336 (1995).

[53] A. Urushibara *et al.*, Phys. Rev. **B51**, 14103 (1995).

[54] Y. Tomioka *et al.*, Phys. Rev. **B53**, 1689 (1996).

[55] H. Hwang *et al.*, Phys. Rev. Lett. **75**, 914 (1995).

[56] Y. Moritomo, H. Kuwahara, and Y. Tokura, J. Phys. Soc. Jpn. **66**, 556 (1997).

[57] P. Radaelli *et al.*, Phys. Rev. **B56**, 8265 (1997).

[58] L. M. Rodriguez-Martinez and J. P. Attfield, Phys. Rev. **B54**, 15622 (1996).

[59] S.-W. Cheong (unpublished).

[60] G. J. Snyder *et al.*, Phys. Rev. **B53**, 14434 (1996).

[61] Y. Tomioka, A. Asamitsu, Y. Moritomo, and Y. Tokura, J. Phys. Soc. Jpn. **64**, 3626 (1995).

[62] H. Yoshizawa, H. Kawano, Y. Tomioka, and Y. Tokura, J. Phys. Soc. Jpn. **65**, 1043 (1996).

[63] E. O. Wollan and W. C. Koehler, Phys. Rev. **100**, 545 (1955).

[64] J. B. Goodenough, Phys. Rev. **100**, 564 (1955).

[65] H. Kawano *et al.*, Phys. Rev. Lett. **78**, 4253 (1997).

[66] T. Akimoto et al., Phys. Rev. **B57**, 5594 (1998).

[67] H. Kuwahara et al., Phys. Rev. **B56**, 9386 (1997).

[68] H. Hwang, S.-W. Cheong, N. Ong, and B. Batlogg, Phys. Rev. Lett. **77**, 2041 (1996).

[69] A. Gupta et al., Phys. Rev. **B54**, 15629 (1996).

[70] P. Raychaudhuri et al., cond-mat/9807084, to be published in Phys. Rev. B1 Feb. 99.

[71] K. Steenbeck et al., Appl. Phys. Lett. **71**, 968 (1997).

[72] N. D. Mathur et al., Nature **387**, 266 (1997).

[73] D. Khomskii and G. Sawatzky, Solid State Comm. **102**, 87 (1997).

[74] N. Hamada, H. Sawada, and K. Terakura, in *Proc. 17th Taniguchi International Conference*, edited by A. Fujimori and Y. Tokura (Springer Verlag, Berlin, 1995).

[75] W. E. Pickett and D. J. Singh, Phys. Rev. **B55**, 8642 (1996).

[76] D. A. Papaconstantopoulos and W. Pickett, Phys. Rev. **B**, 12751 (1998).

[77] Y. Tokura et al., J. Phys. Soc. Jpn. **63**, 3931 (1994).

[78] H. Röder, R. Singh, and J. Zang, Phys. Rev. **B56**, 5084 (1997).

[79] C. W. Searle and S. T. Wang, Canad. J. Phys. **47**, 2703 (1969).

[80] M. F. Hundley et al., Appl. Phys. Lett. **67**, 860 (1995).

[81] A. Asamitsu, Y. Moritomo, and Y. Tokura, Phys. Rev. **B53**, 2952 (1996).

[82] T. G. Perring et al., Phys. Rev. Lett. **77**, 711 (1996).

[83] M. Martin et al., Phys. Rev. **B53**, 14285 (1996).

[84] A. Moudden, L. Vasiliu-Doloc, L. Pinsard, and A. Revcolevschi, Physica B **241-243**, 276 (1998).

[85] L. Vasiliu-Doloc et al. (unpublished).

[86] X. Wang, Phys. Rev. **B57**, 7427 (1998).

[87] Y. Moritomo et al., Phys. Rev. **B56**, 5088 (1997).

[88] E. L. Nagaev, Phys. Rev. **B58**, 816 (1998).

[89] D. Sarma, Phys. Rev. **B53**, 6873 (1996).

[90] Y. Okimoto et al., Phys. Rev. Lett. **75**, 109 (1995).

[91] K. Takenaka et al. (unpublished).

[92] H. Röder, J. Zang, and A. R. Bishop, Phys. Rev. Lett. **76**, 1356 (1996).

[93] D. Louca et al., Phys. Rev. **B56**, 8475 (1997).

[94] P. Dai et al., Phys. Rev. **B54**, R3694 (1996).

[95] S. J. L. Billinge et al., Phys. Rev. Lett. **77**, 715 (1996).

[96] A. J. Millis, B. I. Shrainman, and R. Mueller, Phys. Rev. Lett. **77**, 175 (1996).

[97] A. J. Millis, R. Mueller, and B. I. Shraiman, Phys. Rev. **B54**, 5405 (1996).

[98] E. Müller-Hartmann and E. Dagotto, Phys. Rev. **B54**, 6819 (1996).

[99] R. Allub and B. Alascio, Sol. State Comm. **99**, 613 (1996).

[100] Q. Li, J. Zang, A. R. Bishop, and C. Soukoulis, Phys. Rev. **B56**, 4541 (1997).

[101] L. Sheng, D. Xing, D. Sheng, and C. Ting, Phys. Rev. **B56**, 7053 (1997).

[102] E. Kogan and M. Auslender, cond-mat / 9807069.

[103] J. Lynn et al., Phys. Rev. Lett. **76**, 4046 (1996).

[104] J. D. Teresa et al., Phys. Rev. **B54**, 12689 (1996).

[105] J. Fernandez-Baca et al., Phys. Rev. Lett. **80**, 4012 (1998).

[106] H. Hwang et al., Phys. Rev. Lett. **80**, 1316 (1998).

[107] S. Kaplan et al., Phys. Rev. Lett. **77**, 2081 (1996).

[108] J. H. Jung et al., Phys. Rev. **B57**, 11043 (1998).

[109] M. Quijada et al., cond-mat/9803201.

[110] A. Machida, Y. Moritomo, and A. Nakamura, Phys. Rev. **B58**, 4281 (1998).

[111] R. H. Heffner et al., Phys. Rev. Lett. **77**, 1869 (1996).

[112] S. Yoon et al., Phys. Rev. **B58**, 2795 (1998).

[113] C. H. Booth et al., Phys. Rev. **B57**, 10440 (1998).

[114] T. Perring, G. Aeppli, Y. Moritomo, and Y. Tokura, Phys. Rev. Lett. **78**, 3197 (1997).

[115] M. Jaime and M. Salamon (unpublished).

[116] C. M. Varma (unpublished).

[117] S. Ishihara, J. Inoue, and S. Maekawa, Physica **C 263**, 130 (1996).

PHASE SEPARATION IN MODELS FOR MANGANITES: THEORETICAL ASPECTS AND COMPARISON WITH EXPERIMENTS

E. Dagotto, S. Yunoki, and A. Moreo

National High Magnetic Field Lab and Department of Physics,
Florida State University, Tallahassee, FL 32306, USA

INTRODUCTION

Colossal magnetoresistance in metallic oxides such as $R_{1-x}X_xMnO_3$ (where R = La, Pr, Nd; X = Sr, Ca, Ba, Pb) is attracting considerable attention [1] due to its potential technological applications. A variety of experiments have revealed that oxide manganites have a rich phase diagram [2] with regions of antiferromagnetic (AF) and ferromagnetic (FM) order, as well as charge ordering, and a peculiar insulating state above the FM critical temperature T_c. Recently, *layered* manganite compounds $La_{1.2}Sr_{1.8}Mn_2O_7$ have also been synthesized [3] with properties similar to those of their 3D counterparts. Strong correlations are important for transition-metal oxides, and, thus, theoretical guidance is needed to understand the behavior of manganites and for the design of new experiments.

The appearance of ferromagnetism at low temperatures can be explained using the so-called Double Exchange (DE) mechanism [4, 5]. However, the DE model is incomplete to describe the entire phase diagram observed experimentally. For instance, the electron-phonon coupling may be crucial to account for the insulating properties above T_c [6]. The presence of a Berry phase at large Hund-coupling also challenges predictions from the DE model [7]. In a series of recent papers (Refs. [8, 9, 10, 11]) we have remarked that another phenomena occurring in manganites and not included in the DE description, namely the charge ordering effect, may be contained in a more fundamental Kondo-like model. More precisely, in those papers [8, 9, 10] it was reported the presence of *phase separation* (PS) between hole-undoped antiferromagnetic and hole-rich ferromagnetic regions in the low temperature phase diagram of the one-orbital FM Kondo model. In a recent paper by the same authors a similar phenomenon was also observed using two orbitals and Jahn-Teller phonons [11]. In this case the two phases involved are both spin ferromagnetic and the *orbital* degrees of freedom are responsible for the phase separation. Upon the inclusion of long-range Coulombic repulsion, charge ordering in the form of non-trivial extended structures (such as stripes) could be stabilized, similarly as discussed for the cuprates [12, 13, 14, 15] but now also including ferromagnetic domains. The analysis carried out by our group suggests that phenomena as rich as observed in the high-Tc superconductors may exist in the manganites as well, and hopefully our effort will induce further theoretical and experimental work in this context.

The present paper should be considered as an informal review of the present status of the computational studies of models for manganites that reported the existence of phase separation at low temperatures. It does *not* pretend to be a comprehensive article, and thus we encourage the readers to consult the literature mentioned here to find out additional references. The paper is organized as follows: in the next section the results for the case of the one-orbital Kondo model are reviewed, with emphasis on the phase diagram. Most of the calculations are performed with classical t_{2g}-spins but results for quantum spins are also shown. In addition, models that interpolate between Cu-oxides and Mn-oxides have also been studied, as well as the influence on our conclusions of Coulomb interactions beyond the on-site term. In the following section results recently reported for the case of the two-orbital model with Jahn-Teller phonons are described. Once again the main emphasis is given to the phase diagram. In both of those previous sections the main result is that tendencies to ferromagnetism and phase-separation are in strong competition in these models. Such a phenomenon appears so clearly in all dimensions of interest and for such a wide range of models that it leads us to believe that it may be relevant for experiments in the context of manganites. A discussion of experimental literature that have reported some sort of charge inhomogeneity compatible with phase separation (after Coulomb interactions are added) are briefly described towards the end of the paper. In the last section a summary is provided.

ONE ORBITAL FERROMAGNETIC KONDO MODEL

Results for Classical t_{2g}-Spins

The FM Kondo Hamiltonian [4, 16] is defined as

$$H = -t \sum_{\langle \mathbf{ij} \rangle \sigma} (c^\dagger_{\mathbf{i}\sigma} c_{\mathbf{j}\sigma} + h.c.) - J_H \sum_{\mathbf{i}\alpha\beta} c^\dagger_{\mathbf{i}\alpha} \sigma_{\alpha\beta} c_{\mathbf{i}\beta} \cdot \mathbf{S_i}, \qquad (1)$$

where $c_{\mathbf{i}\sigma}$ are destruction operators for one species of e_g-fermions at site $\mathbf{i}$ with spin σ, and $\mathbf{S_i}$ is the total spin of the t_{2g} electrons, assumed localized. The first term is the e_g electron transfer between nearest-neighbor Mn-ions, $J_H > 0$ is the Hund coupling, the number of sites is L, and the rest of the notation is standard. The density is adjusted using a chemical potential μ. In this section and Refs. [8, 9] the spin $\mathbf{S_i}$ is considered classical (with $|\mathbf{S_i}| = 1$), unless otherwise stated. Although models beyond Eq.(1) may be needed to fully understand the manganites (notably those that include lattice effects as studied later in this paper), it is important to analyze in detail the properties of simple Hamiltonians to clarify if part of the experimental rich phase diagram can be accounted for using purely electronic models.

To study Eq.(1) in the t_{2g}-spin classical limit a Monte Carlo (MC) technique was used: first, the trace over the e_g-fermions in the partition function was carried out exactly diagonalizing the $2L \times 2L$ Hamiltonian of electrons in the background of the spins $\{S_i\}$, using library subroutines. The fermionic trace is a positive function of the classical spins and the resulting integration over the two angles per site parametrizing the S_i variables can be performed with a standard MC algorithm without "sign problems". In addition, part of the calculations of Refs. [8, 9] were also performed with the Dynamical Mean-Field approximation (D=∞) [16], the Density-Matrix Renormalization Group (DMRG), and the Lanczos method. Special care must be taken with the boundary conditions (BC) [17, 18].

Our results are summarized in the phase diagram of Fig.1. In 1D (and also in 2D, not shown) and at low temperatures clear indications of (i) ferromagnetism, (ii) spin incommensurate (IC) correlations, and (iii) phase separation were identified. Results are also available in small 3D clusters and qualitatively they agree with those in Fig.1. The same occurs working at D = ∞ (see Refs. [8, 9]). In 1D we also obtained results with quantum t_{2g}-spins S=3/2 (shown below) which are in good agreement with Fig.1.

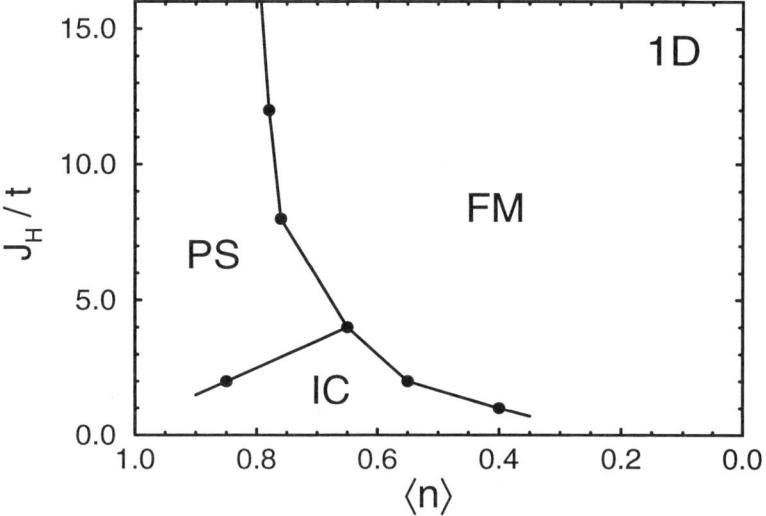

Figure 1: Phase diagram of the FM Kondo model reported in Refs. [8, 9] obtained using numerical techniques. FM, IC, and PS denote regimes with FM correlations, spin incommensurate correlations, and with phase separation between undoped AF and hole-rich FM regions, respectively. The results shown correspond to one dimension and they were obtained with MC simulations at $T = t/75$ using chains with L=20, 30 and 40 sites. This result is taken from Refs. [8, 9].

The boundaries of the FM region of the phase diagram were found evaluating the spin-spin correlation defined as $S(\mathbf{q}) = (1/L) \sum_{\mathbf{j},\mathbf{m}} e^{i(\mathbf{j}-\mathbf{m})\cdot\mathbf{q}} \langle \mathbf{S_j} \cdot \mathbf{S_m} \rangle$. Fig.2 shows $S(\mathbf{q})$ at zero momentum vs. T/t for typical examples in 1D and 2D. The rapid increase of the spin correlations as T is reduced and as the lattice size grows clearly points towards the existence of ferromagnetism in the system [19]. It is natural to assume that the driving force for FM in this context is the DE mechanism. Repeating this procedure for a variety of couplings and densities, the robust region of FM shown in Fig.1 was determined. In the small J_H/t region IC correlations were observed monitoring $S(\mathbf{q})$ [20]. Both in 1D and 2D there is one dominant peak which moves away from the AF location at $\langle n \rangle = 1$ towards zero momentum as $\langle n \rangle$ decreases. In the 2D clusters the peak moves from (π, π) towards $(\pi, 0)$ and $(0, \pi)$, rather than along the main diagonal. Note that our computational study predicts IC correlations only in the small and intermediate J_H/t regime.

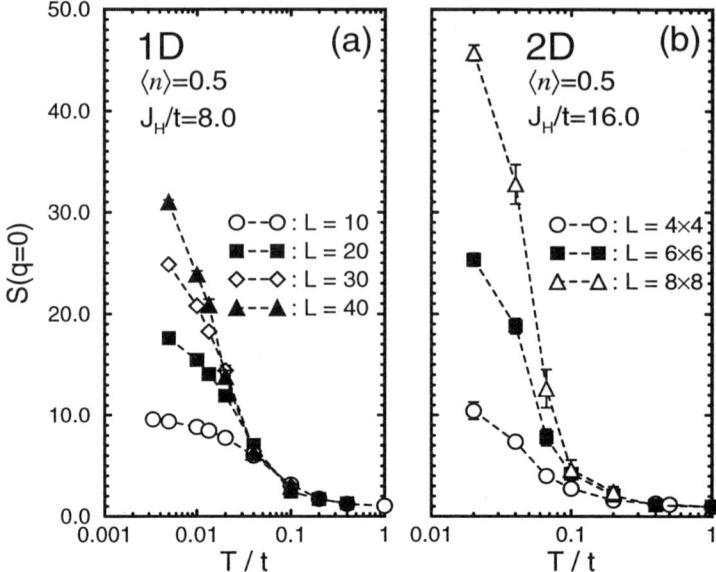

Figure 2: Spin-spin correlations of the classical spins at zero momentum $S(\mathbf{q} = 0)$ vs temperature T (in units of t). MC results for several lattice sizes are shown on (a) chains and (b) 2D clusters. The density and coupling are shown. In (a) closed shell BC are used i.e. periodic BC for $L = 10$ and 30 and antiperiodic BC for $L = 20$ and 40. In (b) open BC are used. Results taken from Ref. [8].

The main result of Refs. [8, 9] is contained in Fig.3 where the computational evidence for the existence of phase separation in dimensions 1, 2, and ∞ is given. The presence of a discontinuity in $\langle n \rangle$ vs μ shows that some electronic densities can not be stabilized by tuning the chemical potential [21]. If the system is nominally prepared with such density it will spontaneously separate into two regions with the densities corresponding to the extremes of the discontinuities of Fig.3. By analyzing these extremes the properties of the two domains can be studied. One region is undoped ($\langle n \rangle = 1$) with strong AF correlations, while the other contains all the holes and the spin-spin correlations between the classical spins are FM (see the inset of Fig.3a. The results are similar in D=2 and infinite). This is natural since holes optimize their kinetic energy in a FM background. On the other hand, at $\langle n \rangle = 1$ the DE mechanism is not operative: if the electrons fully align their spins they cannot move in the conduction band due to the Pauli principle. Then, energetically an AF pattern is formed. As J_H grows, the jump in Fig.3 is reduced and it tends to disappear in the $J_H = \infty$ limit.

Experimentally, PS may be detectable using neutron diffraction techniques, since

$S(q)$ should present a two peak structure, one located at the AF position and the other at zero momentum. Since this also occurs in a canted ferromagnetic state care must be taken in the analysis of the experimental data. Another alternative is that Coulombic forces prevent the macroscopic accumulation of charge intrinsic of a PS regime. Thus, microscopic hole-rich domains may develop in the manganites, perhaps in the form of stripes.

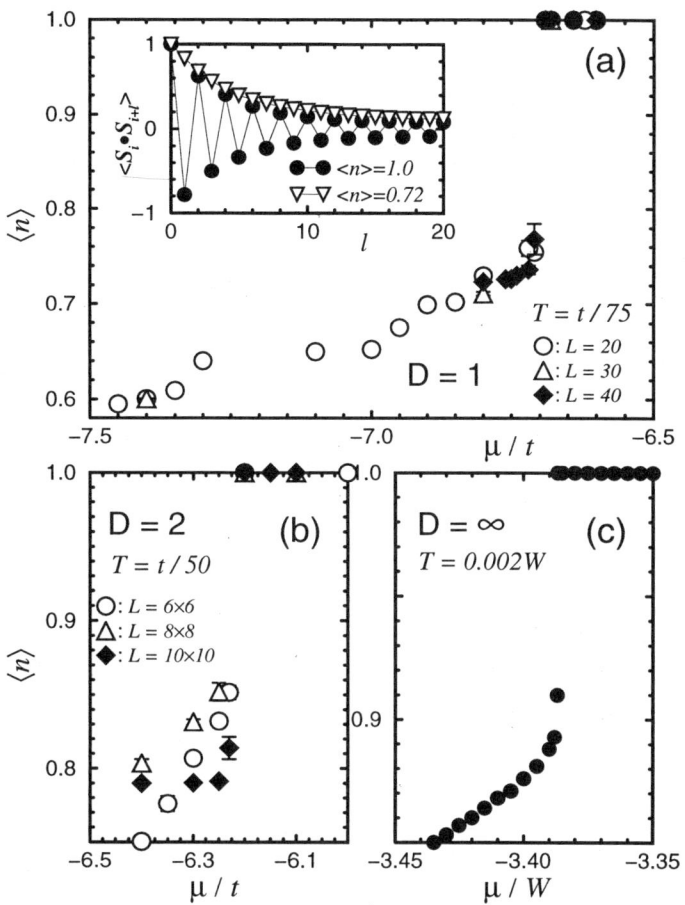

Figure 3: Electronic density $\langle n \rangle$ vs chemical potential μ in (a) D=1, (b) D=2, and (c) D = ∞ clusters. Temperatures are indicated. The coupling is $J_H/t = 8.0$ in (a) and (b) and $J_H/W = 4.0$ in (c) (for the definition of W see Ref. [8]). PBC were used both in D=1 and 2. The discontinuities shown in the figures are indicative of PS. In (a) the inset contains the spin-spin correlation in real space at densities 1.0 and 0.72 showing that indeed PS occurs between AF and FM regions. The results are taken from Refs. [8, 9] where more details can be found.

Although the phase diagram of Fig.1 has PS close to half-filling, actually this phenomenon also occurs at $\langle n \rangle \sim 0$ if an extra direct AF exchange interaction between the localized spins is included [10]. This coupling may be originated in a small hopping amplitude for the t_{2g} electrons. At $\langle n \rangle = 0$, model Eq.(1) supplemented by a Heisenberg coupling J'/t among the localized spins produces an AF phase, as in experiments, which upon electron doping induces a competition between AF (with no e_g-electrons) and FM electron-rich regions, similarly as for $\langle n \rangle = 1$ but replacing holes by electrons. Thus, PS or charge ordering could exist in manganites both at large and small fermionic densities. A careful study of the influence of the J' coupling on the phase diagram of the 1-orbital Kondo model has been recently presented in Ref. [10]

For completeness, upper bounds on the 3D critical temperature T_c were also provided in Refs. [8, 9]. Using MC simulations in principle it is possible to calculate T_c accurately. However, the algorithm used in our studies prevented us from studying clusters larger than 6^3 even at $J_H = \infty$. In spite of this limitation, monitoring the spin-spin correlations in real space allows us to judge at what temperature T^* the correlation length reaches the boundary of the 6^3 cluster. Since the bulk T_c is smaller than T^*, this give us upper bounds for the critical temperature. Following this procedure at $\langle n \rangle = 0.5$, it was found that for $T \sim 0.1t$ robust correlations reach the boundary, while for $T \geq 0.12t$ the correlation is short-ranged. Thus, at this density we estimate that $T_c < 0.12t$. Since results for the e_g electrons bandwidth range from $BW \sim 1\ eV$ [22] to $BW \sim 4eV$ [23], producing a hopping $t = BW/12$ between 0.08 and 0.33 eV, then the estimations for the critical temperature of Refs. [8, 9] range roughly between $T_c \sim 100\ K$ and $400\ K$. This is within the experimental range and in agreement with other results [16, 24]. Then, in principle purely electronic models can account for T_c.

Quantum t_{2g}-Spins: Phase Diagram

In the previous section we have used classical localized spins to simplify the technical aspects of the computational study and since the realistic values of those spins is large (3/2). However, it is interesting to study the phase diagram for the case of quantum t_{2g}-spins at least in one special case, in order to verify that indeed the use of classical spins is a good approximation. Using the Lanczos and DMRG techniques, the phase diagram for the cases of t_{2g}-spins $S = 3/2$ and $1/2$ was reported in Refs. [8, 9] for the case of one dimension, and one of them is shown in Fig. 4. The techniques used work in the canonical ensemble, instead of grand-canonical as the MC simulation of the previous section. The presence of phase separation is investigated here using the compressibility (which if it becomes negative signals the instability of the density under study). Fig. 4 shows that the three regimes found before (namely PS, FM, and IC) are also observed in the quantum case. The results for $S = 3/2$ are even quantitatively similar to those found with the MC technique of the previous section, while those for $S = 1/2$ are only qualitatively similar. The overall conclusion is that the use of classical t_{2g}-spins in the Ferromagnetic Kondo Model was found to be a good approximation, and no important differences in the phase diagram have been observed (at least in this one-dimensional example) when quantum spins are used. Studies in dimensions larger than one would be too involved using quantum localized spins, and thus the similarities between the classical and quantum cases found here help us to investigate those more realistic dimensions.

Models for Other Transition-Metal Oxides

In the previous sections we have found that models for manganites analyzed with reliable unbiased techniques provide a phase diagram where there are two main phases in competition: a ferromagnetic regime and a "phase-separated" regime (actually the latter is not a phase but a region of unstable densities). Such a clear result suggest that

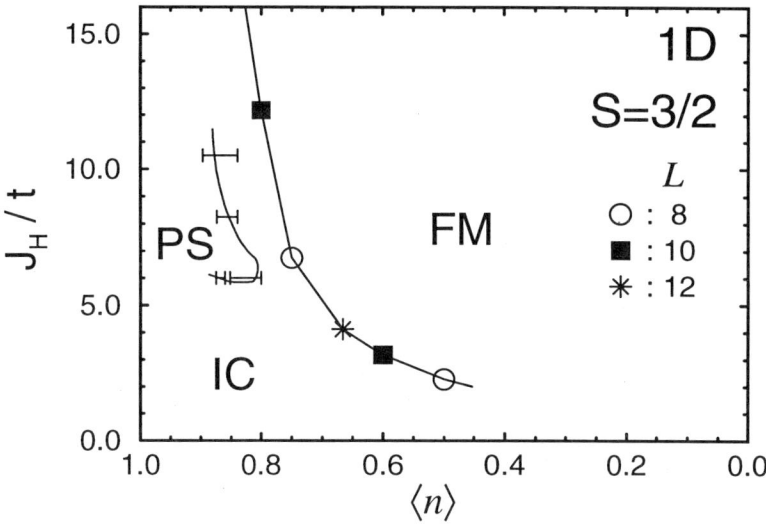

Figure 4: Phase diagram of the Ferromagnetic Kondo model using quantum t_{2g}-spins. The results correspond to a localized spin of value $S = 3/2$ (normalized as $|S| = 1$, see Ref. [9]). The notation is the same as in Fig.1. The techniques used to obtain the results are Lanczos and DMRG. The result is taken from Ref. [9].

the phenomenon maybe more general than expected. Actually recent studies of models for transition-metal oxides in general [17], in the limit of a large Hund-coupling, revealed phase diagrams where indeed FM and PS dominated. Let us consider the evolution of such a phase diagram as we move in the periodic table of elements from copper, to nickel, etc, eventually arriving to manganese. The study of Ref. [17] was carried out in one-dimension for technical reasons, but nevertheless the tendencies observed are strong enough to be valid in other dimensions as well. For the case of copper the model used was the plain "t-J" model, namely a model where we have spins of value $S = 1/2$ and holes which carry no spin. The phase diagram in this case was known for a long time. It contains no indications of ferromagnetism (at least no "clear" indications), and there is phase separation only at large J/t. However, the study of Ref. [17] for Ni-oxides using a t-J-like model with "spins" of value $S = 1$ and "holes" with spin $S = 1/2$ (since they correspond to the removal of one electron of a Ni-site) revealed that both FM and PS are clearly enhanced in the phase diagram with respect to the case of Cu-oxides. The technical details can be found in Ref. [17]. As the spin grows these tendencies become stronger and actually for the case of the Mn-oxides only a small window separates the fully polarized FM and PS regimes. This result suggests that the phenomenon is very general and should be searched for in other transition-metal oxides as well.

Influence of Coulomb Interactions

The accumulation of charge that occurs in the phase separation regime described in the previous section is not stable once Coulombic interactions beyond the on-site term are added to the problem. It is expected that in the presence of these long-range interactions the ground state will have clusters of one phase embedded into the other (droplet regime). However, a computational study including these Coulomb interactions is very difficult. For instance, the analysis of model Eq.(1) would require a Hubbard-Stratonovich decomposition of the fermionic interaction which not only introduces more degrees of freedom, but in addition the infamous "sign problem" which will prevent the study from being carried out at low enough temperatures. Then, only techniques such as Lanczos or DMRG can handle properly this type of models. Recently, the analysis including Coulombic terms started by using the "minimum" possible Hamiltonian, namely the 1-orbital Ferromagnetic Kondo Model with localized spin-1/2 and with the addition of an on-site Coulomb term and a nearest-neighbor (NN) density-density V-coupling that penalizes the accumulation of charge in two NN sites [25]. The results are still being analyzed so here only a qualitative summary will be presented.

The conclusions of the study of Malvezzi et al [25]. thus far are the following: (1) the on-site coupling U apparently does not affect qualitatively the results found at $U = 0$ as long as the Hund-coupling is large; (2) in the regime of phase separation near $\langle n \rangle = 1$ the addition of a V-term produces results compatible with the cluster formation anticipated in the previous paragraph. Actually even at finite V in a regime where the compressibility shows that phase separation no longer exists (i.e. all densities are stable), clear remnants of such a PS regime are found. For instance, at $V = 0$ the spin structure factor $S(q)$ has robust peaks at both $q = 0$ and π. Introducing V, these peaks still exist but they move slightly from the 0 and π positions forming incommensurate structures; (3) in the regime where at $V = 0$ there is ferromagnetism induced by a double-exchange process, the inclusion of a nonzero coupling V transforms the ground state into a FM charge-density-wave (CDW). In this phase the spins are aligned but the charge is not uniformly distributed. The overall phase diagram is rather complicated. We are actively working in this context to analyze the qualitative aspects of such a result. But it is already clear that charge-ordering tendencies are observed at all densities upon introducing V, in agreement with experiments. More detailed results will be presented soon.

TWO ORBITALS AND JAHN-TELLER PHONONS

Phase Diagram

In spite of the rich phase diagram observed in the study of the 1-orbital Ferromagnetic Kondo Model in the previous section and its similarities with experiments, there are aspects of the real manganites that require a more sophisticated approach. For instance, dynamical Jahn-Teller (JT) distortions are claimed to be very important [6], and a proper description of the recently observed orbital order [26] obviously needs at least two orbitals. Such a multi-orbital model with JT phonons is nontrivial, and thus far it has been studied only using the dynamical mean-field approximation [27]. The previous experience with the 1-orbital case suggests that a computational analysis is actually crucial to understand its properties. In addition, it is conceptually interesting to analyze whether the PS described before [8, 9, 10] exists also in multi-orbital models.

The first computational study of a 2-orbital model for manganites including JT phonons was reported recently by the authors in Ref. [11] and here a summary of the main conclusions will be discussed. The results show a rich phase diagram including a novel regime of PS induced by the *orbital*, rather than the spin, degrees of freedom (DOF). The Hamiltonian used in that study had three contributions $H_{KJT} = H_K + H_{JT} + H_{AF}$. The first term is

$$H_K = - \sum_{\langle ij \rangle \sigma ab} t_{ab}(c^\dagger_{ia\sigma} c_{jb\sigma} + h.c.) - J_H \sum_{ia\alpha\beta} \mathbf{S_i} \cdot c^\dagger_{ia\alpha} \sigma_{\alpha\beta} c_{ia\beta}, \qquad (2)$$

where $\langle ij \rangle$ denotes nearest-neighbor lattice sites, $J_H > 0$ is the Hund coupling, the hopping amplitudes t_{ab} are described in Ref. [28], $a, b = 1, 2$ are the two e_g-orbitals, the t_{2g} spins $\mathbf{S_i}$ are assumed to be classical (with $|\mathbf{S_i}| = 1$) since their actual value in Mn-oxides (3/2) is large [29], and the rest of the notation is standard. None of the results described below depends crucially on the set $\{t_{ab}\}$ selected [28]. The energy units are chosen such that $t_{11} = 1$ in the x-direction. In addition, since J_H is large in the real manganites, here it will be fixed to 8 (largest scale in the problem) unless otherwise stated. The e_g-density $\langle n \rangle$ is adjusted using a chemical potential μ.

The coupling with JT-phonons is through [6, 30]

$$H_{JT} = \lambda \sum_{iab\sigma} c^\dagger_{ia\sigma} Q^{ab}_i c_{ib\sigma} + \frac{1}{2} \sum_i ({Q^{(2)}_i}^2 + {Q^{(3)}_i}^2), \qquad (3)$$

where $Q^{11}_i = -Q^{22}_i = Q^{(3)}_i$, and $Q^{12}_i = Q^{21}_i = Q^{(2)}_i$. These phonons are assumed to be classical, which substantially simplifies the computational study. This is a reasonable first approximation towards the determination of the phase diagram of the H_{KJT} model. Finally, a small coupling between the t_{2g}-spins is needed to account for the AF character of the real materials even when all La is replaced by Ca or Sr (fully doped manganites). This classical Heisenberg term is $H_{AF} = J' \sum_{\langle ij \rangle} \mathbf{S_i} \cdot \mathbf{S_j}$, where J' is fixed to 0.05 throughout the paper, a value compatible with experiments [31]. To study H_{KJT} a Monte Carlo (MC) technique similar to that employed in Refs. [8, 9, 10] and in the previous section for the 1-orbital problem was used. Finally, to analyze orbital correlations the pseudopin operator $\mathbf{T_i} = \frac{1}{2} \sum_{\sigma ab} c^\dagger_{ia\sigma} \sigma_{ab} c_{ib\sigma}$ was used, while for spin correlations the operator is standard. The Fourier-transform of the pseudospin correlations is defined as $T(\mathbf{q}) = \frac{1}{L} \sum_{\mathbf{l},\mathbf{m}} e^{i\mathbf{q}\cdot(\mathbf{l}-\mathbf{m})} \langle \mathbf{T_m} \cdot \mathbf{T_l} \rangle$, with a similar definition for the spin structure factor $S(\mathbf{q})$.

Let us first consider the limit $\langle n \rangle = 1.0$, corresponding to undoped manganites. Fig.5 shows $T(q)$ and $S(q)$ at representative momenta $q = 0$ and π vs. λ. For small electron-phonon coupling the results are similar to those at $\lambda = 0.0$, namely a large $S(0)$ indicates a tendency to spin-FM order induced by DE (as in the qualitatively similar 1-orbital problem at $\langle n \rangle = 0.5$ [8]). The small values of $T(q)$ imply that in this regime the orbitals remain disordered. When the coupling reaches $\lambda_{c1} \sim 1.0$, the rapid increase of $T(\pi)$ now suggests that the ground state has a tendency to form a *staggered* (or "antiferro" AF) orbital pattern, with the spins remaining FM aligned since $S(0)$ is large. The existence of this phase was discussed before, but using multi-orbital Hubbard models with Coulomb interactions and without phonons [32]. Our results show that it can be induced by JT phonons as well. As the coupling increases further, another transition at $\lambda_{c2} \sim 2.0$ occurs to a spin-AF orbital-FM state ($S(\pi)$ and $T(0)$ are large). In this region a 1-orbital approximation is suitable.

The three regimes of Fig.5 can be understood in the limit where λ and J_H are the largest scales, and using $t_{12} = t_{21} = 0$, $t_{11} = t_{22} = t$ for simplicity. For parallel spins with orbitals split in a staggered (uniform) pattern, the energy per site at lowest order in t is $\sim -t^2/\Delta$ (~ 0), where Δ is the orbital splitting. For antiparallel spins with uniform (staggered) orbital splitting, the energy is $\sim -t^2/2J_H$ ($\sim -t^2/(2J_H + \Delta)$).

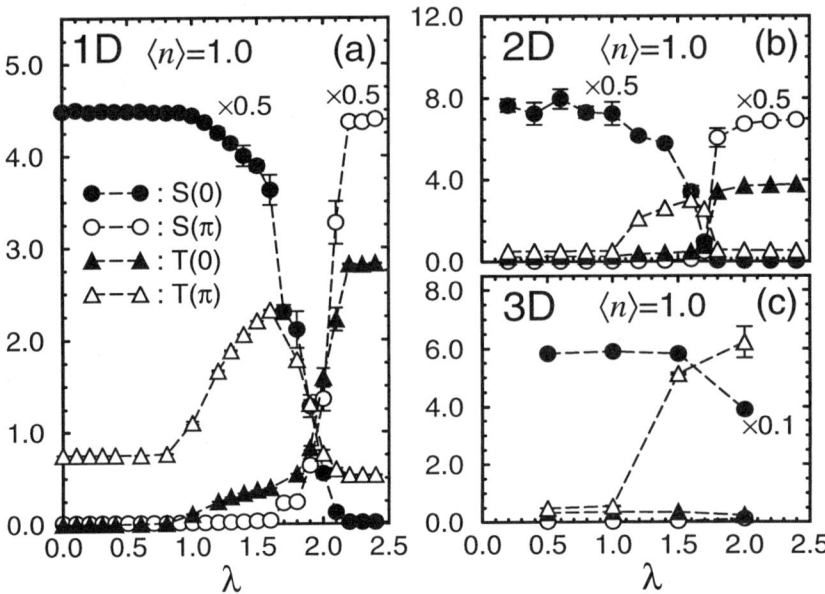

Figure 5: (a) $T(q)$ and $S(q)$ vs λ, working at $\langle n \rangle = 1.0$, $T = 1/75$, $J_H = 8$, $J' = 0.05$, and in 1D with 10 sites. $\{t_{ab}\}$ correspond to set T_1 (see [28]). Results with sets T_2 and T_3 are qualitatively the same; (b) Same as (a) but for a 4^2 cluster, $T = 1/50$, and hopping T_3 (T_4) in the y (x) direction. $q = 0(\pi)$ denotes $(0,0)$ ((π,π)); (c) Same as (a) but for a 4^3 cluster, $T = 1/50$, the 3D hopping amplitudes of Ref. [28], and $J_H = \infty$. $q = 0(\pi)$ denotes $(0,0,0)$ ((π,π,π)). These results were taken from Ref. [11].

Then, when $\Delta < 2J_H$ ("intermediate" λs), a spin-FM orbital-AF order dominates, while as λ grows further a transition to a spin-AF orbital-FM ground state is expected. Note that this reasoning is actually valid in any dimension. To confirm this prediction, results for a 2D cluster in Fig.5b are shown. Indeed the qualitative behavior is very similar to the 1D results. In 3D (Fig.5c), where studies on 4^3 clusters can only be done at $J_H = \infty$ to reduce the number of DOF, at least two of the regimes of Fig.5a have been identified [33].

Let us discuss the transport properties in the three regimes at $\langle n \rangle = 1$. The algorithm used in Ref. [8, 11] allow us to calculate *real-time* dynamical responses accurately, including the optical conductivity $\sigma(\omega > 0)$, since all the eigenvectors in the fermionic sector for a given spin and phonon configuration are obtained exactly. From the sum-rule, e_g kinetic-energy, and the integral of $\sigma(\omega > 0)$, the $\omega = 0$ Drude-weight D_W can be obtained. Carrying out this calculation it was observed that D_W vanishes at λ_{c1} signaling a *metal-insulator* transition (MIT). Here the insulating phase is spin-FM and orbital-AF, while the metallic one is spin-FM and orbital-disordered [34]. The density of states (DOS) for $\lambda > \lambda_{c1}$ was also calculated in Ref. [11] and it presents a clear gap at the Fermi level. The qualitative shape of D_W vs λ on 4^2 and 4^3 clusters was found to be the same as in 1D and, thus, it is reasonable to assume that the MIT exists also in all dimensions of interest.

Consider now the influence of hole doping on the $\langle n \rangle = 1.0$ phase diagram, with special emphasis on the stability of other densities as μ is varied. Fig.6 shows $\langle n \rangle$ vs μ in the intermediate-λ regime. It is remarkable that *two* regions of unstable densities exist at low-T (similar conclusions were reached using the Maxwell's construction [35]). These instabilities signal the existence of PS in the H_{KJT} model. At low-density there is separation between an (i) empty e_g-electron band with AF-ordered t_{2g}-spins and a (ii) metallic spin-FM orbital uniformly ordered phase. This regime of PS is analogous to the spin-driven PS found at low-density in the 1-orbital problem [10]. On the other hand, the unstable region near $\langle n \rangle = 1.0$ is *not* contained in the 1-orbital case. Here PS is between the phase (ii) mentioned above, and (iii) the insulating intermediate-λ spin-FM and orbital-AF phase described in Fig.5 [36]. In 2D the results were found to be very similar (Ref. [11]). The driving force for this novel regime of PS are the *orbital* degree of freedom, since the spins are uniformly ordered in both phases involved. Studying $\langle n \rangle$ vs μ, for $\lambda < \lambda_{c1}$ only PS at small densities is observed, while for $\lambda > \lambda_{c2}$ the PS close to $\langle n \rangle = 1$ is similar to the same phenomenon observed in the 1-orbital problem since it involves a spin-AF orbital-FM phase [8].

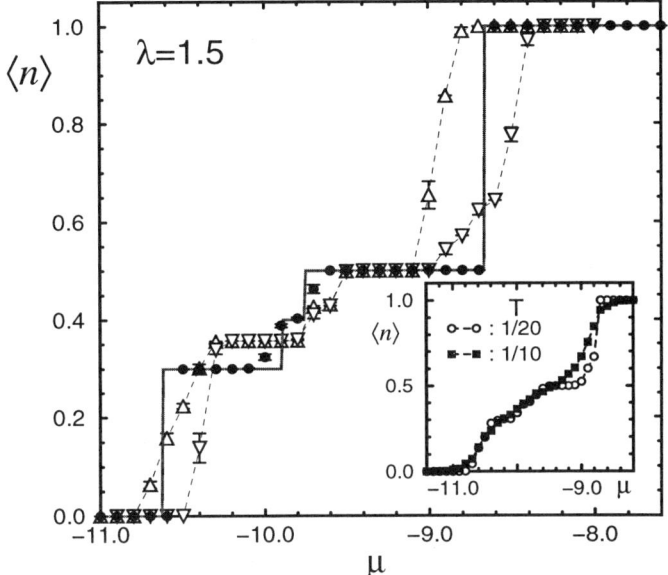

Figure 6: $\langle n \rangle$ vs μ at $\lambda = 1.5$, $L = 10$, and $T = 1/40$ (solid circles). The discontinuities near $\langle n \rangle = 1.0$ and 0.0 show the existence of unstable densities. The solid line is obtained from the Maxwell's construction. The triangles are results also at $\lambda = 1.5$ and $T = 1/40$, but using 14 sites and only 2×10^4 MC sweeps to show the appearance of *hysteresis* loops as in a first-order transition. The inset shows the T-dependence of the results at $L = 10$. These results were taken from Ref. [11].

The phase diagram of the 1D H_{KJT} model is given in Fig.7. The two PS regimes are shown, together with the metallic spin-FM region. This phase is separated into two regions, one ferro-orbital ordered and the other orbitally disordered. The existence of these two regimes can be deduced from the behavior of the pseudospin correlations, the mean value of the pseudospin operators, and the probability of double occupancy of the same site with different orbitals. The results are similar for several $\{t_{ab}\}$ sets [28]. The simulations of Ref. [11] suggest that the qualitative shape of Fig.7 should be valid also in $D = 2$ and 3. Recently we learned of experiments where the low-T coexistence of domains similar to those described in Ref. [11], i.e. one orbitally-ordered and the other FM-metallic, has been observed in Mn-perovskites [37].

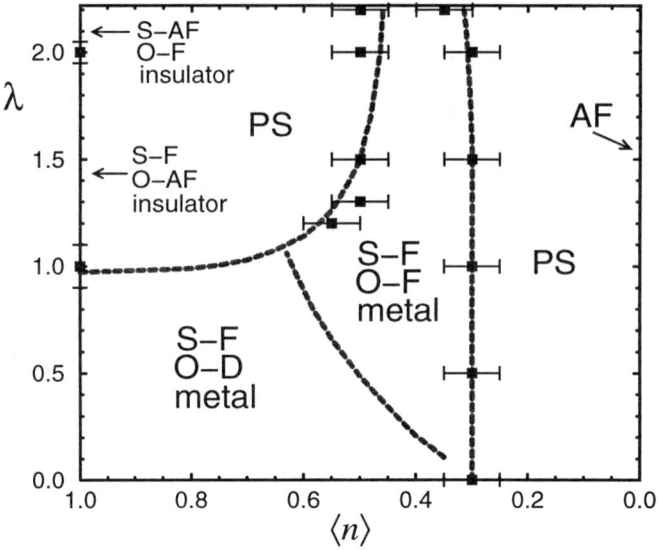

Figure 7: Phase diagram of the H_{KJT} model at low-T, $J_H = 8.0$, $J' = 0.05$, and using set T_1 for $\{t_{ab}\}$. S – F and S – AF denote regimes with FM- and AF-spin orders, respectively. O – D, O – F, and O – AF represent states with disordered, uniform and staggered orbital orders, respectively. PS means phase separation. These results were taken from Ref. [11].

Optical Conductivity

In Ref. [11] results for $\sigma(\omega)$ were also presented. Experimental studies for a variety of manganites such as $Nd_{0.7}Sr_{0.3}MnO_3$, $La_{0.7}Ca_{0.3}MnO_3$, and $La_{0.7}Sr_{0.3}MnO_3$ reported a broad peak at $\omega \sim 1eV$ (for hole doping $x > 0.2$ and $T > T_c^{FM}$) [38, 39]. At room-T there is negligible weight near $\omega = 0$, but as T is reduced the $1eV$-peak shifts to smaller energies, gradually transforming into a Drude response well below T_c. The finite-ω peak can be identified even inside the FM phase. The coherent spectral weight is only a small fraction of the total. Other features at larger energies $\sim 3eV$ involve transitions between the J_H-split bands and the O-ions. In addition, Jung et al. [39] interpreted the $1eV$ feature at room-T as composed of two peaks due to intra- and inter-atomic transitions in JT-distorted environments.

In Fig.8, $\sigma(\omega)$ for the H_{KJT} model is shown at $\langle n \rangle = 0.7$ and several temperatures near the unstable PS region of Fig.7 (weight due to J_H split bands occurs at higher energy). Here the FM spin correlation length grows rapidly with the lattice size for $T^* \leq 0.05t$, which can be considered as the "critical" temperature. Both at high- and intermediate-T a broad peak is observed at $\omega \sim 1$, smoothly evolving to lower energies as T decreases. The peak can be identified well-below T^* as in experiments [38, 39]. Eventually at very low-T, $\sigma(\omega)$ is dominated by a Drude peak. The T-dependence shown in the figure is achieved at this λ and $\langle n \rangle$ by a combination of a finite-ω phonon-induced broad feature that looses weight, and a Drude response that grows as T decreases (for smaller λs, the two peaks can be distinguished even at low-T). The similarity with experiments suggests that real manganites may have couplings close to an unstable region in parameter space. In the inset, D_W vs T is shown. Note that at $T \sim T^*$, D_W vanishes suggesting a MIT, probably due to magneto polaron localization. Results for the 1-orbital case are smoother, with no indications of a singularity. These features are in agreement with experiments, since the manganite "normal" state is an insulator. Work is currently in progress to analyze in more detail this MIT transition.

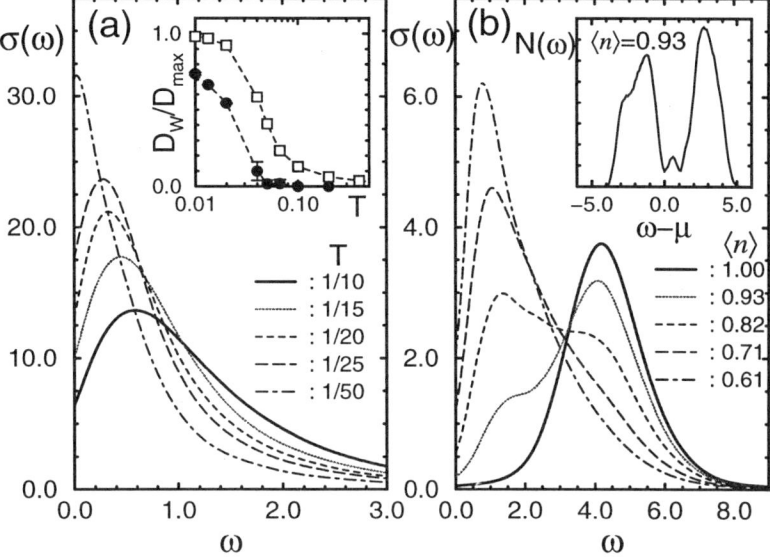

Figure 8: (a) $\sigma(\omega)$ parametric with T, at $\lambda = 1.0$, $\langle n \rangle = 0.7$, and $L = 20$. The inset shows D_W vs T for both the H_{KJT} (soild circles) and the 1-orbital model of Ref. [8] (open squares) (the latter at $\langle n \rangle = 0.65$). D_W is normalized to its maximum value at $T = 0.01$; (b) $\sigma(\omega)$ vs ω parametric with $\langle n \rangle$ at $\lambda = 1.5$, $T = 1/10$, and $L = 16$ (results for $L = 10$ are very similar). The inset shows the lower J_H-split DOS at $\langle n \rangle = 0.93$. Both in (a) and (b) a δ-function broadening $\epsilon = 0.25$ was used. These results were taken from Ref. [11].

A similar good agreement with experiments was observed working in the regime of the orbitally-induced PS but at a temperature above its critical value T_{PS} (roughly $\sim 1/20$, see inset Fig.6). Here the broad feature observed at high-T in Fig.8a moves to higher energies (Fig.8b) since λ has increased. At the temperature of the plot the system is an insulator at $\langle n \rangle = 1$, but as hole carriers are added a second peak at lower energies develops, in addition to a weak Drude peak (which carries, e.g., just 1% of the total weight at $\langle n \rangle = 0.61$). This feature at high-T is reminiscent of recent experimental results [39] where a two-peak structure was observed at room-T and several densities. Similar results were obtained on 4^2 clusters. In Fig.8b the peak at large-ω is caused by phononic effects since its position was found to grow rapidly with λ. It corresponds to intersite transitions between Mn^{3+} JT-split states. The lower energy structure is compatible with a Mn^{3+}-Mn^{4+} transition [40]. The inset of Fig.8b shows the DOS of the system. The two peaks above μ are responsible for the features found in $\sigma(\omega)$. This interpretation is the same as given in Ref. [27] at $D = \infty$.

EXPERIMENTAL CONSEQUENCES OF PHASE SEPARATION

A large number of papers in the context of experimental studies of manganites have reported the presence of some sort of inhomogeneity in the system that is tempting to associate with the phase-separation tendencies observed in our studies. In the next paragraph we will mention some of these results such that the interested reader can have at hand at least part of the relevant references on the subject to decide by him/herself on this issue.

Some of the experiments that have reported results that could be compatible with ours are the following (in a random order): **1.** De Teresa et al. [41] studied $La_{1-x}Ca_xMnO_3$ using small-angle neutron scattering, magnetic susceptibility and other techniques, at x=1/3. The analysis of their data showed the existence of "magnetic clusters" of size approximately 12Å above the ferromagnetic critical temperature; **2.** Hennion et al. [42] recently presented elastic neutron scattering results below T_c at $x = 0.05$ and 0.08 also for $La_{1-x}Ca_xMnO_3$. They interpreted their results as indicative of "magnetic droplets". The density of these droplets was found to be much smaller than the density of holes implying that each droplet contains several holes; **3.** Lynn and collaborators [43] have studied $La_{1-x}Ca_xMnO_3$ at x=1/3 also using neutron scattering. Lattice anomalies and magnetic irreversibilities near T_c were interpreted as evidence of two coexisting distinct phases; **4.** Perring et al. [31] studying $La_{1.2}Sr_{1.8}Mn_2O_7$ with neutron scattering reported the presence of long-lived antiferromagnetic clusters coexisting with ferromagnetic critical fluctuations above T_c; **5.** Allodi et al. [44] using NMR applied to $La_{1-x}Ca_xMnO_3$ at $x \sim 0.1$ observed coexisting resonances corresponding to FM and AF domains; **6.** Cox et al. [45] studying $Pr_{0.7}Ca_{0.3}MnO_3$ with x-ray and powder neutron scattering reported the presence of ferromagnetic clusters; **7.** Bao et al. [46] in their analysis of $Sr_{2-x}La_xMnO_4$ (2D material) found phase separation at small e_g-densities; **8.** Yamada et al. [47] in their study of $La_{1-x}Sr_xMnO_3$ with neutron scattering at x=0.1 and 0.15 interpreted their results as corresponding to polaron ordering; **9.** Roy et al. [48] studying $La_{1-x}Ca_xMnO_3$ near $x = 0.50$ found the coexistence of two carrier types: nearly localized carries in the charge-ordered state and a parasitic population of free carriers tunable by stoichiometry; **10.** Booth et al. [49] also studying $La_{1-x}Ca_xMnO_3$ found that the number of delocalized holes n_{dh} in the ferromagnetic phase changes with the magnetization M as $ln(n_{dh}) \propto -M$. Since the magnetization saturates well below T_c, this implies that there is a range of temperatures where sizable fractions of localized and delocalized holes coexist; **11.** Jaime et al. [50] reported the possibility of polaronic distortions of the paramagnetic phase of La − Ca − Mn − O manganites persisting into the ferromagnetic phase, and analyzed the data with a two-fluid model; **12.** Zhou and Goodenough [51] studying the thermopower and resistivity of $^{16}O/^{18}O$ isotope-exchanged $(La_{1-x}Nd_x)_{0.7}Ca_{0.3}MnO_3$ found indications of a segregation of hole-rich clusters within a hole-poor matrix in the paramagnetic state; **13.** Billinge et al. [52] reported the coexistence of localized and delocalized carriers in a wide range of densities and temperatures below T_c; **14.** Ibarra et al. [53] studied $(La_{0.5}Nd_{0.5})_{2/3}Ca_{1/3}MnO_3$ concluding that insulating charge-ordered and metallic ferromagnetic regions coexist at low temperatures; **15.** Rhyne et al. [54] in their study of $La_{0.53}Ca_{0.47}MnO_3$ with neutron diffraction found two magnetic phases below the transition temperature, one ferromagnetic and the second antiferromagnetic, both persisting down to 10K; **16.** Heffner et al. [55] studying $La_{0.67}Ca_{0.33}MnO_3$ using zero-field muon spin relaxation and resistivity techniques found indications of polarons on the spin and charge dynamics; **17.** Jung et al. [56] have very recently studied the optical conductivities of $La_{7/8}Sr_{1/8}MnO_3$ concluding that there are indications in this system of phase separation.

The common theme of these experiments, and possibly others that have escaped our attention, is that some sort of inhomogeneity appears in the analysis of the data. The phenomenon is apparently more clear at small hole densities x and low temperatures which is precisely the region where our results indicate a strong tendency to phase separate. But also above T_c in the interesting densities around x=1/3 magnetic clusters have been reported. It is nothing but natural to imagine a "smooth" connection between x $\sim$ 0.1 and low temperature with x $\sim$ 1/3 and temperatures above T_c. In this case the large droplets found by Hennion et al. [42] could have evolved as to become the magnetic clusters of De Teresa et al. [41] by a reduction of their size and number of holes inside. We strongly encourage experimental work that could contribute to the answer of this or other important issues around the common theme of possible charge seggregation in real manganites.

SUMMARY

Summarizing, a comprehensive computational study of models for manganites have found that the expected double-exchange induced strong tendencies to ferromagnetic correlations at low temperatures are in competition with a regime of "phase separation". This regime was identified in all dimensions of interest, using one and two orbitals (the latter with Jahn-Teller phonons), and both with classical and quantum localized t_{2g} spins. It also appears in the presence of on-site Coulomb interactions. This robustness of our results suggests that phase separation may also be present in real manganites. In the previous section experimental literature that have reported some form of charge inhomogeneity in the context of the manganites has been briefly reviewed. It is concluded that theory and experiments seem to be in qualitative agreement and phase separation tendencies (which may correspond to the formation of magnetic droplets or even stripes once Coulomb interactions beyond the on-site term are included in the analysis) should be taken seriously. They may even be responsible for the phenomenon of Colossal Magnetoresistance that motivated the current enormous interest in the study of manganites in the first place!.

ACKNOWLEDGEMENTS

E. D. and A. M. are supported by the NSF grant DMR-9520776. S. Y. thanks the NHMFL for support.

REFERENCES

[1] S. Jin et al., Science **264**, 413 (1994); J. M. D. Coey, M. Viret, and S. von Molnar, *Mixed Valence Manganites*, Adv. Phys. (1998), in press.

[2] P. E. Schiffer, A. P. Ramirez, W. Bao, and S-W. Cheong, Phys. Rev. Lett. **75**, 3336 (1995); A. P. Ramirez et al., Phys. Rev. Lett. **76**, 3188 (1996); C. H. Chen and S-W. Cheong, Phys. Rev. Lett. **76**, 4042 (1996); S-W. Cheong and C. H. Chen, in *Colossal Magnetoresistance and Related Properties*, ed. by B. Raveau and C. N. R. Rao (World Scientific).

[3] Y. Moritomo, A. Asamitsu, H. Kuwahara, Y. Tokura, Nature **380**, 141 (1996).

[4] C. Zener, Phys. Rev. **82**, 403 (1951).

[5] P. G. de Gennes, Phys. Rev. **118**, 141 (1960).

[6] A. J. Millis, et al., Phys. Rev. Lett. **74**, 5144 (1995); H. Röder, et al., Phys. Rev. Lett. **76**, 1356 (1996).

[7] E. Müller-Hartmann and E. Dagotto, Phys. Rev. **B 54**, R6819 (1996).

[8] S. Yunoki, J. Hu, A. Malvezzi, A. Moreo, N. Furukawa, and E.Dagotto, Phys. Rev. Lett. **80**, 845 (1998).

[9] E. Dagotto, S. Yunoki, A. Malvezzi, A. Moreo, J. Hu, S. Capponi, D. Poilblanc, and N. Furukawa, Phys. Rev. **B 58**, 6414 (1998).

[10] S. Yunoki and A. Moreo, Phys. Rev. **B 58**, 6403 (1998).

[11] S. Yunoki, A. Moreo, and E. Dagotto, cond-mat/9807149.

[12] V. J. Emery, S. A. Kivelson, and H. Q. Lin, Phys. Rev. Lett. **64**, 475 (1990). See also V. J. Emery, and S. A. Kivelson, Physica **C 209**, 597 (1993).

[13] E. Dagotto, Rev. Mod. Phys. **66**, 763 (1994), and references therein.

[14] J. M. Tranquada et al., Nature **375**, 561 (1995), and references therein.

[15] U. Löw et al., Phys. Rev. Lett. **72**, 1918 (1994); S. Haas et al., Phys. Rev. **B 51**, 5989 (1995).

[16] N. Furukawa, J. Phys. Soc. Jpn. **63**, 3214 (1994).

[17] J. Riera, K. Hallberg, and E. Dagotto, Phys. Rev. Lett. **79**, 713 (1997).

[18] Closed shell BC or open BC are needed to stabilize a ferromagnet. If other BC are used the spin correlations at short distances are still strongly FM (if working at couplings where ferromagnetism is stable), but not at large distances where they become negative. This well-known effect was observed before in K. Kubo, J. Phys. Soc. Jpn. **51**, 782 (1982); J. Zang, et al., J. Phys.: Condens. Matter **9**, L157 (1997); T. A. Kaplan and S. D. Mahanti, *ibid*, L291 (1997); and in Ref. [17]. It does not present a problem in the analysis shown in this paper.

[19] In both 1D and 2D the Mermin-Wagner theorem forbids a nonzero T_c in the model. However, the correlation lengths can be very large even at finite T/t.

[20] IC effects were predicted using a Hartree-Fock approximation (J. Inoue and S. Maekawa, Phys. Rev. Lett. **74**, 3407 (1995)) as an alternative to canted FM [5].

[21] See also E. L. Nagaev, Phys. Status Solidi (b) **186**, 9 (1994); D. Arovas and F. Guinea, cond-mat/9711145; M. Yu. Kagan, et al., cond-mat/9804213; M. Yamanaka, W. Koshibae, and S. Maekawa, preprint, cond-mat/9807173.

[22] Y. Moritomo, A. Asamitsu and Y. Tokura, Phys. Rev. **B 51**, 16491 (1995).

[23] D. D. Sarma et al., Phys. Rev. **B 53**, 6873 (1996).

[24] H. Röder, et al., Phys. Rev. **B 56**, 5084 (1997).

[25] A. Malvezzi, S. Yunoki, and E. Dagotto, in preparation.

[26] Y. Murakami et al., Phys. Rev. Lett. **80**, 1932 (1998).

[27] A. J. Millis et al., Phys. Rev. **B 54**, 5405 (1996).

[28] Most of the work in one-dimension (1D) has been performed using $t_{11} = t_{22} = 2t_{12} = 2t_{21}$ (set T_1), but results have also been obtained with $t_{11} = t_{22}$ and $t_{12} = t_{21} = 0$ (T_2), as well as with the hopping that takes into account the proper orbital overlap, namely $t_{11} = 3t_{22} = \sqrt{3}t_{12} = \sqrt{3}t_{21}$ (T_3) (see S. Ishihara et al., Phys. Rev. **B 56**, 686 (1997)). In two-dimensions (2D), the set T_1 in both directions was used, but also the combination of T_3 in the y-direction and $t_{11} = 3t_{22} = -\sqrt{3}t_{12} = -\sqrt{3}t_{21}$ (T_4) in the x-direction. Finally, in three-dimensions (3D) T_4 was used in the x-direction, T_3 in the y-direction, and $t_{11} = t_{12} = t_{21} = 0$, $t_{22} = 4/3$ (T_5) in the z-direction.

[29] This approximation was shown to be accurate in Ref. [8, 9].

[30] J. Kanamori, J. Appl. Phys. (Suppl.) **31**, 145 (1960).

[31] T. G. Perring et al., Phys. Rev. Lett. **78**, 3197 (1997).

[32] J. B. Goodenough, Phys. Rev. **100**, 565 (1955); K. I. Kugel and D. I. Khomskii, JETP Lett. **15**, 446 (1972); S. Ishihara, J. Inoue and S. Maekawa, Phys. Rev. **B 55**, 8280 (1997); T. Mizokawa and A. Fujimori, Phys. Rev. **B 56**, R493 (1997).

[33] Studies at larger values of λ's are too costly in CPU time.

[34] In Ref. [27], a MIT at $\lambda \sim 1$ was also found but it was not associated with orbital order.

[35] The energies needed for this construction are obtained at low-T from the MC evolution of the density [10].

[36] Slave-fermion mean-field studies of a model with strong Coulomb interactions and no JT phonons by S. Maekawa et al. (private communication) have also recently observed a similar PS pattern.

[37] P. G. Radaelli et al., talk given at the *Second International Conference on Stripes and High Tc Superconductivity*, Rome, June 2-6 (1998). The results of Ref. [11] were presented at the same conference session by one of the authors (A.M.).

[38] M. Quijada et al., cond-mat/9803201. See also S. G. Kaplan et al., Phys. Rev. Lett. **77**, 2081 (1996); Y. Okimoto et al., Phys. Rev. Lett. **75**, 109 (1995) and Phys. Rev. **B 55**, 4206 (1997); T. Ishikawa et al., Phys. Rev. **B 57**, R8079 (1998).

[39] J. H. Jung et al., Phys. Rev. **B 57**, R11043 (1998); K. H. Kim et al., cond-mat/9804167 and 9804284.

[40] Although its dependence with λ was weak.

[41] J. M. De Teresa et al., Nature (London) **386**, 256 (1997).

[42] M. Hennion et al., Phys. Rev. Lett. **81**, 1957 (1998)

[43] J. W. Lynn et al., Phys. Rev. Lett. **76**, 4046 (1996).

[44] G. Allodi et al., Phys. Rev. **B 56**, 6036 (1997).

[45] D. E. Cox et al., Phys. Rev. **B 57**, 3305 (1998).

[46] Wei Bao et al., Solid State Comm. **98**, 55 (1996).

[47] Y. Yamada et al., Phys. Rev. Lett. **77**, 904 (1996).

[48] M. Roy, J. F. Mitchell, A. P. Ramirez, and P. Schiffer, preprint.

[49] C. H. Booth, F. Bridges, G. H. Kwei, J. M. Lawrence, A. L. Cornelius, and J. J. Neumeier, Phys. Rev. Lett. **80**, 853 (1998); Phys. Rev. **B 57**, 10440 (1998).

[50] M. Jaime, P. Lin, M. B. Salamon, P. Dorsey, and M. Rubinstein, preprint.

[51] J.-S. Zhou and J. B. Goodenough, Phys. Rev. Lett. **80**, 2665 (1998).

[52] S. J. L. Billinge, R. G. DiFrancesco, G. H. Kwei, J. D. Thompson, M. F. Hundley, and J. Sarrao, proceedings of the workshop "Physics of Manganites", Michigan State University, July 1988, Eds. T. Kaplan and S. D. Mahanti (this volume).

[53] M. R. Ibarra, G-M. Zhao, J. M. De Teresa, B. Garcia-Landa, Z.Arnold, C. Marquina, P. A. Algarabel, H. Keller, and C. Ritter, Phys. Rev. **B 57**, 7446 (1998).

[54] J. J. Rhyne, H. Kaiser, H. Luo, Gang Xiao, and M. L. Gardel, J. Appl. Phys. **93**, 7339 (1998).

[55] R. H. Heffner et al., Phys. Rev. Lett. **77**, 1869 (1996).

[56] J. H. Jung, K. H. Kim, H. J. Lee, J. S. Ahn, N. J. Hur, T. W. Noh, M. S. Kim, and J.-G. Park, cond-mat/9809107.

TWO FERROMAGNETIC STATES IN MAGNETORESISTIVE MANGANITES
- FIRST ORDER TRANSITION DRIVEN BY ORBITALS -

S. Maekawa, S. Ishihara and S. Okamoto

Institute for Materials Research,
Tohoku University,
Sendai 980-8577, Japan

ABSTRACT

A systematic study of the electronic structure in perovskite manganites is presented. The effective Hamiltonian is derived by taking into account the degeneracy of e_g orbitals and strong electron correlation in Mn ions. The spin and orbital orderings are examined as functions of carrier concentration in the mean-field approximation applied to the effective Hamiltonian. We obtain the first order phase transition between ferromagnetic metallic and ferromagnetic insulating states in the lightly doped region. The transition is accompanied with the orbital order-disorder one which is directly observed in the anomalous X-ray scattering experiments. The present investigation shows a novel role of the orbital degree of freedom on metal-insulator transition in manganites.

1. INTRODUCTION

Perovskite manganites have recently attracted much attention. They exhibit a variety of anomalous phenomena such as colossal magnetoresistance (CMR)[1-4] and charge ordering/melting transitions,[5-6] which depend crucially on carrier-doping. In this paper, we examine a systematic study of the electronic structure and discuss the mechanism of the properties. An effective Hamiltonian is derived by taking into account the degeneracy of e_g orbitals, strong electron correlation and Hund coupling in Mn ions.[7] In the Hamiltonian, charge, spin and orbital degrees of freedom are included on an equal footing. The spin and orbital orderings are examined as functions of carrier concentration in the mean-field approximation applied to the effective Hamiltonian. We obtain the first order phase transition between ferromagnetic metallic and ferromagnetic insulating states in the lightly doped region. The transition is accompanied with the orbital order-disorder one. We show theoretically how to observe the orbital ordering by the anomalous X-ray scattering.

In Sec. 2, the effective Hamiltonian is derived and examined in the mean-field approximation. We obtain the phase diagram as a function of carrier-doping. In Sec. 3, the anomalous X-ray scattering is discussed as a probe to observe the orbital ordering. The first-order phase transition between two ferromagnetic states is discussed in the light of the phase diagram in Sec. 4. In Sec. 5, the summary and conclusion are given. The present investigation shows a novel role of the orbital degree of freedom on metal-insulator transition in manganites.

2. EFFECTIVE HAMILTONIAN AND PHASE DIAGRAM

Let us start with undoped perovskite manganites such as $LaMnO_3$, where each Mn^{3+} ion has one e_g and three t_{2g} electrons. The e_g electron occupies one of the orbitals $d_{3z^2-r^2}$ and $d_{x^2-y^2}$ as shown in Fig. 1, and couples with the localized t_{2g} spins ferromagnetically due to the Hund coupling. The Hund coupling H_K and the antiferromagnetic exchange interaction between nearest-neighbor t_{2g} spins $H_{t_{2g}}$ are given by

$$H_K + H_{t_{2g}} = -K\sum_i \vec{S}_i^{t_{2g}} \cdot \vec{s}_i + J_s \sum_{\langle i,j \rangle} \vec{S}_i^{t_{2g}} \cdot \vec{S}_j^{t_{2g}}, \qquad (1)$$

where both the constants K and J_s are defined to be positive, and $\vec{S}_i^{t_{2g}}$ and $\vec{s}_i$ are the operators for a t_{2g} spin with $S=3/2$ and that of e_g electron with $S=1/2$, respectively. The two e_g orbitals are assumed to be degenerate. The matrix elements of the electron transfer between γ orbital at site i and γ' orbital at nearest neighbor site j, $t_{ij}^{\gamma\gamma'}$, is estimated by the

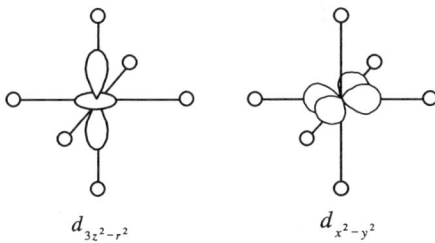

Fig. 1. Schematic illustration of $d_{3z^2-r^2}$ and $d_{x^2-y^2}$ orbitals in an octahedron of O ions.

second-order perturbation with respect to the electron transfer between Mn$3d$ and O$2p$ orbital (t_{pd}). t_{pd} is parameterized as $t_{pd} = \alpha(\gamma)V_{pd\sigma}$, where α is a numerical factor and $V_{pd\sigma}$ is an overlap integral independent of the orbitals. Then, $t_{ij}^{\gamma\gamma'}$ is denoted by $t_{ij}^{\gamma\gamma'} = \alpha(\gamma)\alpha(\gamma')t_0$, where t_0 ($\propto V_{pd\sigma}^2$) is treated as a parameter. The strong intra- and inter-orbital Coulomb interactions U and U', respectively, cause the localization of e_g electrons. We eliminate the doubly occupied configuration in the e_g states and obtain the leading term of the effective Hamiltonian,[7]

$$H_{eff} = \tilde{H}_{e_g} + H_K + H_{t_{2g}}, \tag{2}$$

with

$$\tilde{H}_{e_g} = -2\tilde{J}\sum_{\langle ij \rangle}(\tfrac{3}{4} + \vec{S}_i \cdot \vec{S}_j)(\tfrac{1}{4} - \psi_i^\dagger \hat{\tau}_{ij} \psi_j), \tag{3}$$

where

$$\psi_i^\dagger = (T_i^z, T_i^x), \tag{4}$$

$$\hat{\tau}_{ii\pm x} = \frac{1}{4}\begin{pmatrix} 1 & -\sqrt{3} \\ -\sqrt{3} & 3 \end{pmatrix}, \quad \hat{\tau}_{ii\pm y} = \frac{1}{4}\begin{pmatrix} 1 & \sqrt{3} \\ \sqrt{3} & 3 \end{pmatrix}, \quad \hat{\tau}_{ii\pm z} = \frac{1}{4}\begin{pmatrix} 4 & 0 \\ 0 & 0 \end{pmatrix}. \tag{5}$$

Here, $\tilde{J} = t_0^2/(U'-J')$ with J' (>0) being the intra-orbital exchange interaction between e_g electrons. $\vec{T}_i$ is the pseudo-spin operator for the orbital degree of freedom at site i defined as

$$\vec{T}_i = \frac{1}{2}\sum_{\sigma\gamma\gamma'} \tilde{d}_{i\gamma\sigma}^\dagger \vec{\sigma}_{\gamma\gamma'} \tilde{d}_{i\gamma'\sigma}, \tag{6}$$

where $\vec{\sigma}$ is the Pauli matrix and $\tilde{d}_{i\gamma\sigma} = d_{i\gamma\sigma}(1-n_{i\gamma-\sigma})(1-n_{i-\gamma\sigma})(1-n_{i-\gamma-\sigma})$ with $d_{i\gamma\sigma}$ being the annihilation operator of an electron with spin σ in the orbital γ at site i and $n_{i\gamma\sigma} = d_{i\gamma\sigma}^\dagger d_{i\gamma\sigma}$. The eigenstates of the operator $\vec{T}_i$ correspond to the occupied and unoccupied e_g orbitals. For example, for the $T^z = 1/2$ and $-1/2$, an electron occupies the $d_{3z^2-r^2}$ and $d_{x^2-y^2}$ orbitals, respectively. We note that Eq. (3) does not include T^y.

In the doped manganites such as $La_{1-x}Sr_xMnO_3$, e_g electrons have the kinetic energy,

$$H_t = \sum_{\langle ij \rangle \sigma \gamma \gamma'} (t_{ij}^{\gamma\gamma'} \tilde{d}_{i\gamma\sigma}^{\dagger} \tilde{d}_{j\gamma'\sigma} + h.c.). \tag{7}$$

A similar model Hamiltonian with Eq. (3) has been proposed by Khomskii and Kugel[8] and Castellani, Natoli and Ranninger.[9] However, effects of t_{2g} spins were not introduced in their model. Roth,[10] Cyrot and Lyon-Caen[11] and Inagaki[12] also proposed a model, which coincides with ours if the matrix elements of the electron transfer are assumed as $t_{ij}^{\gamma\gamma'} = t_0 \delta_{\gamma\gamma'}$. In this case, the orbital interaction is isotropic in contrast with Eq. (3).

The Hund coupling K is so strong that e_g and t_{2g} spins at the same site are parallel. As seen in Eq. (3), spins prefer ferromagnetic ordering, whereas orbitals do the alternate ordering which is called the antiferro-type, hereafter. It is known that in the doped manganites, the carrier motion induces spin ferromagnetism due to the double exchange interaction. Therefore, we may expect a rich phase diagram as a function of the parameters $\tilde{J}$, J_s and t_0 as well as carrier concentration.

We obtain the phase diagram at zero temperature as a function of hole concentration and J_s/t_0 in the mean field approximation[13] applied to the effective Hamiltonian. In the spin and orbital structures, four types of the mean field are considered; the ferro (F)-type, and three antiferro-types (layer(A)-type, rod(C)-type and NaCl(G)-type). We introduce the rotating frame in the spin and orbital spaces[14] and describe the states by the rotating angle $\theta_i^{(s)}$ and $\theta_i^{(t)}$, respectively. In the rotating frame, $\langle \tilde{S}_i^z \rangle$ ($=\cos\theta_i^{(s)} \langle S_i^z \rangle - \sin\theta_i^{(s)} \langle S_i^x \rangle$), $\langle \tilde{S}_i^{t_{2g}z} \rangle$, and $\langle \tilde{T}_i^z \rangle$ ($=\cos\theta_i^{(t)} \langle T_i^z \rangle - \sin\theta_i^{(t)} \langle T_i^x \rangle$) are adopted as the mean field order parameters. The kinetic energy term (H_t) is rewritten as $-\sum_{\langle ij \rangle} h_j^{\dagger} h_i \sum_{\sigma} z_{i\sigma}^{(s)*} z_{j\sigma}^{(s)} \sum_{\gamma\gamma'} z_{i\gamma}^{(t)*} t_{ij}^{\gamma\gamma'} z_{j\gamma'}^{(t)} + h.c.$, where h_i is a fermion operator describing the hole carrier and $z_{i\sigma}^{(s)}$ ($z_{i\gamma}^{(t)}$) is the element of the unitary matrix for the rotation in the spin (orbital) frame. We assume $\langle h_i^{\dagger} h_i \rangle = x$ and represent $\langle z_{i\sigma}^{(s)*} z_{j\sigma}^{(s)} \rangle$ and $\langle z_{i\gamma}^{(t)*} t_{ij}^{\gamma\gamma'} z_{j\gamma'}^{(t)} \rangle$ by the rotating angle. By minimizing the energy, the phase diagram at $T=0$ is obtained.

The results are shown in Fig. 2. The value of J_s/t_0 in the manganites which we are interested in is estimated to be 0.001~0.01 from the Néel temperature of $CaMnO_3$.[15] Therefore, let us consider the case with $J_s/t_0 = 0.004$. At the hole concentration $x=0.0$, the A-type antiferromagnetic spin ordering is realized. With increasing x, the ferromagnetic spin state appears where the orbitals show the ordering given in Fig. 3 (a). This state is called F_1. With further increasing x, we find the phase separation between two ferromagnetic spin states F_1 and F_2. The state F_2 shows the orbital ordering given in Fig. 3 (b) which provides the gain of the kinetic energy of hole carriers. Since the state has more holes than F_1, the ferromagnetism in the state F_2 is caused by the double exchange interaction. On the other hand, the state F_1 has less holes and the ferromagnetism is due to the superexchange

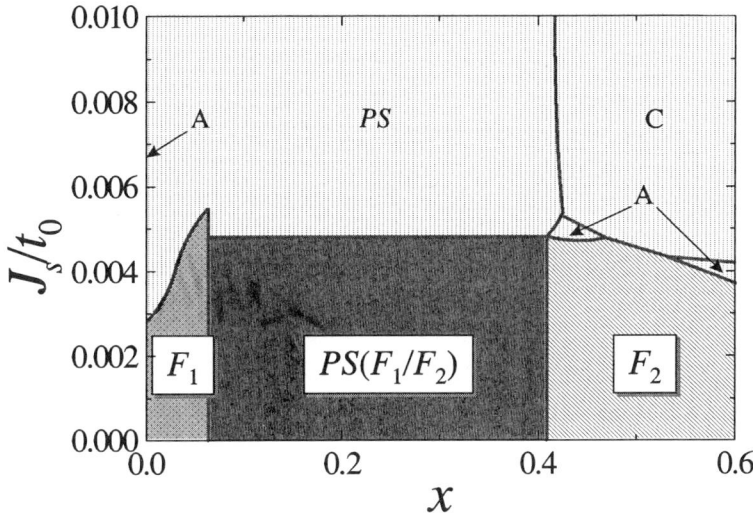

Fig. 2. Theoretical phase diagram at zero temperature calculated in the mean field approximation as a function of the carrier concentration (x) and the antiferromagnetic superexchange interaction (J_s) between localized t_{2g} spins. J_s is normalized by t_0 which is the electron transfer intensity between neighboring e_g orbitals in Mn ions, and the realistic value of J_s/t_0 is estimated to be of the order of 0.001 for the Néel temperature of $CaMnO_3$. The two kinds of ferromagnetic phases (F_1 and F_2) with different orbital structures are shown and phase separation region $PS(F_1/F_2)$ where F_1 and F_2 coexist with different volume fractions appears between them. A and C imply the layer-type and rod-type antiferromagnetic phases, respectively, and PS implies the phase separated region between ferromagnetic and antiferromagnetic phases, and the two antiferromagnetic phases.

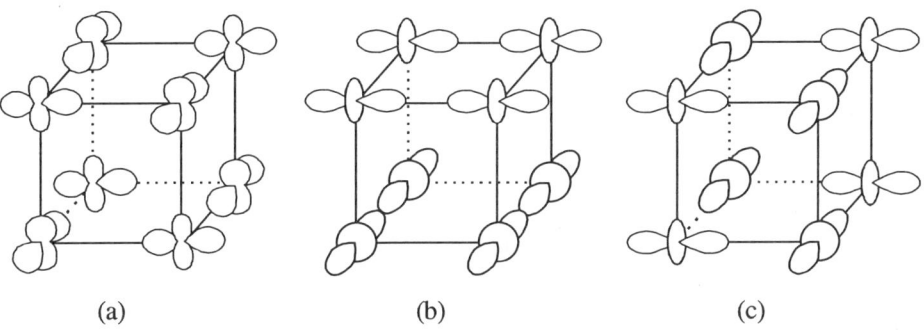

Fig. 3. Schematic illustration of the orbital orderings obtained by the theoretical calculation. (a) ferromagnetic insulating state (F_1) and (b) ferromagnetic metallic state (F_2) given in Fig. 2. The orbital structure in $LaMnO_3$ is shown in (c) for comparison.

interaction induced by the orbital antiferro-type ordering. In Fig. 4, the total energy is plotted as a function of x. The two minima correspond to F_1 and F_2. For example, the compound with $x=0.2$ shows the phase separation, and 60 % and 40 % of the sample are F_1 with $x=0.06$ and F_2 with $x=0.41$, respectively. Recently, the first-order phase transition between two ferromagnetic spin states has been experimentally observed. The details will be discussed in Sec. 4. Fig. 2 also shows the phase separation between A-type and C-type antiferromagnetic states when J_s/t_0 is large. Such a phase separation may be realized in the materials with small hopping parameter t. The two ferromagnetic states have also been shown by Maezono et al.[16] The results are in accord with ours although the theoretical method is different each other. Yunoki et al.[17] have proposed the phase separation between two ferromagnetic states due to the Jahn-Teller coupling without electron correlation.

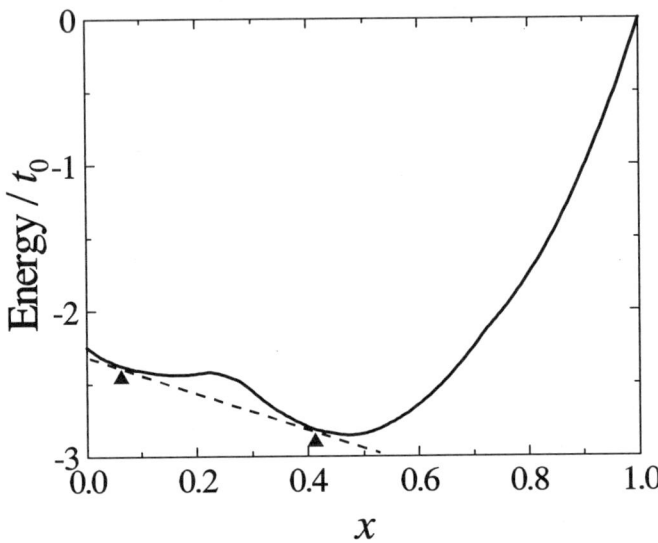

Fig. 4. Total energy of a function of hole concentration x. Two minima correspond to the F_1 and F_2 states.

3. ORBITAL ORDERING AND ANOMALOUS X-RAY SCATTERING

As discussed in the previous section, the orbital ordering plays a crucial role in the magnetic and electronic properties in manganites. However, the direct observation of the orbitals was limited experimentally. Recently, Murakami et al. have applied the anomalous X-ray scattering in order to detect the orbital ordering in single layered manganites

$La_{0.5}Sr_{1.5}MnO_4$[18]. They focused on a reflection at $(3/4,3/4,0)$ point and obtained a resonant-like peak near the K-edge of a Mn^{3+} ion below about 200K. They further observed the unique polarization dependence which is attributed to the tensor character of the anomalous scattering factor. When all Mn^{3+} ions are equivalent, the reflection at $(3/4,3/4,0)$ is forbidden. Therefore, an appearance of the intensity implies that two kinds of orbital are alternately aligned in the MnO_2 plane (antiferro-type). The experimental results also imply that the dipole transition between Mn 1s and Mn 4p orbitals causes the scattering. The experimental method was extended to $La_{1-x}Sr_xMnO_3$ with $x=0.0$[19] and 0.12[20]. In this section, we study theoretically the anomalous X-ray scattering in relation to its role as a detector of the orbital ordering in manganites.[21]

The structure factor of the X-ray scattering is expressed as a sum of the normal and anomalous part of the atomic scattering factor. The normal part is given by the Fourier transform of the charge density ρ_i in the i-th atom, $f_{0i} = \langle f | \rho_i(\vec{K} = \vec{k}'' - \vec{k}') | 0 \rangle$, where $|0\rangle$ ($|f\rangle$) is the initial (final) electronic state with energy ε_0 (ε_f), and $\vec{k}'$ and $\vec{k}''$ are the momenta of incident and scattered photons, respectively. The anomalous part is derived by the interaction between electronic current and photon and is expressed as

$$\Delta f_{i\alpha\beta} = \frac{m}{e^2} \sum_l \left\{ \frac{\langle f | j_{i\alpha}(-\vec{k}')|l\rangle\langle l|j_{i\beta}(\vec{k}'')|0\rangle}{\varepsilon_0 - \varepsilon_l - \omega_{k''} - i\delta} + \frac{\langle f|j_{i\beta}(\vec{k}'')|l\rangle\langle l|j_{i\alpha}(-\vec{k}')|0\rangle}{\varepsilon_0 - \varepsilon_l + \omega_{k'} - i\delta} \right\}, \qquad (8)$$

where $|l\rangle$ is the intermediate electronic states with energy ε_l and δ is a dumping constant. The current operator $j_{i\alpha}(\vec{k})$ describes the dipole transition between Mn 1s and 4p orbitals at site i coupled with photon with polarization in the α direction. The contribution from the quadrupole transition is neglected because the inversion symmetry is preserved in the system which we are interested in.

As mentioned above, the anomalous scattering is dominated by the Mn 1s→4p E1 transition. In this case, how does the 3d orbital ordering reflect on the anisotropy of the anomalous scattering factor? In order to study the problem, we consider the electronic structure in a MnO_6 octahedron, since the local electronic excitation dominates Δf_i. Then, we find that the electron hybridization do not result in the anisotropy of the scattering factor, since the hybridization between the Mn 3d and O 2p orbitals and between the Mn 4p and O 2p ones are decoupled. One of the promising origins of the anisotropy of the scattering factor is the Coulomb interaction between Mn 3d and 4p electrons. The electron-electron interaction in the orbital ordered state breaks the cubic symmetry and thus lifts the degeneracy of Mn 4p orbitals.

The interaction between Mn 3d and 4p electrons is represented as

$$V(3d_{\gamma_{\theta\pm}}, 4p_\gamma) = F_0 + 4F_2 \cos\left(\theta \pm m_\gamma \frac{2\pi}{3}\right), \qquad (9)$$

where $m_x=+1$, $m_y=-1$, and $m_z=0$, and $|3d_{\gamma_{\theta+}}\rangle = \cos\frac{\theta}{2}|3z^2-r^2\rangle + \sin\frac{\theta}{2}|x^2-y^2\rangle$ and $|3d_{\gamma_{\theta-}}\rangle$ is its counterpart. F_n is the Slater integral between $3d$ and $4p$ electrons. The explicit formula of F_n is given by $F_0 = F^{(0)}$ and $F_2 = \frac{1}{35}F^{(2)}$ with $F^{(n)} = \int dr\, dr'\, r^2 r'^2 R_{3d}(r)^2 R_{4p}(r')^2 \frac{r_<^n}{r_>^{n+1}}$, where $r_<$ ($r_>$) is the smaller (larger) one between r and r'. When $d_{3z^2-r^2}$ orbital is occupied ($\theta=0$), the energy in the $4p_z$ orbital is higher than that of the $4p_x$ ($4p_y$) orbital by $6F_2$. As a result, $(\Delta f_i)_{xx(yy)}$ dominates the anomalous scattering near the edge in comparison with $(\Delta f_i)_{zz}$.

The inter atomic Coulomb interaction between Mn $4p_y$ electron and O $2p_{\gamma_{\theta-}}$ hole also provides an origin of the anisotropy of the scattering factor through the Mn $3d$-O $2p$ hybridization. The interaction is represented by $V(2p_{\gamma_{\theta-}}, 4p_y) = -\varepsilon + \frac{\varepsilon\rho^2}{5}\cos(\theta + m_y\frac{2\pi}{3})$, where the definition of m_y is the same as that in Eq. (9). $\varepsilon=Ze^2/a$ and $\rho=\langle r_{4p}\rangle/a$, where $Z=2$, a is the Mn-O bond length, and $\langle r_{4p}\rangle$ is the average radius of Mn $4p$ orbital. Although the above two interactions cooperate to bring about the anisotropy of the scattering factor, it seems likely that the magnitude of $V(2p_{\gamma_{\theta-}}, 4p_y)$ is much reduced by the screening effects in comparison with $V(3d_{\gamma_{\theta+}}, 4p_y)$.

Being based on the Hamiltonian, the imaginary part of the scattering factor is calculated by the configuration interaction method. The calculated $(\Delta f_i)_{\alpha\alpha}$ near the K-edge is shown in

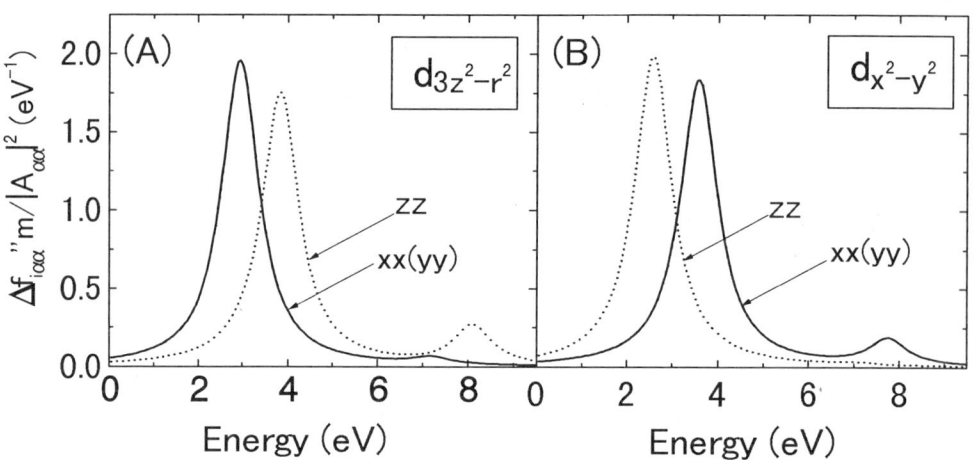

Fig. 5. The imaginary part of the scattering factor $[(\Delta f_i'')_{xx(zz)} m/\pi |A_{x(z)}|^2]$ in the case where the following orbital is occupied: (a) $\theta=0$ ($d_{3z^2-r^2}$) and (b) $\theta=\pi$ ($d_{x^2-y^2}$). The straight and broken lines show $(\Delta f_i'')_{xx}$ and $(\Delta f_i'')_{zz}$, respectively. The origin of the energy is taken to be arbitrary. Here, m is the mass of an electron and $A_{x(z)}$ is the coupling constant between electron and photon with $x(z)$ polarization of the electric field.

Fig. 5 (a), where the $d_{3z^2-r^2}$ orbital is occupied. It is noted that the edge of the lowest main peak corresponds to the Mn K-edge. The detailed structure away from the edge may become broad and be smeared out in the experiments by overlapping with other peaks which are not included in the calculation. In the figure, the scattering intensity is governed by $(\Delta f_i)_{xx}$. Owing to the core hole potential, main and satellite peaks are attributed to the transition from the ground state, which is mainly dominated by the $|3d^1_{\gamma_{\theta+}}\rangle$ state, to the $|\underline{1s}\,3d^1_{\gamma_{\theta+}}\,3d^1_{\gamma_{\theta-}}\,4p^1_{x(z)}\,\underline{2p_{\gamma_{\theta-}}}\rangle$ and $|\underline{1s}\,3d^1_{\gamma_{\theta+}}\,4p^1_{x(z)}\rangle$ excited states, respectively, where the underlines show the states occupied by holes, although the two excited states strongly mix with each other. Therefore, the anisotropy in the main peak is caused by $V(3d_{\gamma_{\theta+}},4p_\gamma)$ through the Mn 3d-O 2p hybridization. As a comparison, the results in the case where the $d_{x^2-y^2}$ orbital is occupied are shown in Fig. 5 (b). In the figure, the anisotropy near the edge is entirely opposite to that in Fig. 5 (a); i.e., the scattering factor near the edge is governed by $(\Delta f_i'')_{zz}$, owing to the positive value of $V(3d_{x^2-y^2},4p_x)-V(3d_{x^2-y^2},4p_z)$.

4. FIRST ORDER TRANSITION BETWEEN TWO FERROMAGNETIC STATES

Recently, the first-order phase transition between ferromagnetic metallic and ferromagnetic insulating states has been discovered in La$_{1-x}$Sr$_x$MnO$_3$ with $x\sim0.12$.[20] In this section, we discuss the transition in the light of the phase diagram given in Sec. 2. Let us first review the experimental data. The electrical resistivity $\rho(T)$ shows an anomalous temperature dependence below the Curie temperature $T_C=170$ K. It is metallic between T_C and $T_L=145$ K. However, as temperature decreases below T_L, ρ increases rapidly and the crystal structure changes to the less distorted (pseudo cubic) O^* phase from the distorted orthorhombic O' phase.[22] A first-order transition occurs by applying a magnetic field between T_C and T_L. Discontinuous jump in both ρ and the magnetization curve are brought about at the critical field $H_C(T)$. The striction ΔL defined as $\Delta L = L(T)-L(140K)$ tends to zero above $H_C(T)$. With decreasing temperature, $H_C(T)$ decreases and at T_C it goes to zero.

The temperature dependence of lattice constant and magnetic Bragg reflection intensity were observed in the neutron scattering experiments. The results are plotted in Fig. 6. The ferromagnetic order parameter jumps at T_C simultaneously with the O' to O^* phase transition upon cooling. Superlattice reflections such as $(h,k,l+1/2)$ were also observed, indicating the lattice modulation due to the charge ordering, which is consistent with the previous results.[23]

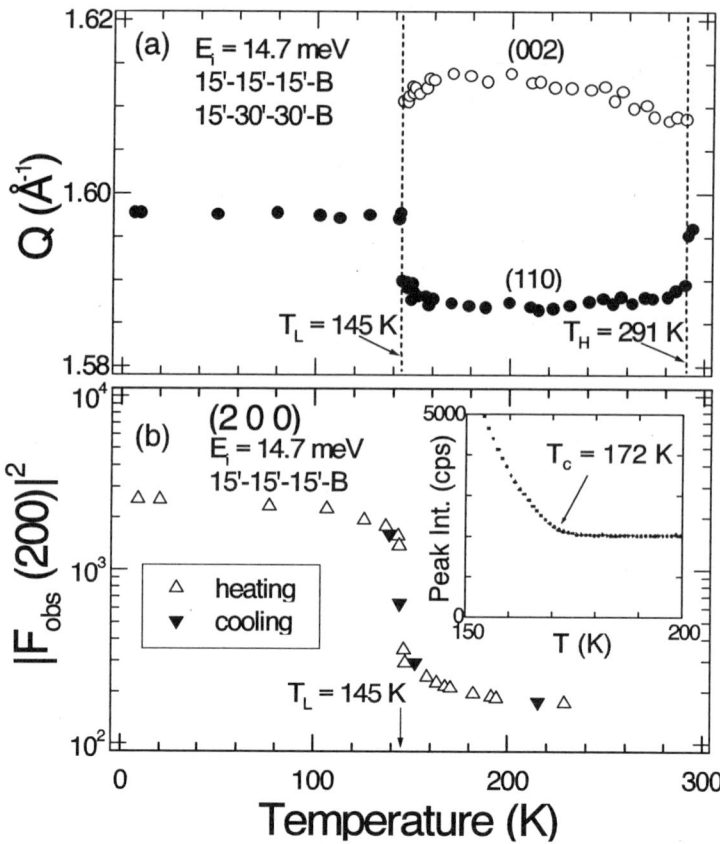

Fig. 6. Temperature dependence of (a) lattice parameter, and (b) integrated intensity of (2 0 0) ferromagnetic Bragg reflection measured with 14.7 meV neutrons.[20] Between T_L (=145 K) and T_H (=291 K), the crystal structure is determined to be O' (orthorhombic). Out side of these temperatures, O^* (pseudo cubic). T_c is determined to be 170 K for $La_{0.88}Sr_{0.12}MnO_3$ which is consistent with the magnetization measurement.

The two ferromagnetic phases have different orbital structure. The resonance-like peak appears at the $(0,3,0)$ reflection in the anomalous X-ray scattering experiments, when the photon energy is tuned at the K-edge (6.552 KeV) in a Mn^{3+} ion as shown in Fig. 7. Besides the energy scan, the azimuthal scan around the scattering vector shows the angle dependence of two-fold sinusoidal symmetry, which gives rise to a direct evidence of the antiferro-type orbital ordering as that in the undoped system $LaMnO_3$.[19] We stress the fact that the intensity appears only below $T_L(O^*$ phase), as shown in Fig. 7, where the notable lattice distortion does not exist. Therefore, the antiferro-type orbital ordering is not the one associated with

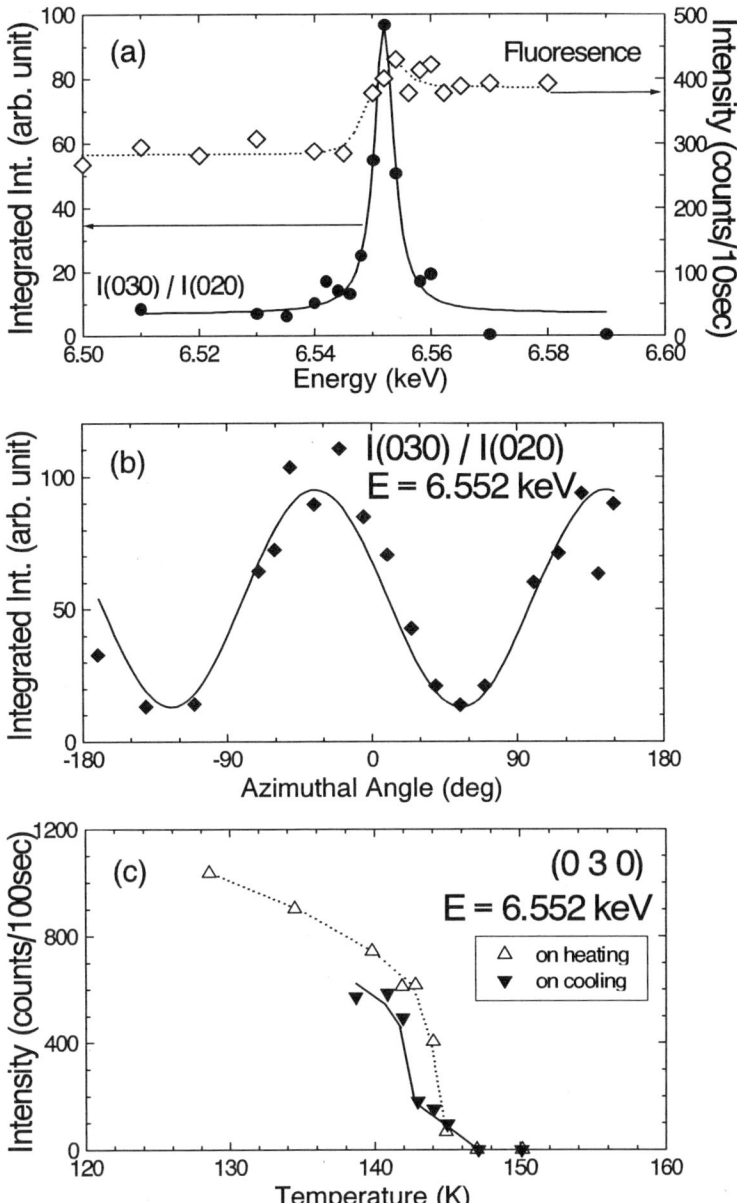

Fig. 7. (a) Energy dependence of intensities in the anomalous X-ray scattering experiments at the orbital ordering reflection (0 3 0) at T=12 K in $La_{0.88}Sr_{0.12}MnO_3$.[20] The resonant energy is determined to be 6.552 KeV. The dashed curve represents fluorescence showing the resonant energy corresponding to the K-edge of Mn cation. (b) The azimuthal angle dependence of orbital ordering reflection (0 3 0). The solid line is two-fold squared sine curve of angular dependence. (c) Temperature dependence of peak intensities of orbital ordering reflection (0 3 0).

the Jahn-Teller type lattice distortion. Note that the spin-wave dispersion in $La_{0.88}Sr_{0.12}MnO_3$ is nearly isotropic, which is entirely different from the two-dimensional relation in $LaMnO_3$, due to the antiferro-type orbital ordering of $d_{3x^2-r^2}/d_{3y^2-r^2}$, which is shown in Fig. 3 (c). Therefore, we anticipate that $La_{0.88}Sr_{0.12}MnO_3$ should have a different orbital state, e.g. the hybridization of $d_{z^2-x^2(y^2-z^2)}$ and $d_{3x^2-r^2(3y^2-r^2)}$.

We consider that the two ferromagnetic states observed in this compound correspond to F_1 and F_2 in the phase diagram in Fig. 2, and the first-order transition occurs between them by applying magnetic field and/or changing temperature. At high temperatures, the F_2 phase is favorable because the entropy promotes the orbital disordering and carrier mobility. At low temperatures, on the other hand, the F_1 phase becomes dominant and occupies the large volume fraction in the system. The stabilization of F_1 phase by an applied magnetic field is also explained as follows: (1) both ferromagnetic ordering and antiferro-type orbital ordering are confirmed to be cooperatively stabilized (2) the magnetic moment is enlarged by changing the dominant magnetic coupling from the double exchange interaction to the superexchange interaction. The first-order phase transition from ferromagnetic metallic to ferromagnetic insulating states is ascribed to the simultaneous transition of orbital order-disorder states. The phase separation occurs between these two ferromagnetic phases and the insulating phase dominates the system with increasing the magnetic field and/or decreasing temperature. The present investigations show a novel role of the orbital degree of freedom in the metal-insulator transition as a hidden parameter, unambiguously for the first time.

5. CONCLUSION

We have studied the magnetic and electronic structure in perovskite manganites taking into account the degeneracy of e_g orbitals and strong electron correlation in Mn ions. The spin and orbital orderings were examined as functions of carrier concentration in the mean-field approximation applied to the effective Hamiltonian, which describes the low energy states. We obtained the phase separation between ferromagnetic metallic and ferromagnetic insulating states in the lightly doped region. The two ferromagnetic states have been discovered experimentally in $La_{1-x}Sr_xMnO_3$ with $x \sim 0.12$. The ferromagnetic metallic state is due to the double exchange interaction, whereas the ferromagnetic insulating state is caused by the superexchange interaction coupled with the orbital degree of freedom.

The orbital degree of freedom was considered to be a hidden parameter until recently, since the direct observation was much limited experimentally. We have studied theoretically the anomalous X-ray scattering in relation to its role as a probe of the orbital ordering in manganites. We conclude that the orbital degree of freedom is not a hidden parameter but is examined together with spin and charge degrees of freedom of electrons in manganites.

ACKNOWLDGEMENTS

The work described in Sect. 4 was done in collaboration with Y. Endoh, K. Hirota, T. Fukuda, H. Kimura, H. Nojiri, and K. Kaneko at Tohoku University and Y. Murakami at KEK. We thank them for providing the experimental data prior to publication and valuable discussions. This work was supported by Priority Areas Grants from the Ministry of Education, Science and Culture of Japan, and CREST (Core Research for Evolutional Science and Technology Corporation) Japan. The numerical calculation was performed at the supercomputer facilities of Institute for Materials Research, Tohoku University and Institute of Solid State Physics, University of Tokyo. S. O. acknowledges the financial support of JSPS Research Fellowship for Young Scientists.

REFERENCES

1. K. Chahara, T. Ohono, M. Kasai, Y. Kanke, and Y. Kozono, Appl. Phys. Lett. **62**, 780 (1993).
2. R. von Helmolt, J. Wecker, B. Holzapfel, L. Schultz and K. Samwer, Phys. Rev. Lett. **71**, 2331 (1993).
3. Y. Tokura, A. Urushibara, Y. Moritomo, T. Arima, A. Asamitsu, G. Kido, and N. Furukawa, Jour. Phys. Soc. Jpn. **63**, 3931 (1994).
4. S. Jin, T. H. Tiefel, M. McCormack, R. A. Fastnacht, R. Ramesh, and L. H. Chen, Science, **264**, 413 (1994).
5. Z. Jirak, S. Krupicka, Z. Simsa, M. Dlouha, and S. Vratislav, J. Mag. Mag. Mat. , **53**, 153 (1985).
6. Y. Tomioka, A. Asamitsu, Y. Moritomo, and Y. Tokura, Jour. Phys. Soc. Jpn. **64**, 3626 (1995).
7. S. Ishihara, J. Inoue, and S. Maekawa, Physica C **263**, 130 (1996), and Phys. Rev. B **55**, 8280 (1997).
8. K. I. Kugel, and D. I. Khomskii, ZhETF Pis. Red. **15**, 629 (1972). (JETP Lett. **15**, 446 (1972)); D. I. Khomskii, and K. I. Kugel, Sol. Stat. Comm. **13**, 763 (1973).
9. C. Castellani, C. R. Natoli, and J. Ranninger, Phys. Rev. **B18**, 4945, (1978); T. M. Rice,

in: *Spectroscopy of Mott Insulators and Correlated Metals, Springer Series in Solid-State Science, Vol. 119, p. 221*, A.Fujimori and Y.Tokura ed., Springer, Berlin, (1998).
10. L. M. Roth, Phys. Rev. **149**, 306, (1966).
11. M. Cyrot, and C. Lyon-Caen, Le Jour. de Physique **36**, 253 (1975).
12. S. Inagaki, Jour. Phys. Soc. Jpn. **39**, 596 (1975).
13. S. Ishihara, S. Okamoto, and S. Maekawa, J. Phys. Soc. Jpn. 66, 2965 (1997).
14. S. Ishihara, M. Yamanaka, and N. Nagaosa, Phys. Rev. B **56**, 686 (1997).
15. J. B. Goodenough, in: *Progress in Solid State Chemistry*, H. Reiss ed., Pergamon, London, (1971).
16. R. Maezono, S. Ishihara, and N. Nagaosa, Phys. Rev. B **57,** R13993 (1998).
17. S. Yunoki, A. Moreo, and E. Dagotto, (unpublished, cond-mat/9807149)
18. Y. Murakami, H. Kawada, H. Kawata, M. Tanaka, T. Arima, H. Moritomo and Y. Tokura, Phys. Rev. Lett. **80**, 1932 (1998).
19. Y. Murakami, J. P. Hill, D. Gibbs, M. Blume, I. Koyama, M. Tanaka, H. Kawata, T. Arima, T. Tokura, K. Hirota, and Y. Endoh, Phys. Rev. Lett. **81**, 582 (1998).
20. Y. Endoh, K. Hirota, Y. Murakami, T. Fukuda, H. Kimura, N. Nojiri, K. Kaneko, S. Ishihara, S. Okamoto, and S. Maekawa , (unpublished).
21. S. Ishihara, and S. Maekawa, Phys. Rev. Lett. **80**, 3799 (1998).
22. H. Kawano, R. Kajimoto, M. Kubota, and H. Yoshizawa, Phys. Rev. B. **53**, R14709 (1996).
23. Y. Yamada, O. Hino, S. Nohdo, R. Kanao, T. Inami, and S. Katano, Phys. Rev. Lett. **77**, 904 (1996).

MAGNETISM AND ELECTRONIC STATES OF SYSTEMS WITH STRONG HUND COUPLING

K. Kubo,[1] D. M. Edwards,[2] A. C. M. Green,[2] T. Momoi[1] and H. Sakamoto[1]

[1]Institute of Physics, University of Tsukuba, Tsukuba,
Ibaraki 305-8571, Japan
[2]Department of Mathematics, Imperial College, London SW7 2BZ,
UK

INTRODUCTION

The fascinating physics of the perovskite manganites[1] is governed by electrons which hop among or are localized on Mn^{3+} and Mn^{4+} ions. It was pointed out already in 1951 by Zener[2] that the strong Hund coupling between e_g and t_{2g} electrons is essential in understanding the ferromagnetism[3] caused by substitution of triply valent La with doubly valent elements such as Ca, Sr or Ba in $LaMnO_3$. The two-fold degeneracy of e_g orbitals causes other important effects in the physics of manganites. The existence of the orbital degrees of freedom leads to orbital ordering in the insulating phases.[4] Various charge and orbital ordered states are experimentally observed.[5] Another important effect is the Jahn-Teller distortion, which causes strong correlations between lattice distortions and orbital and magnetic interactions.[6-8] All these effects are entangled and give rise to the rich phase diagrams and peculiar transport phenomena of the manganites. Though it might be necessary to take into account all of these effects at the same time in order to explain experimental results quantitatively, it is useful to extract one of these important factors and study its effect in detail, in order to understand the physics of manganites in a profound way. In this paper we aim to clarify the effects of the Hund coupling by employing simplified models. First we study the ground state phase diagram of a doubly degenerate Hubbard model. Our main concern is the effectiveness of the Hund coupling on the ferromagnetism. We compare the result in one dimension with that in infinite dimensions, and examine common features and differences between them. The second model we study is a simple ferromagnetic Kondo lattice model or double exchange model. We investigate one-particle states in this model using a single-site approximation and calculate the electrical resistivity. We treat the localized spins as quantum mechanical ones and study the quantum effects on the electronic states.

Doubly Degenerate Hubbard Model

We consider in this section a doubly degenerate Hubbard model described by the Hamiltonian

$$H = -t \sum_{\substack{m=1,2 \\ \sigma=\uparrow,\downarrow}} \sum_{\langle i,j\rangle \in \text{N.N.}} (c^\dagger_{im\sigma} c_{jm\sigma} + h.c.) + U \sum_{i,m} n_{im\uparrow} n_{im\downarrow} + U' \sum_{i,\sigma,\sigma'} n_{i1\sigma} n_{i2\sigma'}$$
$$- J \sum_{i,\sigma,\sigma'} c^\dagger_{i1\sigma} c_{i1\sigma'} c^\dagger_{i2\sigma'} c_{i2\sigma} - J' \sum_{i} (c^\dagger_{i1\uparrow} c^\dagger_{i1\downarrow} c_{i2\uparrow} c_{i2\downarrow} + h.c.), \qquad (1)$$

where $c_{im\sigma}$ ($c^\dagger_{im\sigma}$) denotes the annihilation (creation) operator of the electron at site i with orbital m (=1 or 2) and spin σ. The number operators are denoted by $n_{im\sigma}$. Hoppings of electrons are assumed to occur between the same orbitals of nearest neighbor sites. In real systems, there are off-diagonal hoppings and also hopping integrals are anisotropic for e_g orbitals, that is, they are dependent on the directions of hoppings. This anisotropy may have an important effect on the orbital and antiferromagnetic ordering in manganites.[9, 10] We take here, however, the simplest model which can take account of the effects of orbital degeneracy and Hund coupling. The interaction terms in eq. (1) originate from the Coulomb interaction between electrons at the same site. The last term, which transfers two electrons on one orbital to the other, is often neglected. But this term should be properly considered, since it enhances local quantum fluctuations and the coefficient J' is equal to that of the Hund coupling J if we assume the orbital wave functions are real. The interaction parameters satisfy the relation

$$U = U' + 2J \qquad (2)$$

for e_g orbitals, and we assume this relation in the following.

The Hamiltonian (1) produces various magnetic correlations and effective ferromagnetic interactions between electrons on different sites. Let us first consider two electrons on one site. The spin-triplet states, where two electrons occupy different orbitals, are stabilized by the Hund coupling and have the lowest energy $U' - J$. There are three spin-singlet states; one with energy $U' + J$ where two electrons occupy different orbitals, and the other two with energies $U - J'(= U' + J)$ and $U + J'(= U' + 3J)$ where electrons occupy the same orbital. An effective spin interaction between neighboring sites is derived from this one-site spectrum in the strong correlation regime ($U' > J \gg t$). Let us consider two nearest neighbor sites each of which is occupied by a single electron in this regime. When two electrons with parallel spins sit on different orbitals, virtual hoppings of the electrons between two sites lower the energy by $-2t^2/(U' - J)$. When two electrons have antiparallel spins, the energy is lowered by $-2t^2 U/(U^2 - J'^2)$ or $-2t^2 U'/(U'^2 - J^2)$ depending on whether they are on the same or different orbitals. Hence there is an effective interaction between neighboring sites which favors ferromagnetic spin alignment but alternating alignment of orbital degrees of freedom.[11] In a system with quarter-filled bands, i.e. for $n \equiv N_e/N = 1$, both ferromagnetic long-range order (LRO) and alternating orbital order are expected to coexist in the ground state.[11–15] Here N_e and N denote the total number of electrons and sites, respectively. The second-order perturbation from the atomic limit ($t = 0$) leads to the following effective Hamiltonian for spin operators $\mathbf{S}_i$ and pseudo-spin operators $\boldsymbol{\tau}_i$[12, 25]:

$$\begin{aligned} H_{\text{eff}} = -t^2 \sum_{\langle i,j \rangle} \Big[&\frac{4U}{U^2 - J'^2}\Big(\frac{1}{4} + \tau_i^z \tau_j^z\Big)\Big(\frac{1}{4} - \mathbf{S}_i \cdot \mathbf{S}_j\Big) \\ &- \frac{2J'}{U^2 - J'^2}(\tau_i^- \tau_j^- + \tau_i^+ \tau_j^+)\Big(\frac{1}{4} - \mathbf{S}_i \cdot \mathbf{S}_j\Big) \\ &+ \frac{2U'}{U'^2 - J^2}\Big\{\frac{1}{4} - \tau_i^z \tau_j^z - 2(\boldsymbol{\tau}_i \cdot \boldsymbol{\tau}_j - \tau_i^z \tau_j^z)\Big(\frac{1}{4} + \mathbf{S}_i \cdot \mathbf{S}_j\Big)\Big\} \\ &+ \frac{2J}{U'^2 - J^2}\Big\{\tau_i^z \tau_j^z - \boldsymbol{\tau}_i \cdot \boldsymbol{\tau}_j + 2\Big(\frac{1}{4} - \tau_i^z \tau_j^z\Big)\Big(\frac{1}{4} + \mathbf{S}_i \cdot \mathbf{S}_j\Big)\Big\}\Big]. \end{aligned} \quad (3)$$

In strongly correlated systems with the filling $1 < n < 2$, each lattice site is either singly- or doubly-occupied, and electrons hop from doubly-occupied sites to singly-occupied ones. Doubly-occupied sites are almost necessarily in spin-triplet states due to the Hund coupling and the hopping probability is largest between pairs of sites with parallel spins. As a result the kinetic energy is lowered by ferromagnetic spin correlations. This mechanism favoring ferromagnetism is quite similar to that in the double exchange model of electrons, where electrons interacting with localized spins have lower kinetic energy when spins are aligned parallel.[2, 16] In the following we call this mechanism which favors ferromagnetism the "double exchange mechanism" even when we are not treating localized spins. In the case with less-than-quarter filling ($n < 1$), the "double exchange mechanism" may not work for $U' - J \gg t$. Nevertheless the Hund coupling may lead to ferromagnetism even for $n < 1$, if $t/(U' - J)$ is not too small. The effective ferromagnetic interaction described by eq. (3) between nearest neighbor electrons may have a sizable effect and cause metallic ferromagnetism. Van Vleck argued that this mechanism may be operative in realizing ferromagnetism in Ni.[17]

Though the mechanism favoring ferromagnetism can be understood qualitatively as above, it is far from trivial whether ferromagnetic long-range order occurs in bulk systems. In the following, we present a numerical study of the model in one and infinite dimensions.

One-Dimensional Model

There are rigorous proofs for the ferromagnetic ground state of the one dimensional model in strong coupling limits.[18–20] These proofs are valid in different limits of strong coupling. For the strong Hund coupling case ($J \to \infty$ and $U \to \infty$), existence of ferromagnetism is proved for arbitrary $U'(> 0)$ in $1 < n < 2$,[18, 19] and also for $0 < n \leq 1$ in the special limit $J = U' \to \infty$ and $U \to \infty$.[19] Shen obtained a rather general result that the ground state is fully spin-polarized for any n between 0 and 2 except for 1 if $U = \infty$, and $U'(> 0)$ and $J = J'(> 0)$ are finite.[20] (We note that this result cannot be applied naively to our case which assumes the relation (2).)

So far several numerical studies were done, and ferromagnetism was found for densities near quarter filling.[21, 19, 22–24] These studies were done by diagonalizing relatively small systems with sizes up to 12, and size dependence was not studied yet. We thus need to study systems with larger sizes and examine size effects to obtain conclusive results. We note also that most previous studies did not take into account the J'-term, and assumed relations between U, U' and J which differ from eq. (2).

We report in the following a study of finite-size chains with up to 16 sites applying the exact diagonalization method as well as the density matrix renormalization group (DMRG) method. We employ open boundary conditions, since the periodic boundary condition causes very large size dependence in one dimension (e.g. even-odd oscillations). Remarkably we found little size dependence due to the use of the open boundary conditions. Details of this study will be published elsewhere.[26]

First, we show the ground-state phase diagram for $n = 1$ in Fig. 1a. We obtained the ground state with full spin polarization for $J \simeq U'(\gtrsim 5t)$. The appearance of ferromagnetism in the $J < U'$ region can be well understood with the effective Hamiltonian (3). This ferromagnetic ground state has a strong alternating correlation in the orbital degrees of freedom. This is also consistent with the argument from the effective Hamiltonian. In the perfect ferromagnetic ground state, the orbital degrees of freedom have isotropic (Heisenberg) antiferromagnetic interaction with pseudo-spin and hence the alternating correlation decays in a power form. The phase boundary for $J < U'$ approaches an asymptote $J = \alpha U'$ with $\alpha \simeq 0.35$ for large U'. The asymptote corresponds to the ground-state phase boundary of the effective Hamiltonian.[26] The paramagnetic state for $0 \lesssim J < \alpha U'$ may have the same properties as the ground state of the $SU(4)$ model at $J = 0$.[30–32] The ferromagnetic phase extends to the parameter region $J > U'$, as well. Though this region $J > U'$ is unrealistic, it is of interest from the viewpoint of triplet superconductivity. An attractive force acts between electrons with parallel spins due to the Hund coupling and the present model might have some relevance to experimental results on organic superconductors.[27] For $J > U' \gg t$ two electrons are paired to form a hard-core boson with spin unity. The perturbation due to the hopping term in eq. (1) leads to the effective Hamiltonian for these bosons, which includes hopping, repulsion and antiferromagnetic spin interaction between nearest neighbor bosons. The effective Hamiltonian does not favor ferromagnetism and indeed the ferromagnetic phase does not extend to a region with large $J - U'$ in our numerical calculations. The slope of the phase boundary approaches unity for large U'. We note that only the systems with N_e = even are used in determining the phase diagram. In fact we found a large difference in the phase boundaries for $J > U'$ between systems with even N_e and odd N_e. We consider that the results for small odd N_e are strongly affected by the existence of an unpaired electron and are not useful for extracting bulk properties.

Next we show the ground state phase diagram for $n = 0.5$ in Fig. 1b. The ferromagnetic phase expands compared to that in the quarter-filled case both for $J > U'$ and $J < U'$, and the size dependence is very weak. It is remarkable that ferromagnetism is realized for rather weak Hund coupling, that is, $J \simeq 2t$ for $U' \simeq 5$. Recently Hirsch[24] argued that the Hund coupling is not effective enough to realize ferromagnetism in systems with low density ($n < 1$) and that ferromagnetic exchange interactions between different sites are necessary to explain ferromagnetism in low density systems like Ni. From the present results, we expect that a moderate Hund coupling realizes ferromagnetism in a bulk system in one dimension. This behavior should be compared with the result for infinite dimensions where we could not find ferromagnetism for $n < 1$. (See next section.)

Finally, as an example of the case with $n > 1$, we show the result for $n = 1.25$ in Fig. 1c. In this case ferromagnetism appears in a wider region than in the quarter filled case especially for small J. The lower phase boundary apparently approaches the line $J = 0$ for large U'. This enhanced stability of the ferromagnetic state may be understood as the result of the "double exchange mechanism". On the other hand the phase boundary for $J > U'$ is almost same as that for $n = 1$.

For all densities we found ferromagnetism on the line $J = U'$ with strong $J(\gtrsim 5t)$. This result is consistent with the rigorous result in the limit $J = U' \to \infty$ by Kusakabe and Aoki.[19] We note that all the ferromagnetic ground states obtained above are fully

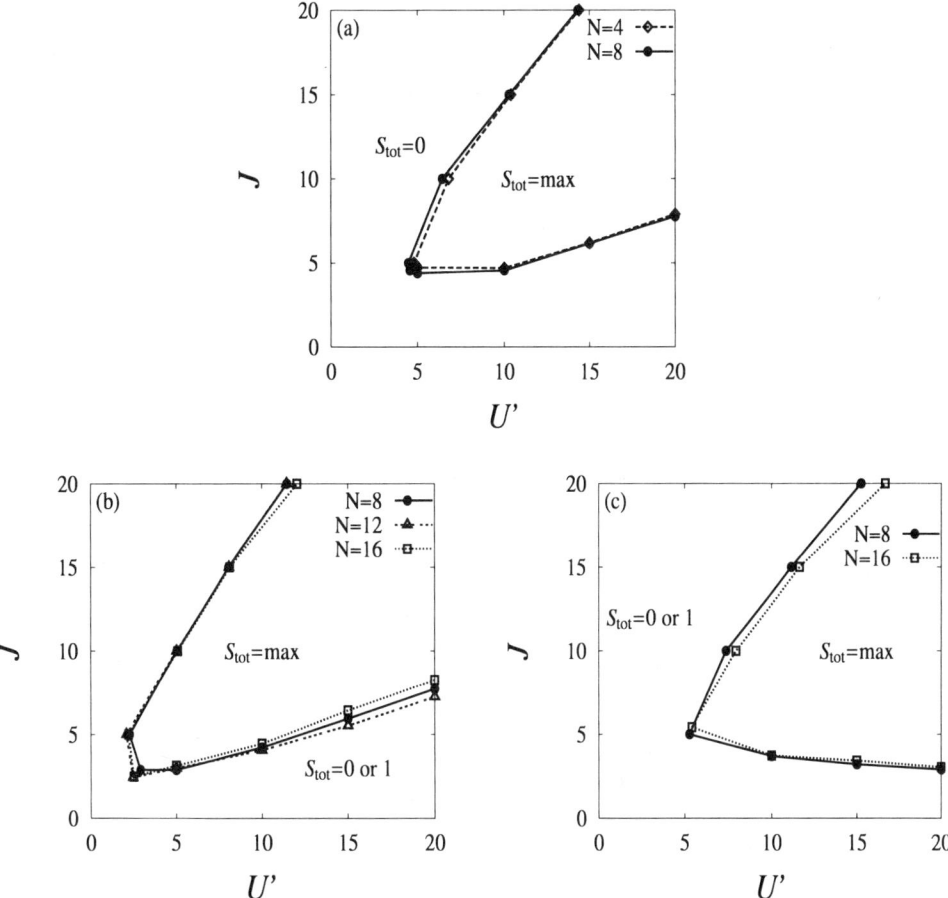

Figure 1: Ground-state phase diagrams of the 1D doubly degenerate Hubbard model for the filling $n = 1$(a), 0.5(b) and 1.25(c). We assume t to be unity.

polarized. Since the ferromagnetic state is fully polarized the spin degrees of freedom are completely suppressed. The orbital degrees of freedom in the ground state are mapped into the usual spin degrees of freedom in the single-band Hubbard model with the interaction parameter $U' - J$.[20] Then we learn that, for $U' - J > 0$, the pseudo-spin (orbital) correlation function decays with a power law as $\cos(|i-j|n\pi) \cdot |i-j|^{-\sigma}$. On the other hand, for $U' - J < 0$, it shows an exponential decay but the pair-pair correlation function of pseudospin-singlet (spin-triplet) pairs decays with a power law, which is a sign of quasi-long-range order of the triplet superconductivity.

Infinite Dimensional Model

Next we discuss the model (1) on a hypercubic lattice in infinite dimensions.[25] We scale the hopping integrals between nearest neighbor sites as $t = \tilde{t}/2\sqrt{d}$ in d-dimensions and consider the limit $d = \infty$. Then the density of states (DOS) of each energy band has the Gaussian form $D(\varepsilon) = \exp(-\varepsilon^2/\tilde{t}^2)/\tilde{t}\sqrt{\pi}$. We assume $\tilde{t} = 1$ in the following. In this limit we can treat quantum fluctuations completely by taking local interactions into account and spatial correlations can be neglected.[28] The system is described in terms of a one-site effective action which is determined self-consistently. Generally one must rely on numerical methods to solve the effective action. In this study we approximated the action by that of a two-channel impurity model with finite (n_s) number of levels in each channel, and solved the impurity model by exact diagonalization. The energy levels and mixing parameters of the impurity model were determined self-consistently. We searched for ground states which are uniform in space as well as those with two-sublattice structures. Numerical calculations were done for $n_s = 5$ or 6. We mostly studied the system with $n_s = 5$ and confirmed phase boundaries by using the system with $n_s = 6$. We found that the results do not depend much on n_s. We studied the ground state mainly at the fillings $n = 1, 1.2,$ and 0.8, controlling the chemical potential.

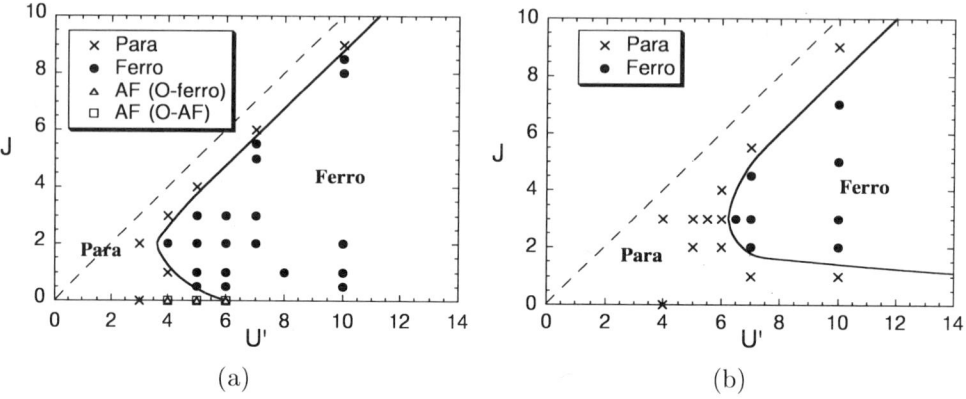

Figure 2: Ground state phase diagram of the $D = \infty$ doubly degenerate Hubbard model for $n = 1$(a) and 1.2(b).

In the quarter-filling ($n = 1$) case we found paramagnetic and ferromagnetic ground states for $0 < J < U'$ as is shown in Fig. 2a. Near the phase boundary two solutions coexist and their energies cross over. We selected the ground state by comparing energies and determined the phase diagram. The paramagnetic state obtained is spatially uniform and metallic. On the other hand the ferromagnetic state has a two-sublattice structure with alternating orbital order, and is insulating. We found a narrow paramagnetic region for $J \simeq U'$. Therefore the ferromagnetic phase seems to be confined within the region $J < U'$, though we did not study the case with $J > U'$ in $d = \infty$. The ferromagnetic ground state appears even for $J \simeq 0$ for $U' \gtrsim 6t$. At $J = 0$ the model possesses an $SU(4)$ symmetry. We found the coexistence of several ground states in this case.

We show the phase diagram for $n = 1.2$ in Fig. 2b. At this filling we obtain a metallic ferromagnetic phase and a metallic paramagnetic one, both of which are spatially uniform. Both states have no orbital ordering. The area of the ferromagnetic phase is reduced in the phase diagram compared to that of the insulating ferromagnetic phase at $n = 1$. As a hole-doped case, we studied the ground state for $n = 0.8$. At this filling we found only metallic ground states which are uniform in space and could not find any magnetically ordered phase for $U' \leq 20t$.

We show the variation of the number density as a function of the chemical potential

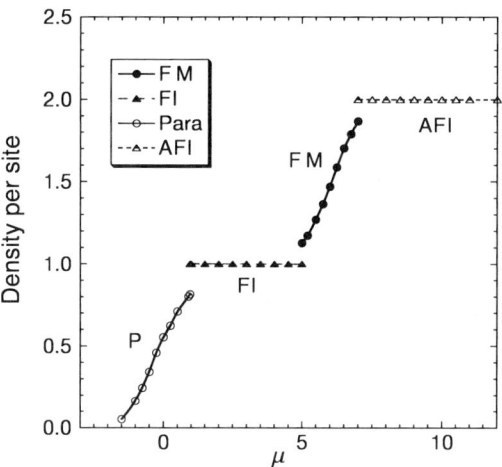

Figure 3: The number density as a function of the chemical potential μ for $U' = 10$ and $J = 4$. The flat parts indicate insulating states.

for $U' = 10$ and $J = 4$ in Fig. 3. For these parameter values we have the paramagnetic metallic ground state for $0 < n < 0.82$. For $n = 1$ the ground state is a ferromagnetic insulator. The ferromagnetic metal is stable for $1.14 < n < 1.86$. The antiferromagnetic insulator is realized for $n = 2$. It is interesting that there are small jumps of n on both sides of quarter-filling. One is from $n = 0.82$ to 1 and the other is $n = 1$ to 1.14. There is another jump close to half-filling, i.e. between $n = 1.86$ and 2. They imply that phase separation occurs for n in these intervals. Occurrence of phase separation was found also in the double exchange model.[29]

We show the kinetic and interaction energy per site as a function of U' on the line $J = 0.4U'$ for $n = 1.2$ in Fig. 4. For these parameters the ground state is ferromagnetic for $U' \gtrsim 6$. The kinetic energy increases linearly with U' for small U' where the ground state is paramagnetic. Then it starts to saturate and stays almost constant in the ferromagnetic phase. The interaction energy increases linearly with U' but its slope decreases as an effect of (local) correlations. The slope slightly increases again in the ferromagnetic region and the potential energy is nearly $(n - 1)(U' - J) = 0.12U'$ for large U'. The above result clearly shows that the ferromagnetism is caused by reduction of kinetic energy rather than interaction energy. That means that the "double exchange mechanism" is the cause of ferromagnetism for this density.

We have seen above that the Hund coupling is effective both in one and infinite dimensions. Especially for $n > 1$ the ferromagnetic ground state is realized in quite a large parameter region in both dimensions. This result suggests that the "double exchange mechanism" is quite effective in realizing the ferromagnetic state for $1 < n < 2$. We may expect that the situation is similar in two and three dimensions. For $n = 1$, ferromagnetism accompanied by alternating orbital order is realized for $J < U'$. We found that the state is destabilized for weak J in one dimension. This result may be understood as lower dimensionality stabilizing the paramagnetic liquid state and destabilizing alternating orbital order. (In fact we have only quasi-long-range orbital order in one dimension.) The paramagnetic state for $J = 0$ is known to be an $SU(4)$ singlet state.[30–32]

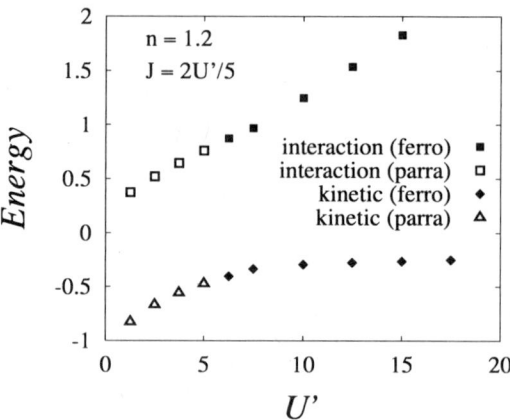

Figure 4: The kinetic and the potential energy per site vs U' for $n = 1.2$ on the line $J = 0.4U'$.

We found that the phase diagram is strongly dependent on the dimensionality for $n < 1$. Though ferromagnetism is realized in a large parameter region in one dimension, we could not find it in infinite dimensions. Although our study does not exhaust the whole parameter region, it seems likely that there is no ferromagnetism for $n < 1$ in $d = \infty$. There may be a general tendency for ferromagnetism in a low density system to be stabilized in one dimension. In a 1D Hubbard model with nearly flat bands low density was found to be favorable for ferromagnetism.[33] This tendency may be understood as a result of the diverging DOS at the zone boundary in one dimension. Since we have quite different results for $d = 1$ and $d = \infty$ for $n < 1$, a study of the ground states in two and three dimensions is desirable in order to answer the important question as to whether Hund coupling is effective in realizing ferromagnetism in low density systems like Ni. It should be noted that Ni has a fcc lattice structure, and its DOS has a sharp peak near the edge, which is similar to the one-dimensional one.

ELECTRONIC STATES IN THE DOUBLE EXCHANGE MODEL

In this section we consider the so-called double exchange model (DEM), which is composed of electrons in a single conduction band and localized spins of magnitude S at all lattice sites. The electrons and localized spins interact through intraatomic Hund coupling. The DEM may be the simplest lattice model for electrons in manganites. If we assume a single orbital instead of the doubly degenerate e_g orbitals and regard three electrons occupying t_{2g} orbitals as a localized spin in Mn^{3+} ions, we obtain the DEM with $S = 3/2$. We may consider also the DEM for arbitrary S. For example, we may consider a level separation Δ between two orbitals in the doubly degenerate Hubbard model. If both U' and Δ are much greater than t, then the model reduces to the DEM with $S = 1/2$ for $n > 1$. The DEM is described by the following Hamiltonian

$$H = -t \sum_{\langle i,j \rangle \in \text{n.n.}, \sigma} (c_{i\sigma}^{\dagger} c_{j\sigma} + \text{h.c.}) - J \sum_{i,\sigma,\sigma'} \boldsymbol{S}_i \cdot \boldsymbol{s}_{\sigma\sigma'} c_{i\sigma}^{\dagger} c_{j\sigma'}, \qquad (4)$$

where $\boldsymbol{s} = \frac{1}{2}(\sigma^x, \sigma^y, \sigma^z)$ and σ^α denotes the Pauli matrix. The parameter J in (4) corresponds to $2J$ in (1). Direct interactions between localized spins as well as the coupling between electronic and lattice degrees of freedom are neglected. For doped LaMnO$_3$ typical values of the conduction band width $2W$ and $J(2S+1)/2$ are thought to be $1 \sim 2$ eV[34] and $2 \sim 3$ eV, respectively. For large JS the spin of an electron is always coupled parallel with the localized spin and forms a total spin of size $S + 1/2$. Since the original hopping term in (4) conserves the spin of the electron, the hopping probability between two neighboring sites is effectively reduced when localized spins at these sites are not parallel to each other. The factor of the reduction is given by $\cos\theta/2$ if the localized spins are classical and make the angle θ between them.[16] In the paramagnetic state the localized spins are oriented randomly and as a result the conduction band is narrowed due to the reduction of the hopping integrals as well as decoherence effects due to scattering. Band narrowing increases the kinetic energy of the paramagnetic state and favors the ferromagnetic state. Ferromagnetism due to this "double exchange" mechanism was studied earlier.[35, 36] The electronic states of the model with classical localized spins were studied by use of dynamical mean field theory by Furukawa.[37] Classical localized spins are not affected when they scatter conduction electrons. Quantum mechanical spins may be flipped during the scattering processes. The effect of this "spin exchange scattering" was studied earlier using the coherent potential approximation (CPA).[38–40] The electronic states of the model with quantum spins in one and two dimensions were recently studied by using numerical methods extensively.[41, 42]

The CPA theory mentioned above treated a single electron in a system with randomly oriented localized spins and did not take into account the presence of other electrons. As a result the theory is valid only in the low density limit, but in this limit it gives a qualitatively correct description of the change in the electronic states due to interactions with localized spins. The single conduction band is modified by the interactions and the density of states splits into two bands for $JS \gtrsim W$. The lower band corresponds to electronic states with electron spins parallel to the localized spins (we call them "parallel electrons") and the upper band to those of "antiparallel electrons". The relative weights of the lower and the upper bands are $(S+1)/(2S+1)$ and $S/(2S+1)$, corresponding to the total spin of a site $S + 1/2$ and $S - 1/2$, respectively. If we naively consider these bands as a rigid one-particle density of states, the half-filled system ($n = N_e/N = 1$) cannot be an insulator since the lower band is not full. Surely the half-filled system should be an insulator if $JS \gg W$. We need a theory which realizes the insulating half-filled system for $JS \gg W$ in order to discuss the transport properties of the DEM.

Let us first consider the atomic limit, i.e. the case with $t = 0$. The energy spectrum of the Green function in this limit is composed of four levels. The lowest level at $\omega = -J(S+1)/2$ corresponds to the process of creating a parallel electron at a site which is already occupied by an antiparallel one. The second one at $-JS/2$ comes from creating a parallel electron at an empty site. The levels at $JS/2$ and $J(S+1)/2$ correspond to creating an antiparallel electron at an occupied and unoccupied site, respectively. The spectral weights of these four levels are $(Sn - 2\langle \boldsymbol{S}_i \cdot \boldsymbol{s}_i \rangle)/(2S+1)$, $[(S+1)(2-n) + 2\langle \boldsymbol{S}_i \cdot \boldsymbol{s}_i \rangle]/(2S+1)$, $[(S+1)n + 2\langle \boldsymbol{S}_i \cdot \boldsymbol{s}_i \rangle]/(2S+1)$ and $[S(2-n) - 2\langle \boldsymbol{S}_i \cdot \boldsymbol{s}_i \rangle]/(2S+1)$, respectively, where $\boldsymbol{s}_i$ denotes the electron spin operator at the

i-th site. When the hopping term is turned on, these discrete levels will broaden and compose four separate bands for $t \ll J$. They broaden as t increases and will finally merge into a single band for $t \gg J$. We note that the first and third levels vanish in the low density limit ($n = 0$) and that the weights of the two bands of the spectrum in the earlier one-electron CPA theory reproduce those of the second and fourth levels correctly. As for the metal-insulator transition, we note that the two lower levels are fully occupied at $n = 1$ in the exact atomic limit Green function. Therefore, we can expect a Green function to reproduce the correct insulating behavior at $n = 1$ for $JS \gg W$, if it reduces to the exact one in the atomic limit. We report below on the derivation of a Green function in the one-site approximation which gives the correct atomic limit and also reduces to the one-electron CPA result in the low density limit. We calculate the resistivity ρ in the framework of this approximation. A brief account of this work has already appeared and full details will be published elsewhere.[43]

In a single-site approximation the Green function $G_{\boldsymbol{k},\sigma}(\omega)$ is written as

$$G_{\boldsymbol{k},\sigma}(\omega) = \{\tilde{G}_\sigma(\omega)^{-1} - (\epsilon_{\boldsymbol{k}} - J_\sigma(\omega))\}^{-1}, \tag{5}$$

where

$$J_\sigma(\omega) = \omega - \Sigma_\sigma(\omega) - \tilde{G}_\sigma(\omega)^{-1}. \tag{6}$$

Here $\epsilon_{\boldsymbol{k}}$ and $\Sigma_\sigma(\omega)$ denote the free band energy and the self energy, respectively. The local Green function $\tilde{G}_\sigma(\omega)$ is related to the self-energy through the DOS of the free band energy $D_0(\epsilon_{\boldsymbol{k}})$ as

$$\tilde{G}_\sigma(\omega) = \int \frac{D_0(x)\mathrm{d}x}{\omega - \Sigma_\sigma(\omega) - x}. \tag{7}$$

In order to close the above equations we need another equation for $\tilde{G}_\sigma(\omega)$. We follow the equation of motion of the Green function along the line of Hubbard.[44] We can close the equations for arbitrary S in the paramagnetic state. In the paramagnetic state we can omit the spin suffices and the self-consistent equation for $\tilde{G}(\omega)$ is written as

$$\tilde{G}(\omega) = \sum_{\alpha=\pm} \frac{(E(\omega) + \alpha J/2)(n/2)_\alpha + \alpha J/2 \langle \boldsymbol{S}_i \cdot \boldsymbol{s}_i \rangle}{(E(\omega) - \alpha SJ/2)(E(\omega) + \alpha J(S+1)/2)}, \tag{8}$$

where $E(\omega) \equiv \omega - J(\omega)$ and $(n/2)_\alpha \equiv \delta_{\alpha-} + \alpha n/2$. The above set of equations reduces to that of the one-particle CPA at $n = 0$. We show in Fig. 5 the one-particle density of states for $J = 4W$ and $S = 3/2$ obtained from above set of equations by using the elliptic DOS given by

$$D_0(\epsilon) = 2/(\pi W^2)\sqrt{W^2 - \epsilon^2}. \tag{9}$$

At $n = 0$ the spectrum is composed of two bands centered at $\omega = -JS/2$ and $J(S+1)/2$, respectively. As n increases the third band centered at $\omega = JS/2$ emerges and correspondingly the weight of the band at $\omega = J(S+1)/2$ decreases. There should be a fourth band at $\omega = -J(S+1)/2$ as well, but its weight is very small for $JS \gtrsim W$, since $\langle \boldsymbol{S}_i \cdot \boldsymbol{s}_i \rangle \simeq nS/2$ (In the calculation shown in Fig. 5 we approximated as $\langle \boldsymbol{S}_i \cdot \boldsymbol{s}_i \rangle = nS/2$ for simplicity). The third band grows with n and finally it takes over the second one at $n = 1$. At $n = 1$ the lower two bands (the lowest band in Fig. 5) are completely filled and the system becomes a Mott insulator at $T = 0$ as is expected. These three bands should be observable by photoemission experiments on manganites. We note that the

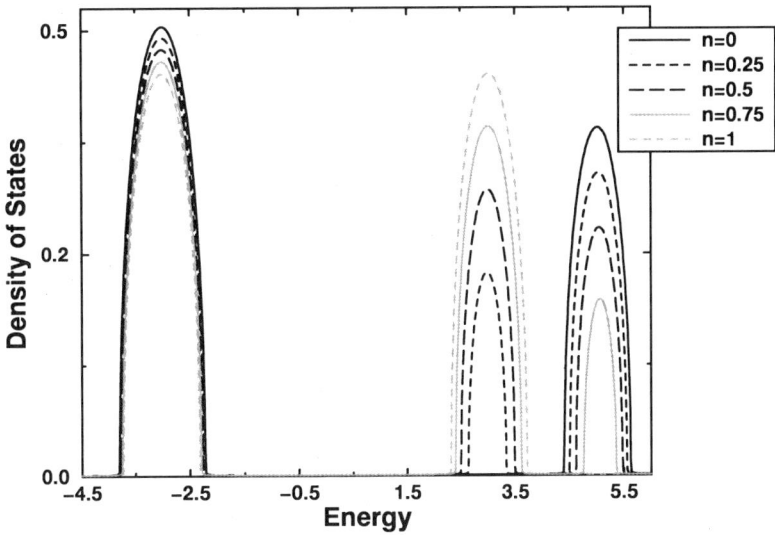

Figure 5: The density of states obtained from the approximate Green function in the paramagnetic state for $J = 4W$ and $S = 3/2$ at the filling $n = 0$, $n = 0.25$, $n = 0.5$, $n = 0.75$ and $n = 1.0$.

position of the band at $\omega \simeq JS/2$ will be shifted to $\omega \simeq JS/2 + U$ in the presence of the intraatomic Coulomb repulsion U. For $J = \infty$ the lowest band has width

$$2\bar{W} = 2W\sqrt{(S + 1 - n/2)/(2S + 1)} \qquad (10)$$

for the elliptic DOS. The band narrowing factor is a minimum at $n = 1$ with the value $1/\sqrt{2}$ independent of S. It should be noted that this factor may depend considerably on the choice of D_0 as was found for one-particle CPA.[38]

We calculate the DC resistivity ρ by using the above paramagnetic Green function in the Kubo formula. Vertex corrections do not enter in the calculation of the conductivity in the single-site approximation. The expression of the static resistivity at $T = 0$ is simplified by assuming the cubic tight-binding form of the hopping term in H. We find

$$\rho^{-1} = \frac{2\pi e^2}{3a\hbar} \int \epsilon d\epsilon \phi(\epsilon) D_0(\epsilon), \qquad (11)$$

where $\frac{d\phi(\epsilon)}{d\epsilon} = A_{\boldsymbol{k}}(\mu)^2|_{\epsilon_{\boldsymbol{k}}=\epsilon}$ and $A_{\boldsymbol{k}}(\mu)$ is the spectral weight function of the Green function $G_{\boldsymbol{k},\sigma}(\omega)$ at $\omega = \mu$. This expression is evaluated using the elliptical approximation to the DOS, both in $D_0(\epsilon)$ itself and in $\phi(\epsilon)$ via the Green function calculated above. In Fig. 6 we show the resistivity obtained for $J = \infty$ as a function of n for various values of S. We used 5Å for the lattice constant a. Note that the correct insulating behavior is obtained for $n = 0$ and $n = 1$. We find that ρ hardly depends on J for $JS \gtrsim 5W$. The resistivity ρ is of order of mΩcm for $0.1 \lesssim n \lesssim 0.9$ and this is much smaller than typical experimental values for doped LaMnO$_3$ except for the case of La$_{1-x}$Sr$_x$MnO$_3$

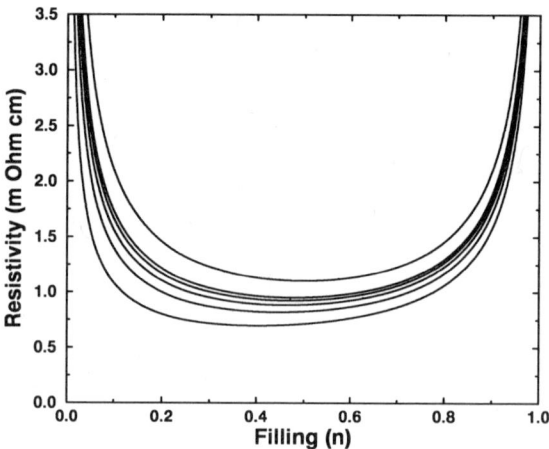

Figure 6: Resistivity at $T = 0$ (the paramagnetic state is assumed) is depicted versus n for $J = \infty$. Data for $S = 1/2 \sim 5/2$ and ∞ obtained by using the elliptic DOS are shown. Resistivity increases with increasing S.

with $x(= 1 - n) \simeq 0.3$.[45] The result shows also too weak a dependence on n compared to the experimental result. Furukawa[37] calculated ρ by using the Lorentzian DOS for $S = \infty$ and obtained good agreement with experimental result at $x = 0.2$. We also calculated ρ by using the Lorentzian DOS defined as

$$D_0(\epsilon) = W/[\pi(\epsilon^2 + W^2)] \tag{12}$$

to calculate $\phi(\epsilon)$ but retaining the elliptical approximation to $D_0(\epsilon)$ so that the integral in eq. (11) converges. The results are shown Fig. 7. They show much stronger dependence on n than those obtained for the elliptic DOS, and increase very rapidly when the fermi level approaches the band edge. The magnitude of ρ at $x \simeq 0.2$ is several tens of mΩcm, which is of the same order as the experimental data. However, the elliptic DOS is considered to be more realistic than the Lorentzian one, since the Lorentzian DOS gives divergent second moments. Hence the good agreement with experiment obtained for the Lorentzian DOS may be an artifact. Our results suggest that scattering by random localized spins is not enough to explain the correct order of magnitude of the resistivity in manganites. The present result is obtained by treating dynamical aspects of the scatterings approximately. Effects of finite temperature and short-range correlations between localized spins are also neglected, since our calculation has assumed complete Fermi degeneracy and completely random configurations of localized spins. These effects may modify the above result to some extent, but we do not expect that they will change the order of magnitude of the resistivity. Therefore some other effects should be taken into account to explain experimental results.[7, 46, 47]

We also studied the magnetic properties of the system by using the Green function. The self-consistent equation for the general magnetization for $S = 1/2$ was obtained as

$$\tilde{G}_\uparrow(\omega) = \tag{13}$$

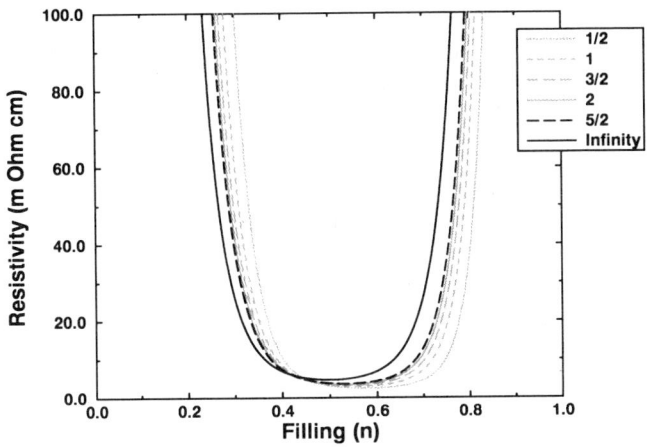

Figure 7: Resistivity at $T=0$ for Lorentzian DOS.

$$\sum_{\alpha=\pm} \frac{\langle n_{i\downarrow}^\alpha \rangle \left(E_\uparrow(\omega) E_\downarrow^\alpha(\omega) - J^2/8 \right) - (J/2) \left(\langle S_i^z n_{i\downarrow}^\alpha \rangle E_\downarrow^{-\alpha}(\omega) - \alpha \langle S_i^- s_i^+ \rangle E_\uparrow^{-\alpha}(\omega) \right)}{E_\uparrow^{-\alpha}(\omega) \left(E_\uparrow^\alpha(\omega) E_\downarrow^\alpha(\omega) - J^2/4 \right)},$$

where $E_\sigma^\alpha(\omega) = \omega - J_\sigma(\omega) + \alpha J/4$. We calculated the magnetic susceptibility χ for $J=\infty$ by including a magnetic field and expanding $\langle S_i^z + s_i^z \rangle$ about the paramagnetic state. We find that χ never diverges at a finite temperature for any $0 < n < 1$, i.e. there is no ferromagnetic transition. At $n=0$ the correct Curie law $\chi = (g\mu_B)^2 \tilde{S}(\tilde{S}+1)/(3k_B T)$ with $\tilde{S} = 1/2$ was obtained. On the other hand for $n=1$ and $J=\infty$ χ correctly obeys the Curie law with $\tilde{S}=1$ at high temperatures but it obeys the law with $\tilde{S}(\tilde{S}+1) = 2/15$ at low temperatures.

In the DEM with $J=\infty$ the ground state is proven to be ferromagnetic in one dimension for any $0 < n < 1$.[18] At present no reliable study of the ground state phase diagram of the DEM with $S=1/2$ seems to be available in higher dimensions than one. In three dimensions a high-temperature series expansion analysis suggests a finite Curie temperature for all electron density between 0 and 1,[50] though the fully polarized state is not stable for $0.12 < n < 0.45$.[48] It is reasonable to expect that the ferromagnetic ground state is stable in three dimensions in some density region. Therefore we consider that our approximate Green function fails to reproduce the low temperature properties of the model correctly. It is known that the analogous CPA in the Hubbard model does not give ferromagnetism at any density[49] or the correct Curie law at $n=1$.[51] Our present approximation suffers from a similar failure and an improved treatment is necessary in order to discuss the magnetic properties. We believe, however, that the present theory gives a qualitatively correct picture of the system in the paramagnetic state and that the paramagnetic resistivity obtained above is of the correct order of magnitude. We need an improved approximation to the Green function to study the magnetic properties of the system. A similar approach to that of Kawabata[51] for the strongly correlated Hubbard model might be useful.

SUMMARY

We discussed two topics on the Hund coupling in lattice systems employing simplified models. First we examined the effectiveness of the Hund coupling in realizing ferromagnetism in the doubly degenerate Hubbard model. In quarter-filled systems the insulating ferromagnetic state accompanied by alternating orbital order was found stable. In more-than-quarter filling case metallic ferromagnetism is stabilized by the "double exchange mechanism". The above results are common to one and infinite dimensions and we expect them to hold in general dimensions. In less-than-quarter filled systems the ferromagnetic ground state is stable in one dimension but not in infinite dimensions. To study this case in two and three dimensions is an interesting future problem.

Secondly we examined the electronic states and the resistivity in the double exchange model by using the one-particle Green function. The splitting and narrowing of the one-particle spectrum due to the Hund coupling were clarified in the framework of a single-site approximation. The resistivity due to the scattering by random localized spins was shown to be too small to explain the experimental results of doped manganites. The present approximation failed to give the ferromagnetic state and we need an improved treatment to study the properties at low temperatures.

ACKNOWLEDGMENTS

We thank T.A. Kaplan, S.D. Mahanty, P. Horsch and N. Furukawa for useful discussions. K. K. and T. M. were supported by the JSPS Grant No. 09640453. K. K. was supported by the EPSRC Grant No. GR/L90804, and A. C. M. G. by an EPSRC studentship.

References

[1] See papers in this volume and references therein.

[2] C. Zener, Phys. Rev. **81**, 440 (1951).

[3] G.H. Jonker and J.H. Van Santen, Physica, **16** 337 (1950); J.H. Van Santen and G.H. Jonker, Physica, **16** 550 (1950).

[4] Y. Murakami, H. Kawada, H. Kawata, M. Tanaka, T. Arima, Y. Moritomo and Y. Tokura, Phys. Rev. Lett. **80**, 1932 (1998).

[5] Y. Tomioka, in this volume and references therein.

[6] J.B. Goodenough, A. Wold, R.J. Arnott and N. Menyuk, Phys. Rev. **124**, 373 (1961).

[7] A.J. Millis, P.B. Littlewood, B.I. Shraiman, Phys. Rev. Lett. **74**, 5144 (1995).

[8] G. Zhao, K. Conder, H. Keller and K.A. Müller, Nature **381** 676 (1996)

[9] S. Ishihara, J. Inoue and S. Maekawa, Phys. Rev. B **55**, 8280 (1997).

[10] R. Shiina, T. Nishitani and H. Shiba, J. Phys. Soc. Jpn. **66**, 3159 (1997).

[11] L.M. Roth, Phys. Rev. **149**, 306 (1966).

[12] K.L. Kugel' and D.I. Khomskii, Sov. Phys. -J.E.T.P. **37** 725 (1973).

[13] M. Cyrot and C. Lyon-Caen, J. Phys. C: Solid State Phys. **6**, L247 (1973).

[14] S. Inagaki and R. Kubo, Int. J. Magn. bf 4, 139 (1973).

[15] S. Inagaki, J. Phys. Soc. Jpn. **39**, 596 (1975).

[16] P.W. Anderson and H. Hasegawa, Phys. Rev. **100**, 675 (1955).

[17] J.M. Van Vleck, Rev. Mod. Phys. **25**, 220 (1953).

[18] K. Kubo, J. Phys. Soc. Jpn. **51**, 782 (1982).

[19] K. Kusakabe and H. Aoki, Physica B **194-196**, 217 (1994).

[20] S. Q. Shen, Phys. Rev. B **57**, 6474 (1998).

[21] W. Gill and D.J. Scalapino, Phys. Rev. B **35**, 215 (1987).

[22] K. Kusakabe and H. Aoki, Mol. Cryst. Liq. Cryst. **233**, 71 (1993).

[23] J. Kuei and D.J. Scalapino, Phys. Rev. B **55**, 14968 (1997).

[24] G. Hirsch, Phys. Rev. B **56**, 11022 (1997).

[25] T. Momoi and K. Kubo, Phys. Rev. B **58**, R567 (1998).

[26] H. Sakamoto, T. Momoi and K. Kubo, in preparation.

[27] For example, L.J. Lee, M.J. Naughton, G.M. Danner and P.M. Chaikin, Phys. Rev. Lett. **78**, 3555 (1997).

[28] See for example, A. Georges, G. Kotliar, W. Krauth and M.J. Rozenberg, Rev. Mod. Phys. **68**, 13 (1996).

[29] S. Yunoki, J. Hu, A.L. Malvezzi, A. Moreo, N. Furukawa and E. Dagotto, Phys. Rev. Lett. **80**, 845 (1998).

[30] B. Sutherland, Phys. Rev. B **12**, 3795 (1975).

[31] Y.Q. Li, M.M. Ma, D.N. Shi and F.C. Zhang, Phys. Rev. Lett. **81**, 3527 (1998).

[32] Y. Yamashita, N. Shibata and K. Ueda, Phys. Rev. B **58**, 9114 (1998).

[33] H. Sakamoto and K. Kubo, J. Phys. Soc. Jpn **65**, 3732 (1996).

[34] S. Satpathy, Z.S. Popović and F.R. Vulkavić, Phys. Rev. Lett. **76**, 960 (1996).

[35] P.G. de Gennes, Phys. Rev. **118**, 141 (1960).

[36] K. Kubo and N. Ohata, J. Phys. Soc. Jpn. **33**, 21 (1972).

[37] N. Furukawa, J. Phys. Soc. Jpn. **63**, 3214 (1994); ibid **64**, 2734 (1995).

[38] K. Kubo, J. Phys. Soc. Jpn. **33**, 929 (1972).

[39] K. Kubo, J. Phys. Soc. Jpn. **36**, 32 (1974).

[40] M. Takahashi and K. Mitsui, Pys. Rev. B **54**, 11298 (1996).

[41] P. Horsch, J. Jaklič and F. Mack, condmat/9708007.

[42] E. Dagotto, S. Yunoki, A.L. Malvezzi, A. Moreo, J. Hu, S. Capponi, D. Poilblanc and N. Furukawa, Phys. Rev. B **58**, 6414 (1998).

[43] D.M. Edwards, A.C.M. Green, and K. Kubo, Physica B, in press (1998).

[44] J. Hubbard, Prc. Roy. Soc. **281**, 401 (1964).

[45] A. Urushibara, Y. Moritomo, T. Arima, A. Asamitsu, G. Kido and Y. Tokura, Phys. Rev. B **51**, 14103 (1995).

[46] H. Röder, J. Zang and A.R. Bishop, Phys. Rev. Lett. **76**, 1356 (1996).

[47] A.J. Millis, B.I. Shiraiman and R. Mueller, Phys. Rev. Lett. **77**, 175 (1996).

[48] R.E. Brunton and D.M. Edwards, J. Phys. Condens. Mattter **10**, 5421 (1998).

[49] H. Fukuyama and H. Ehrenreich, Phys. Rev. B **7**, 3266 (1973).

[50] H. Röder, R.R.P. Singh and J. Zang, Phys. Rev. B **56**, 5084 (1997).

[51] A. Kawabata, Prog. Theor. Phys. **54**, 45 (1975).

DENSITY FUNCTIONAL STUDIES OF MAGNETIC ORDERING, LATTICE DISTORTION, AND TRANSPORT IN MANGANITES

W. E. Pickett,[†] D. J. Singh,[‡] and D. A. Papaconstantopoulos[‡]

[†]Department of Physics
University of California
Davis CA 95616

[‡]Complex Systems Theory Branch
Naval Research Laboratory
Washington DC 20375

INTRODUCTION

The observation of colossal magnetoresistance (CMR) in the perovskite manganite system, typified by $La_{1-x}D_xMnO_3$ where D is a divalent cation, in the region of $x \approx \frac{1}{3}$ has led to vigorous study of this system across the entire range of x. There are now reviews[1,2] that present the general behavior and give an indication of the complexities. The general phase diagram evolves from an insulating antiferromagnet (AF) at $x \sim 0$ to a metallic ferromagnet (FM) in the $0.2 < x < 0.45$ region. Around $x = 0.5$ charge, spin, and even orbital ordering (selective occupation of only one of the two e_g states that are degenerate in a cubic field) frequently occurs. At $x=1$ the system reverts to an AF insulating phase.

The general behavior just described makes this system a good candidate for a model Hamiltonian treatment, and there has been extensive work especially on the "double exchange" model since it was suggested by Zener.[3] The double exchange Hamiltonian for a crystal, also known as the ferromagnetic Kondo lattice, describes band electrons coupled to a lattice of classical local spins by a FM "Hund's rule" coupling. The band electrons are majority Mn e_g electrons or possibly minority t_{2g} electrons, and the local spins arise from majority Mn t_{2g} electrons. After much work, the most extensive being due to Furukawa,[4] it appears that this double exchange model does contain a basic mechanism for large negative magnetoresistance.

It is however also clear as one surveys the complexities of the system that much more is happening than is described by the double exchange model. There is a strong coupling between the lattice structure, the magnetic order, and the electronic properties. Structural distortions and magnetic (dis)order are now accepted to be necessary for any serious description of the properties of the manganites, which include structural transitions driven by the application of a magnetic field, and insulator-metal transitions resulting from the exposure to light.[5] There are sometimes substantial changes in behavior at fixed carrier concentration, as when La is substituted by a rare earth ion with a different size. Electronic structure work has shown, rather unexpectedly, that the f states of La have an important influence on the structural distortions in these materials[6] in spite of the fact that any real $4f$ character of the carriers in these compounds is negligible for most purposes. With all the newer phenomena that have been discovered, it may even be argued that the magnetoresistance is not the most interesting feature of this system. The 'really colossal' magnetoresistance, originally thought to be linked to the intrinsic insulator-to-metal transition, now appears to be an extrinsic effect, strongly dependent on the microstructure. Clean crystals and films of (La,Ca)MnO$_3$ show the I-M transition and CMR behavior near the Curie temperature T_C, while (La,Sr)MnO$_3$ shows a metallic resistivity[7] (dρ/dT >0) above T_c and only a cusp in the resistivity at T_c, and much smaller MR around T_C.

This complex behavior calls for the application of band theoretical studies to establish the general characteristics of the electronic structure. Although addressing the effects of non-zero temperature have not yet been attempted, it is possible to learn a great deal about many phenomena with ground state (zero temperature) first principles density functional methods. In the following sections the main results of our studies[8-13] will be described. A substantial amount of related work has been done by Terakura's group,[6,14,15] and there is generally good agreement between the groups on the predictions of LDA band theory for the manganites. In this paper we emphasize in particular (1) the stoichiometric $x = 0$ and $x = 1$ end point compounds, (2) the effects of distortions, both at $x = 0$ and $x = \frac{1}{3}$, (3) the low temperature FM metallic phase at $x = \frac{1}{3}$, with emphasis on Fermi surface characteristics, and (4) the effects of La^{3+}/D^{2+} disorder, based upon a tight-binding coherent potential approximation calculation.

APPLICABILITY OF LDA BAND THEORY: $x = 0$ AND $x = 1$

The initial electronic structure work addressed the end point compounds LaMnO$_3$ and CaMnO$_3$, which are AF insulators. Since the applicability of local density approximation results to some classes of magnetic oxides is limited, such as in the transition metal monoxides, our early work compared the energy, magnetic moments, and electronic structure of non-magnetic, FM, and both A-type and G-type AF ordered states. This work established that the local density description of the magnetic order and the electronic structure is quite realistic for both of the end point compounds LaMnO$_3$ and CaMnO$_3$.

As anticipated, the Mn ion has the charge that is expected on the basis of valence counting, and the Mn ion is strongly magnetic: either FM or AF phases are stable relative to the unpolarized state. For CaMnO$_3$, the spherical d^3 ions lead to the observed structure, which has no Jahn-Teller distortion of the Mn-O bonds and is nearly cubic. The G-type AFM state, a rock-salt arrangement in which each up spin is surrounded by six down spins, is more stable than the obvious competing states. In addition, the band structure is insulating. The calculated gap, 0.42 eV for the cubic structure, is a few times smaller than the experimental value, but LDA calculations underestimate the bandgap, even in the simpler covalent semiconductors. Finally, the calculated Mn moment of almost 3 μ_B is typical of a d^3 ion, reduced slightly by hybridization with the O p orbitals.

The case of LaMnO$_3$ is more interesting. In the cubic structure the FM phase is calculated to be more stable than the AF phases, and its band structure is metallic. It is only because of the strong distortion[8] pictured in Fig. 1 that (i) the A-type AFM phase becomes most stable, and (ii) a gap opens in the band structure. As a result, in the observed Pnma crystal structure, the A-type AFM insulating phase (FM layers in the $a-b$ plane stacked antiferromagnetically along the c axis is most stable. This result was confirmed by Solovyev, Hamada and Terakura,[14] who also found that the distortion impacts the magnetocrystalline anisotropy and the exchange interactions in this compound.

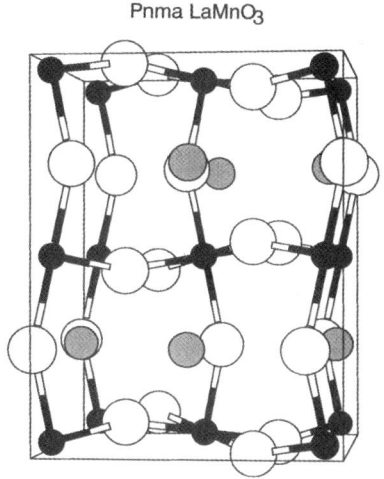

Figure 1. Ball and stick figure of the observed crystal structure (space group Pnma) of LaMnO$_3$. This cell involves a $\sqrt{2} \times \sqrt{2} \times 2$ enlargement of the primitive perovskite cell. Small black spheres are Mn, large white spheres are oxygen, and medium sized gray spheres are lanthanum. The distortion relative to cubic perovskite is described in the text.

The distortion in LaMnO$_3$ is rather complex, consisting of two rotations of MnO$_6$ octahedra followed by a distortion of the octahedra. The first rotation is one around the $\hat{c}$ axis. Since the octahedra are linked at all vertices, a positive rotation of one octahedron results in a negative rotation of the four neighboring octahedra in the $a-b$ plane, resulting in a $\sqrt{2} \times \sqrt{2}$ doubling of the cell. In addition, the octahedra are tipped away from the $\hat{c}$ axis. Due again to the linking the adjacent layers tip in the opposite sense, and this further doubles the cell. Finally, there is a Jahn-Teller internal distortion of the octahedra. The Mn-O bondlengths increase along one direction, lying in the $a-b$ plane, and increase in the other two directions. Again, because of the connectivity of the octahedra and the constraints of a periodic lattice, the direction of stretching of an octahedron alternates within the $a-b$ plane. This arrangement corresponds to "orbital ordering", in which the degeneracy of the e_g orbitals is lifted and $d_{x^2-z^2}$ and $d_{y^2-z^2}$ orbitals are occupied alternately. Since the cell was already doubled by the rotation, this distortion does not increase the cell further.

The distortion also has the effect of increasing the moment on the Mn ion by approximately 5%, bringing it into even better agreement with the experimental value obtained by fitting neutron diffraction data. The induced moment on the O ion in transition metal ions is getting increased attention, and Pierre et al[16] have reported the observation of a positive polarization on the O ion in La$_{0.8}$Ca$_{0.2}$MnO$_3$. The moment on the Mn ion that they infer ($\sim 0.1\mu_B$) is close to that from the band calculations. The calculated band gap of 0.12 eV again is small compared to the observed gap (1 eV), but the description of structural, magnetic, and electronic character seems to be basically correct.

Coulomb Correlation Effects

The question of correlation effects beyond the LDA treatment was been addressed by two groups. Satpathy, Popović, and Vukajlović[17] applied the LDA+U method assuming the Mn on-site Coulomb repulsion U is large (U ≈ 8-10 eV). This treatment does not seem to improve the predictions. The resulting magnetic moments of Mn are overestimated (3.3 μ_B for $CaMnO_3$ in particular is hard to rationalize on chemical grounds). A charge transfer picture of the excitation spectrum also was obtained for $LaMnO_3$, whereas most current indications are that holes doped into this materials primarily change the Mn valence.

A hybrid LDA+U method was applied to the compounds $LaMO_3$, M = Ti-Cu, by Solovyev et al.[15] in which the t_{2g} electrons are treated more as localized electrons whereas the e_g electrons are itinerant (in LDA and conventional LDA+U all electrons are treated identically). Although the resulting picture differs significantly from LDA for most of the compounds, because the two types of d electrons respond and screen differently, for the specific case of $LaMnO_3$ the results were very similar to LDA predictions. In addition, Sarma et al.[18] found that photoemission spectra agree with the LDA density of states rather well. Therefore, except for some corrections around the gap energy to describe excitations accurately, there does not appear to be any great need for correlation corrections beyond LDA for the end-point AF insulators.

MAGNETOSTRUCTURAL COUPLING AT $x = \frac{1}{3}$
Virtual Crystal Approximation

The primary interest in the manganites is for non-integer values of x, and particularly for the CMR region $x \approx \frac{1}{3}$. Studies of ordered supercells corresponding to $x = \frac{1}{4}$ and $x = \frac{1}{3}$ revealed several interesting possibilities of unusual magnetic and electronic behavior.[8] Since the La^{3+} and D^{2+} ions are disordered in samples, apparently randonly, we have implemented for our initial studies a scheme to model the disordered $x = \frac{1}{3}$ FM phase as an ordered crystal. The method is an example of the virtual crystal treatment: we use an effective nuclear charge of $56.667 = \frac{2}{3}Z_{La} + \frac{1}{3}Z_{Ba}$ on the A site, and include a corresponding number of electrons to give a neutral cell. We denote this average atom by Lb, and the corresponding crystal by $LbMnO_3$. This virtual crystal model amounts to neglecting the A-site disorder, while treating the electronic concentration and the electronic and magnetic structure self-consistently within LDA. Superficially, this treatment appears to model only $(La,Ba)MnO_3$. However, since the primary effect of the divalent cation is simply in altering the crystal structure (volume and distortions) to accommodate its size, we expect that this virtual crystal treatment will work almost as well also for Ca or Sr substitution if the calculation is carried out in the appropriate crystal structure.

There are two very important features of the doping regime including $x = \frac{1}{3}$ (a fairly wide regime, which may range from $x \sim 0.15$ to near $x = 0.5$): FM order becomes stable accompanied by metallic character, and the system is either almost half-metallic, or perhaps truly half-metallic in the T→0 limit. LDA calculations, both in the ideal cubic structure and in a distorted structure with rotated octahedra (see below), lead to a *nearly* half-metallic situation, where the minority conduction band is slightly occupied (corresponding Fermi energy of ∼ 0.1 eV). There are at least two reasons why the system could become truly half-metallic. First, in insulators LDA calculations underestimate the bandgap, and for the same reason – discontinuity of the kinetic energy across the bandgap – LDA may underestimate the bandgap of the insulating spin direction in a half-metallic FM. So far we know of no empirical value of the minority gap that would allow us to test this question. Secondly, the relative position of the majority and minority bands in ferromagnetic materials is somewhat sensitive to the exchange-correlation potential. When the Fermi level in a half-metal falls near a band edge in the insulating spin channel, this sensitivity can become important. It has been reported[19] that including gradient corrections as done in the Generalized Gradient Approximation of Perdew and coworkers[20] increases the exchange splitting enough to result in a half-metallic result band structure an ordered compound $La_{0.5}Ca_{0.5}MnO_3$ whereas LDA gives only a nearly half-metallic result, as for $x = \frac{1}{3}$. A similar shift of the relative positions of the majority and minority bands at $x = \frac{1}{3}$ would result in a half-metallic system.

Numerous workers have suggested that this half-metallicity is intimately connected to the colossal magnetoresistance phenomenon. It should be kept in mind that the band structure results apply only at low temperatures, and the question of half-metallicity was still completely open until recently. However, Park et al.[21] have now reported spin-resolved photoemission spectra for single domain films with surfaces that seem to be very similar to bulk material in their properties. The spin dependence at low temperature indeed looked entirely representative of half-metallic behavior. This experiment probably would not see the small amount of minority electrons above the minority gap if the minority conduction band is slightly occupied.

Effects of Distortions From Cubic Structure

To address the effect of structural distortion on the electronic characteristics near the Fermi level E_F of the majority bands at $x = \frac{1}{3}$, we performed relaxations of the oxygen positions within the constraints of the Pnma space group, which is the observed space group symmetry of AF $LaMnO_3$. For both $x = 0$ and $x = \frac{1}{3}$ we obtained no Jahn-Teller distortion (variation in Mn-O bondlengths by ~5% or more) when the magnetic order is constrained to be FM. Since in the average structure there is no Jahn-Teller distortion observed either for (La,Ca)MnO_3 (Ref. 22) or for (La,Ba)MnO_3 (Ref. 23) we have confined our study to the effect of rotations of the MnO_6 octahedra. One important aspect of these rotations is expected to be a decrease in the effective Mn-Mn hopping amplitude, hence a narrowing of bandwidths.

We first comment briefly on $LaMnO_3$, and refer to Ref. 11 for figures and more details. In this fictitious, FM undoped system, the rotations lead to strong decrease in the DOS in a limited region that falls right at the Fermi level. The rotation is fairly large, with Mn-O-Mn bond angles of 165° but all Mn-O bond lengths in the range 2.00-2.01 Å. The Drude plasma energy $\hbar\Omega_p$, given by the Fermi surface quantities from the expression

$$\hbar^2 \Omega_{p,xx}^2 = 4\pi e^2 N_\uparrow(E_F) v_{F\uparrow x}^2 \qquad (1)$$

for the x direction, and similarly for the other directions. Ω_p drops by a factor of eight as a result of the distortion, reflecting a drastic change in the transport behavior.

The CMR Regime

The $x = \frac{1}{3}$ case is more relevant. For the cubic idealization which is nearly half-metallic, the majority and minority bands are pictured in Fig. 2. The majority e_g bands overlap the Fermi level, and they also overlap the O p bands which lie mostly below the Mn d bands. The minority e_g bands lie about 2.5 eV higher than the majority, leaving a 1.5 eV gap. The Fermi level lies just above this gap, about 0.15 eV into the minority e_g conduction bands. This small minority occupation is discussed at more length below.

The majority bands have N(E_F)=0.53 eV^{-1}, $v_{F\uparrow x}$=4.2×10^7 cm/s, and a magnetic moment of 3.40 μ_B. We studied the distorted structure in which all Mn-O bondlengths were equal, and all Mn-O-Mn bond angles were 160°±1°. The structure we used is quite similar to the structure determined by Dai et al.[24] for $La_{0.65}Ca_{0.35}MnO_3$. Similarly to the $x = 0$ case, the distortion resulted in band narrowings (see Fig. 3) and new gaps away from the Fermi level. However, within a few tenths of eV of the Fermi level, the changes are small. $N(E_F)$ increases to 0.68 eV^{-1} and $v_{F\uparrow x}$ decreases to 3.1×10^7 cm/s, and the result is that $\hbar\Omega_{p,xx} = 1.3$ eV is only 30% less than the undistorted value. This indicates a relatively small change in the conductivity, considering the strong electron-lattice coupling and the large distortion. Although there have now been a few optical studies for $x = \frac{1}{3}$ materials, we are aware of only one that presents a Drude plasma energy. For $La_{2/3}Ca_{1/3}MnO_3$, Boris et al.[25] give $\hbar\Omega_{p,xx} = 3.25$ eV. However, this value is for T=78 K, and their data was still showing considerable temperature dependence at this temperature. Sample strains and anisotropy have also been suggested as reasons why the low frequency data in these metallic samples often do not appear to be Drude-like. For these reasons we do not expect this value to be representative of the T→0 value.

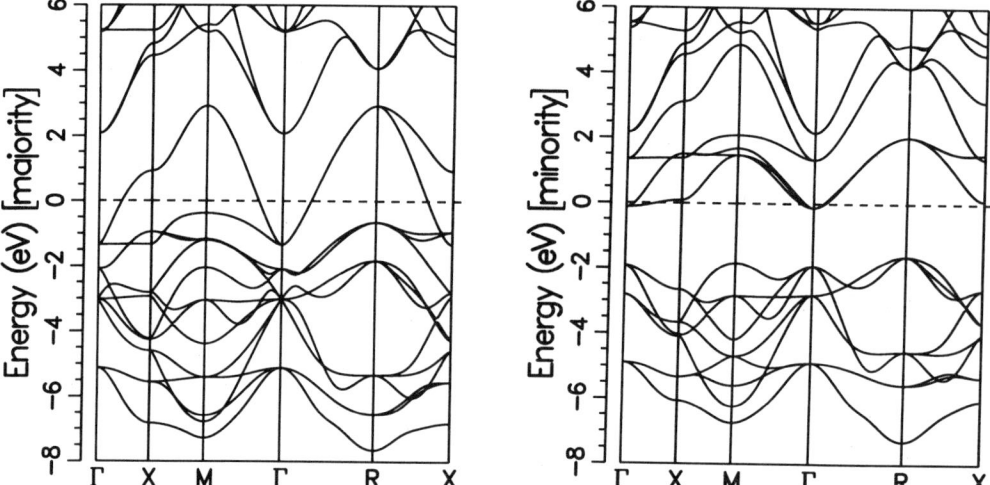

Figure 2. Majority (left) and minority (right) band structures along high symmetry directions for the ferromangetic phase of $x = \frac{1}{3}$ "LbMnO$_3$," from full potential LAPW calculations. The dashed line denotes the Fermi level.

At first look, it seems that these results imply that Jahn-Teller distortions do not occur in the FM phase and that the effects of the rotations of octahedra may not be very important, except to give quantitative changes in thermodynamic and transport coefficients. It must be noted however that these results apply to only to a perfectly ordered, static lattice at full magnetization. While this may be a reasonable picture of the state of affairs in the low temperature limit, the primary interest is at temperatures from $\sim T_C/2$ to above T_C. Dynamical effects at finite temperature give rise to magnetic, structural and electronic fluctuations that surely will complicate the phenomena considerably. It is still important, however, to understand the T→ 0 case and be able to describe it quantitatively.

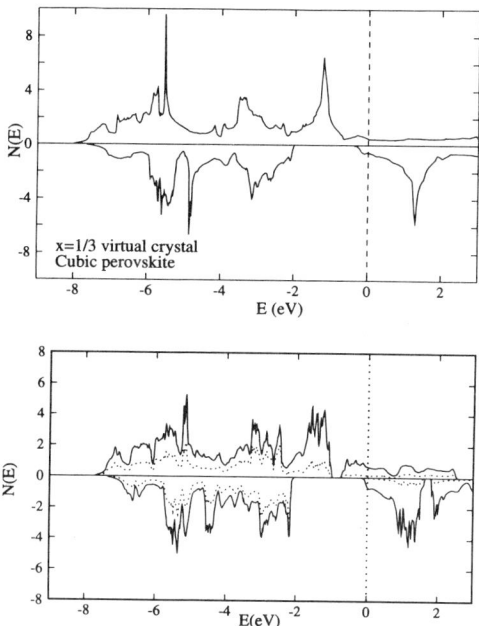

Figure 3. Densities of states for the majority electrons in ferrromagnetic "LbMnO$_3$," for the (top) cubic structure and (bottom) for the distorted Pnma structure. The consequences of these changes are described in the text.

FERMI SURFACE PROPERTIES FOR $x = \frac{1}{3}$

The Fermi surface we calculate for the majority bands of the $x = \frac{1}{3}$ FM material with undistorted cubic structure is simple and yet striking. It consists of a Γ-centered spheroid, which is in fact nearly perfectly spherical, and a zone corner (R) centered cube, which has very flat faces with rounding occurring only very near the edges and corners. These surfaces are pictured in Fig. 4. It is in fact not uncommon to have flat portions of constant energy surfaces in perovskite compounds – the strongly directional bonding leads to certain d bands that are strongly dispersive due to large $dp\sigma$ overlap with oxygen, and also lead to other d bands that are dispersionless because the d orbitals are oriented perpendicular to the corresponding Mn-O direction and have vanishing hopping amplitude. Flat energy surfaces result from bands that lack of dispersion in directions perpendicular to the surface normal.

For the undistorted structure, the Fermi velocities on each of the surfaces are nearly the same, around $(7$-$7.5) \times 10^7$ cm/s. The total density of states, 80% of which arises from the cube, is 0.45 eV^{-1}, and leads to a Drude plasma energy of 1.9 eV. As mentioned above, crystalline distortion consisting of rotations of the octahedra reduce the velocities to around 5×10^7 cm/s and increase N(E$_F$) by 50%, so $\hbar\Omega_p$ is decreased by 30%, to 1.3 eV. Note that these numbers are smaller than what might have been expected from such metallic bands because of the half-metallic nature, *i.e.* there is only one spin channel contributing, so Ω_p is $\sqrt{2}$ smaller than it would be for a normal metallic system.

The cube Fermi surface has consequences that are likely to be observable, perhaps even strikingly so. The flat parallel faces lead to strong nesting, arising from the large phase space for two types of scattering events. The phase space for scattering events through wavevector $\vec{Q}$, with initial and final states both at the Fermi energy (taken as zero) is given by

$$\xi(Q) = \sum_k \delta(\varepsilon_k)\delta(\varepsilon_{k+Q}) \quad (2)$$

Contributions to $\xi(Q)$ are large when the velocities at k and k+Q are small and especially when they are collinear (antiparallel *or* parallel).

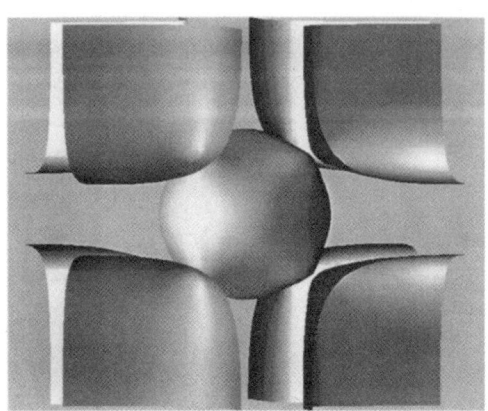 

Figure 4. Fermi surfaces of the majority electrons of "LbMnO$_3$" ($x = \frac{1}{3}$). The sphere (left) is centered at Γ, surrounded by eight cube surfaces. The cube (right) is centered at the R point.

One large contribution from the cube is the usual nesting, involving scattering between two parallel faces of the cube. Because $2k_F$ lies outside the first Brillouin zone, the effect will be observed at a reduced wavevector $2k_{F,red} = 2k_F - 2\pi/a$. The scattering need not lie precisely along a Cartesian direction, it is sufficient that one Cartesian component of the scattering wavevector is equal to $2k_{F,red}$. The resulting planes at which Kohn anomalies should be visible in the phonon spectrum are the planes at $Q_j = \pm 2k_{F,red}$, for $j = x, y$, or z. The size and flatness of the cube edges suggest the Kohn anomalies should be readily observable, as long as the Fermi surface is not washed out by disorder (see the final section). Scattering processes involving the sphere are less important, both because it contains an order of magnitude fewer states, and because its three dimensional character does not focus any scattering events significantly. The associated Kohn anomalies are known to be weak. Thus we do not discuss the spherical Fermi surface further.

The second type of scattering event is less widely known, and consists of a "skipping" event in which the carrier scatters but remains on the same face of the cube. This scattering requires one component of the scattering vector $\vec{Q}$ to vanish. These vectors lie in the (100) planes of the Brillouin zone. We find the phase space for these skipping events to be particularly large.[12] To observe the corresponding Kohn anomalies, the phonon dispersion curves should be scanned across a (100) plane. In fact, if there is a sharp dip in ω_Q across these planes, (*i.e.* if the electron-phonon matrix elements are large), it may be confusing to try to map out the phonon dispersion curves along the <100> directions as is commonly done: a small misalignment may sample different regions of the cusp in ω_Q near the (100) plane.

Such scattering events also contribute to the resistivity. However, a scattering event only contributes to the resistivity to the extent that it degrades the current. The corresponding phase space is described by a quantity $\xi_{tr}(Q)$, which differs from $\xi(Q)$ given above only by having a factor $(\vec{v}_k - \vec{v}_{k+Q})^2/2v_F^2$ within the summation. This weighting factor averages to unity over the zone, so it merely distributes the relative importance of the scattering events for transport compared to their importance in Kohn anomalies. One striking feature is that the velocity factor knocks out the skipping events, so that they contribute nothing to resistivity (because the velocity is unchanged). Hence the normal nesting events will be much more important in determining the resistivity (they *reverse* the velocity of the carrier). Plots of ξ and ξ_{tr} have been presented elsewhere.[12]

TIGHT BINDING REPRESENTATION OF THE BANDS

Ferromagnetic Bands at $x = \frac{1}{3}$

A tight binding parametrization of the band structure makes possible the testing of simple ideas and the implementation of complex procedures, such as calculations of transport properties, that may be impossible otherwise in a complex crystal. We have performed an extensive fit[13] to the energy bands of FM LbMnO$_3$, both majority and minority, using an extensive basis set of Mn d and s orbitals, O p orbitals, and La s and d orbitals. Although the majority and minority bands can be fit separately, we have chosen a more physically based method, and one that allows generalization more easily.

Our choice was to fit *both* majority and minority band structures with a single set of spin-independent hoppping amplitudes, and to include the spin splitting only with an on-site exchange splitting Δ_{ex} on the Mn ion. The exchange splitting was allowed to be different for t_{2g} and e_g states, with the result $\Delta_{ex}^{t_{2g}} = 3.44$ eV, $\Delta_{ex}^{e_g} = 2.86$ eV. This latter quantity is the nearest thing in the band structure to the "Hund's coupling" used in model Hamiltonian treatments to force the conduction e_g electrons to tend to align with the 'core' t_{2g} electrons.

The TB parameters have been presented and discussed fully elsewhere.[13] We mention the main features of the fit. The oxygen p_σ orbitals were allowed a different site energy from the p_π orbitals, consistent with the O site symmetry; this crystal field splitting is analogous to the e_g, t_{2g} difference on Mn. $\varepsilon_{p\sigma}$ turns out to be 1.5 eV below $\varepsilon_{p\pi}$, reflecting the need for this freedom to obtain the best fit. The average separation between Mn d and O p states was found to be $\varepsilon_d - \varepsilon_p = 2.60$ eV. As mentioned above, the photoemission data for LaMnO$_3$ seems to be at least roughly consistent with this description, *i.e.* LaMnO$_3$ is not a charge transfer insulator like the insulating layered cuprates.

The quality of the tight binding fit can be assessed by comparing the resulting bands in Fig. 5 with the *ab initio* LAPW bands shown in Fig. 2. Differences are slight, and especially the important bands near the Fermi level are represented well. The total majority and minority densities of states and their decompositions into partial O p and Mn e_g and t_{2g} contributions are shown in Fig. 6. They track the first principles results very well, with r.m.s. error of the Mn d band region being 0.1 eV. Note that, although the t_{2g} majority states are all occupied, the bandwidth is 6 eV and the uppermost part of these states is within 0.5 eV of the Fermi level. This indicates that these electrons, often referred to as 'core' electrons whose spin is treated classically and whose charge response is neglected, may contribute an important polarizability to the system. Also, a substantial fraction (~25%) of the 'itinerant' e_g majority states have 6 eV binding energy, due to the bonding combination arising from the strong $dp\sigma$ coupling. Thus the electronic structure and its response to perturbations is considerably more complex than that described by the ferromagnetic Kondo lattice Hamiltonian.

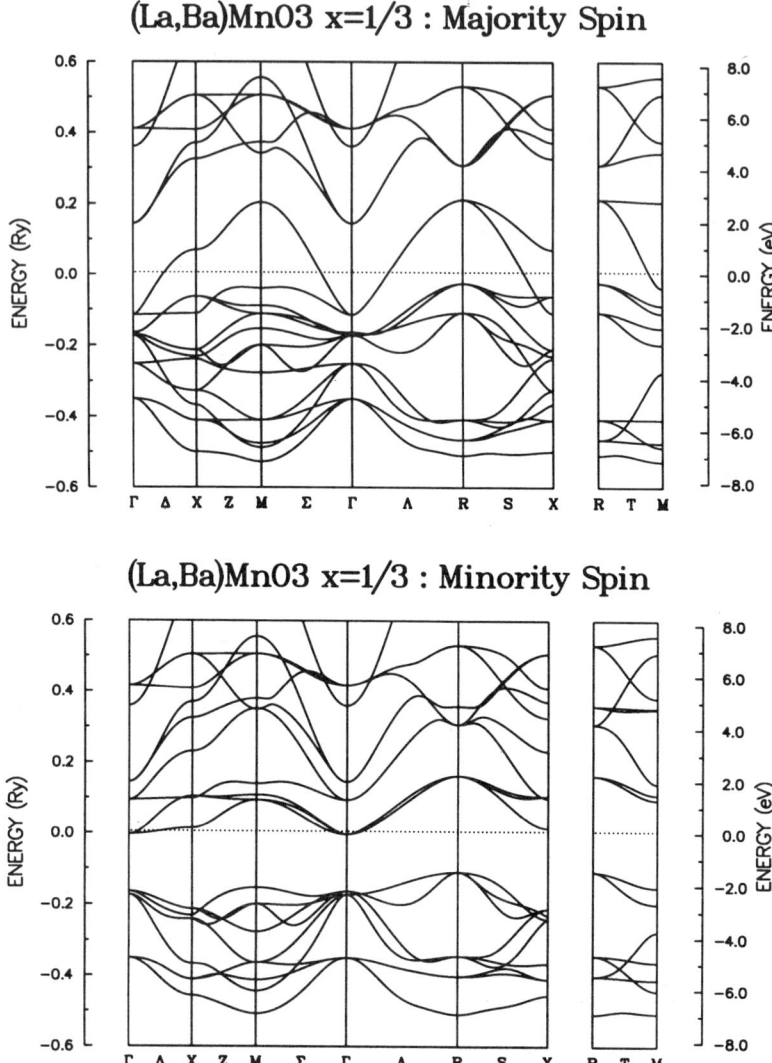

Figure 5. Majority (top) and minority (bottom) band structures resulting from the tight binding fit. These bands can be compared with those shown in Fig. 2 to assess the goodness of the fit.

Simple Fit to the Majority e_g Bands

The majority e_g bands, which provide the conduction states, separate from the t_{2g} bands but overlap with the O p bands. Along the symmetry directions, however, the two e_g bands are easily identified, so a two band model of these conduction states can be constructed. Between e_g states (these will be Mn-centered Wannier functions that involve strong O $p\sigma$ character) only the $dd\sigma \equiv t_\sigma$ and $dd\delta \equiv t_\pi$ overlaps are non-zero. Using Table I of Slater and Koster[26] with the cubic perovskite structure, that is, a simple cubic array of Mn ions, the e_g bands become

$$\varepsilon_k^\pm = \varepsilon_k \pm D_k, \tag{3}$$

$$\varepsilon_k = \bar{\varepsilon} + (t_\sigma - t_\delta)(c_x + c_y + c_z), \tag{4}$$

$$D_k = -(t_\sigma - t_\delta)(c_x^2 + c_y^2 + c_z^2 - c_x c_y - c_y c_z - c_z c_x)^{1/2}. \tag{5}$$

Here $\bar{\varepsilon}$ is a constant and $c_x = cos(k_x a)$ etc. With the choice of sign made here, t_σ will be negative and $t_\sigma - t_\delta$ is assumed to be negative also. This choice of sign is consistent with our tight binding fit of the full band structure.[13]

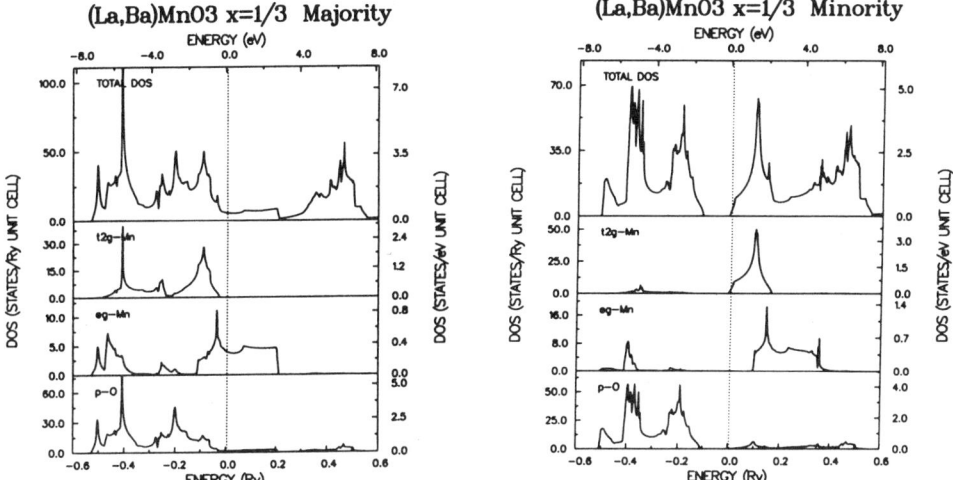

Figure 6. Total and atom projected densities of states for majority (left) and minority (right) states obtained from the tight binding fit to the ferrromagnetic bands of "LbMnO3." The exchange splitting of the Mn e_g and t_{2g} states are apparent from their different positions with respect to the Fermi level.

The e_g bandwidth W is given by the difference in the eigenvalues at Γ and the R point $(1,1,1)\pi/a$: $W = 6(|t_\sigma + t_\delta|)$. The dispersion along Γ-X is equal to $4t_\delta$. Since the majority bands overlap the O p bands, which are not included in this two-band fit, we use the average Γ-X dispersion for majority and minority, giving $4t_\delta = -(0.00 + 0.25)/2$ eV, or $t_\delta \approx -0.03$ eV. The majority bandwidth $W = 6(|t_\sigma + t_\delta|) = 4.30$ eV then gives $t_\sigma = -0.69$ eV. Setting the calculated Fermi level the zero of energy is done by making $\bar{\varepsilon} = 0.81$ eV. t_δ in such a model is unimportant unless there is a reason to "broaden" the van Hove step singularities that arise if t_δ is identically zero.

COHERENT POTENTIAL APPROXIMATION STUDY OF CATION DISORDER

The TB fit to the band structure enables us to consider treating the effects of chemical disorder in LbMnO3 using the tight binding coherent potential approximation (CPA). Related studies of the effects of substituting a D^{2+} ion for La has been reported by Butler, Zhang and MacLaren[27] and by Tobola, Kaprzyk, and Pierre.[28] In their work they considered the direct effect of A site substitution within the CPA, which determines an effective A site cation with the same average scattering of electrons as the disordered alloy. Since we expect the carriers to have very little amplitude on the A site, this direct effect of disorder should be small. Butler *et al.* reported substantial Ca character at and below the Fermi level from their calculation. Such character is not present in the density of states of (La,Ca)MnO3 supercells,[8] and its origin remains unclear.

Model for the Disorder

We treat an indirect effect of the A site disorder, but one that affects the site where the carriers reside: the fact that a neighboring tripositive ion contributes differently to the Mn d site energy than a dipositive ion does. In the CPA work of Butler *et al.*, the La^{3+} and D^{2+} ions on the A site are replaced with an effective ion, and the site energy of all Mn ions is the same. We anticipated that the local Mn environment effect from neighboring positive cations would be of greater importance to the carriers, since they have most of their amplitude on the Mn sites.

We assume that the differences in the onsite energy of the Mn d state, ε, is proportional to the differences in the A site charge on the first and second neighbor shells. Specifically, we take

$$\delta\varepsilon_o = \delta\Big[\sum_{2shells}\frac{Z_i e^2}{\epsilon(R_i - R_o)}\Big], \tag{6}$$

where Z_i is the valence charge of the cation at the site R_i, which is in the first or second neighbor shell of the Mn ion at R_o. The value of ϵ, the background dielectric constant, was determined by fitting the above point charge expression to self-consistent supercell calculations to be $\epsilon \approx 10$. Boris *et al.*[25] have reported from optical studies a background dielectric constant at 78 K of ϵ=7.5, not far from our estimate.

This expression gives the shift $\delta\varepsilon$ of a specific Mn site from the crystal average. For each local configuration (the number of tripositive and dipositive ions in the first and second neighbor shells), we multiply the number of such configurations by the probability that the configuration will occur in an $x = 1/3$ alloy to obtain the probability of that site energy. Since there are nine different first-shell configurations and seven different second-shell configurations, the system we apply the CPA to is a "63 component alloy" in which each 'component' is a Mn ion with a distinct site energy. The distribution of site energies has an approximate Gaussian form[10] with full width 0.4 eV. With such a broad range of Mn d site disorder, and an overlap of the Mn minority t_{2g} band edge from the Fermi level of only 0.15 eV, one might expect that the Fermi level would lie well below the mobility edge in the minority band, thus rendering the minority carriers localized and leaving an effectively, if not purely, half-metallic system.

CPA Results

The CPA calculation leads to a less specific result. The main result of the calculation is the Mn d site self-energy $\Sigma(\varepsilon)$, one each for e_g and t_{2g} character in both the majority and minority bands. The real part of the self-energy gives the site energy shift arising from the disorder, while the imaginary part gives the disorder broadening, which can be interpreted as an inverse lifetime. The self-energy also leads to a spectral density A(k,ε) that is the alloy analog of the delta function $\delta(\varepsilon - \varepsilon_k)$ at the band energy for a periodic crystal.

The real part of the self-energy is always less than 20 meV in magnitude and does not have serious consequences. The disorder broadening described by the imaginary part is the most interesting aspect. For the minority t_{2g} band that just crosses the Fermi level, $|Im\Sigma(\varepsilon_F)| \approx 10$ meV. The quantum mechanical nature of the electrons has averaged over the disorder and resulted in a disorder broadening an order of magnitude smaller in scale. The CPA calculations do not provide any mobility edge, so it is unclear whether all of the minority carriers will be localized. Wang and Zhang[29] have developed a model of electron transport for "nearly half-metallic" FM, where it is assumed that there are non-conducting minority states at the Fermi level in addition to the large density of conducting majority states. They predict a T^α, α 1.5 or 2, temperature dependence of the resistivity at low temperature due to spin-flip processes.

The disorder broadening leads to a tail in the density of states that decreases in Lorentzian fashion into the gap. The calculated behavior is illustrated in Fig. 7. There is a small but appreciable density of states even in the center of the 1.5 eV gap. This disorder-induced band tail is certain to be present, so the conductivity limiting mechanism considered by Wang and Zhang will be present regardless of the position of the Fermi level with respect to the virtual crystal band edge.

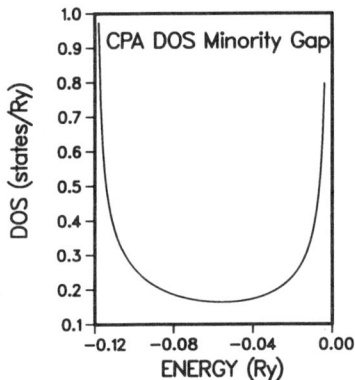

Figure 7. The minority density of states in the band gap region, from the CPA calculation. The decrease toward the center of the gap is roughly Lorentzian in shape. The zero of energy is at the Fermi level of the LAPW bands to which the tight binding fit was done. The Fermi level for the fit bands is at 0.0055 Ry, reflecting the degree of error in the tight binding fit.

The disorder broadening in the majority conducting band has direct experimental consequences. We obtain $|Im\Sigma(\varepsilon_F)| = 8$ meV for the e_g carriers. Equating this with $\hbar/\tau_o$, where τ_o is the elastic scattering time due to the cation disorder, we obtain from the Fermi velocity $v_F \sim 7\text{-}7.5 \times 10^7$ cm/s the mean free path $\ell = v_F \tau_o = 600$ Å. Using our value of $\hbar\Omega_p = 1.9$ eV from above, we obtain a residual resistivity

$$\rho_o = \frac{4\pi}{\Omega_p^2 \tau_o} = 15 \ \mu\Omega \ cm. \qquad (7)$$

If $\hbar\Omega_p$ were 1.3 eV, as we obtained for the distorted structure, then $\rho_o = 32 \mu\Omega$ cm. The value of ρ_o is the prediction of the intrinsic residual resisitivity for $x = \frac{1}{3}$ samples. Values as low as 35 $\mu\Omega$ cm have been reported.[30]

SUMMARY

Current indications are that the local density approximation leads to a reasonable picture of the electronic structure, magnetoelectronic coupling, and magnetostructural coupling in the $La_{1-x}D_xMnO_3$ system for $x = 0, x = 1$, and in the T→0 FM regime for $x \sim \frac{1}{3}$ for the ordered situations that have benn studied. The very important question of spin disorder has not yet been addressed, although that appears to be within the scope of the theory.

The Jahn-Teller distortion is reproduced well in AF $LaMnO_3$. However, for (fictitious) FM $LaMnO_3$, for FM $La_{2/3}D_{1/3}MnO_3$, and for (fictitious) AF $La_{2/3}D_{1/3}MnO_3$, there is no tendency toward John-Teller distortion in the Pnma space group of $LaMnO_3$ and of many $x \sim \frac{1}{3}$ samples. We suspect that the numerous experimental indications of Jahn-Teller-like motions at intermediate and high temperature are closely related to spin disorder, and that the two types of disorder are intimately coupled.

The LDA gives the following picture of the FM $x = \frac{1}{3}$ phase at low temperature. It is half-metallic, or very nearly so, and the metallic majority bands should lead to good metallic behavior for the intrinsic resistivity: $\rho_o \approx 10 - 30 \mu\Omega$ cm. The predicted plasma energy, which should be observable in the far-IR, is ~ 1.5 eV. The majority Fermi surface for $x \approx \frac{1}{3}$ includes an R-centered cube with strikingly flat faces. Since electron-phonon coupling in these materials is very strong, Kohn anomalies should be readily apparent in the phonon dispersion curves. We have provided a simple but good fit to the majority e_g bands. However, other (O p) bands approach within 1 eV of the Fermi level, so this fit is of limited use for the calculation of the optical conductivity even for T→0.

The treatment of spin and structural disorder is now required. The former is implicit in the FM Kondo lattice Hamiltonian, while the latter has been emphasized in several recent models.[31-33] Our experience with the (lack of) Jahn-Teller distortions in the periodic $x = \frac{1}{3}$ system suggest one feature that has been neglected so far: Jahn-Teller distortions around Mn ions are not independent. The interconnectedness of the MnO_6 octahedra imply that a Jahn-Teller distortion is strongly inhibited unless neighboring octahedra also distort. This coupling may mean that the statistical mechanics of internal distortions is substantially different from a model of disconnected distortions. Our tight binding model of the band structure may facilitate a viable treatment of correlated distortions and of disordered spins[34] as well.

Acknowledgements

This work was supported by the Office of Naval Research.

REFERENCES

1. A. P. Ramirez, J. Phys.: Condens. Matt. **9**:8171 (1997).
2. M. Imada, A. Fujimori, and T. Kokura, Rev. Mod. Phys. (in press).
3. C. Zener, Phys. Rev. B **82**:403 (1951).
4. N. Furukawa, J. Phys. Soc. Japan **64**:3164, 2734 (1995)
5. M. Fiebig, K. Miyano, Y. Tomioka, and Y. Tokura, Science **280**:1924 (1998).
6. H. Sawada, Y. Morikawa, K. Terakura, and N. Hamada, Phys. Rev. B bf 56:12154 (1997).
7. A. Urushibara, Y. Moritomo, T. Arima, A. Asamitsu, G. Kido, and Y. Tokura, Phys. Rev. B **51**:14103 (1995).
8. W. E. Pickett and D. J. Singh, Europhys. Lett. **32**:759 (1995).

9. W. E. Pickett and D. J. Singh, Phys. Rev. B **53**:1146 (1996).
10. W. E. Pickett and D. J. Singh, Phys. Rev. B **55**:R8642 (1997).
11. D. J. Singh and W. E. Pickett, Phys. Rev. B **57**:88 (1998).
12. W. E. Pickett and D. J. Singh, J. Magn. Magn. Mater. **172**:237 (1997).
13. D. A. Papaconstantopoulos and W. E. Pickett, Phys. Rev. B **57**:12751 (1998).
14. I. Solovyev, N. Hamada and K. Terakura, Phys. Rev. Lett. **76**:4825 (1996).
15. I. Solovyev, N. Hamada and K. Terakura, Phys. Rev. B **53**:7158 (1996).
16. J. Pierre, B. Gillon, L. Pinsard, and A. Revcolevschi, Europhys. Lett. **42**:85 (1998).
17. S. Satpathy, Z. S. Popović, and F. R. Vukajlović, Phys. Rev. Lett. **76**:960 (1996).
18. D. D. Sarma, N. Shanthi, S. R. Barman, N. Hamada, H. Sawada, and K. Terakura, Phys. Rev. Lett. **75**, 1126 (1995).
19. P. K. de Boer and R. A. de Groot, Comput. Mat. Sci. **10**:240 (1998).
20. J. P. Perdew, J. A. Chevary, S. H. Vosko, K. A. Jackson, M. R., Pederson, D. J. Singh, and C. Fiolhais, Phys. Rev. B **46**:6671 (1992).
21. J.-H. Park, E. Vescovo, H.-J. Kim, C. Kwon, R. Ramesh, and T. Venkatesan, Nature **392**:794 (1998).
22. Q. Huang, A. Santoro, J. W. Lynn, R. W. Erwin, J. A. Borchers, J. L. Peng, K. Ghosh, and R. L. Greene, Phys. Rev. B **58**:2684 (1998).
23. B. Dabrowski, K. Rogacki, X. Siong, P. W. Klamut, R. Dybzinski, J. Shaffer and J. D. Jorgensen, Phys. Reb. B **58**:2716 (1998).
24. P. Dai, J. Zhang, H. A. Mook, S. H. Liou, P. A. Dowben, and E. W. Plummer, Phys. Rev. B **54**:R3694 (1996).
25. A. V. Boris *et al.*, "Infrared Studies of a $La_{0.67}Ca_{0.33}MnO_3$ Single Crystal: Optical Magnetoconductivity in a Half-Metallic Ferromagnet," preprint (cond-mat/9808049).
26. J. C. Slater and G. F. Koster, Phys. Rev. **94**:1498 (1954).
27. W. H. Butler, X.-G. Zhang, and J. M. MacLaren, in *Magnetic Ultrathin Films, Multilayers, and Surfaces*, MRS Symposium Proceedings No. 384, edited by E. E. Marinero, B. Heinrich, W. F. Egelhoff, Jr., A. Fert, H. Fujimori, G. Guntherodt, and R. L. White (Materials Research Society, Pittsburgh,1995), pp.439-443.
28. J. Tobola, S. Kaprzyk, and J. Pierre, Acta Physica Polonica A **92**:461 (1997).
29. X. Wang and X.-G. Zhang, unpublished.
30. H. Y. Hwang et al., Phys. Rev. Lett. **77**:2041 (1996).
31. A. J. Millis, R. Mueller, and B. I. Shraiman, Phys. Rev. B **54**:5405 (1996).
32. J. D. Lee and B. I. Min, Phys. Rev. B **55**:12454 (1997).
33. Q. Li, J. Zang, A. R. Bishop, and C. M. Soukoulis, Phys. Rev. B **56**:4541 (1997).
34. W. E. Pickett, J. Korean Phys. Soc.(Proc. Suppl.) **29**:S70 (1996).

OPTICAL CONDUCTIVITY OF DOPED MANGANITES: COMPARISON OF FERROMAGNETIC KONDO LATTICE MODELS WITH AND WITHOUT ORBITAL DEGENERACY

Frank Mack and Peter Horsch

Max-Planck-Institut für Festkörperforschung,
Heisenbergstr. 1
D-70569 Stuttgart, Germany

INTRODUCTION

The colossal magnetoresistance (CMR) of manganese oxides[1] and their transport properties are usually studied in the framework of the double exchange (DE) Hamiltonian or the more general ferromagnetic Kondo lattice model (KLM)[2]. The essence of these models is the high-spin configuration of e_g-electron and t_{2g}-core electron spins due to a strong ferromagnetic Kondo exchange interaction $K \sim 1\text{eV}$. The kinetic energy in the partially filled e_g band is lowered when neighboring spins are aligned leading to a low-temperature ferromagnetic phase, while the high-T paramagnetic phase is disordered with a high resistivity. In these considerations the orbital degeneracy of the Mn e_g orbitals is usually neglected. Although this model provides an explanation of the phases necessary for a qualitative understanding of CMR, it has been stressed that the double exchange mechanism is not sufficient for a quantitative description.[3]

Doped manganites are characterized by strong correlations and the complex interplay of spin-, charge-, and orbital-degrees of freedom as well as the coupling to the lattice, e.g. via Jahn-Teller coupling.[4,5] This complexity is directly evident from the large number of phases in a typical phase diagram. To quantify the different mechanisms it is helpful to analyze experiments where for certain parameters one or the other degree of freedom is frozen out. Important experiments in this respect are the very detailed investigations of the optical conductivity of $La_{1-x}Sr_xMnO_3$ by Okimoto et al..[6,7] These experiments (see Fig. 1) show (a) a pseudogap in $\sigma(\omega)$ for temperatures above the Curie temperature T_c (paramagnetic phase) with $\sigma(\omega)$ essentially linear in ω, and (b) the evolution of a broad incoherent distribution in the range $0 \leq \omega < 1.0$ eV below T_c, which still grows at temperatures below $T_c/10$, where the magnetization is already close to saturation. Such a temperature dependence of $\sigma(\omega)$ over a wide energy region is quite unusual as compared with other strongly correlated electron systems near a metal-insulator phase boundary.[6] Interestingly there is in addition a narrow

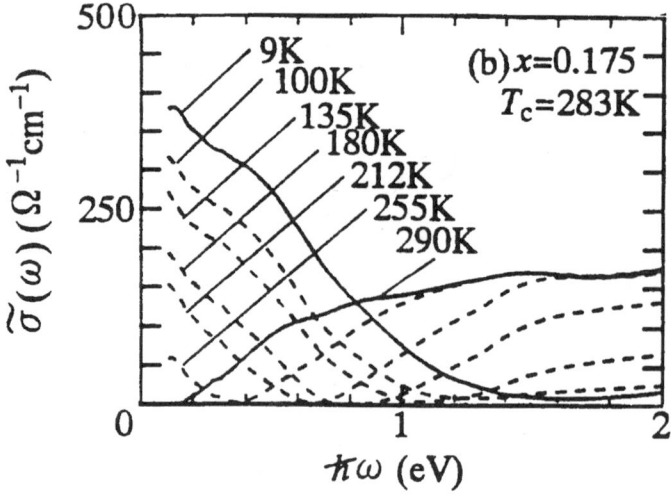

Figure 1: Temperature dependence of the optical conductivity of $La_{1-x}Sr_xMnO_3$ for $x = 0.175$ (reproduced from Okimoto et al.). A temperature independent background has been subtracted from the experimental data.

Drude peak with width of about 0.02 eV and little weight superimposed to the broad incoherent spectrum. That is, the motion of charge carriers is dominated by incoherent processes but there is also a coherent channel leading to a narrow Drude peak. In view of the large energy scale of 1 eV it is plausible that the orbital degeneracy is the source of this incoherent motion, since the spin degrees of freedom are essentially frozen out. Our study of the degenerate Kondo lattice model will support this point of view. Further optical studies for various 3D manganites[8-10] as well as for the bilayer system $La_{1.2}Sr_{1.8}Mn_2O_7$[11] confirm the presence of the large incoherent absorption in the ferromagnetic state.

Although the importance of these experiments was recognized immediately, the few theoretical studies[12-14] were confined to simplified models or approximations, thereby ignoring important aspects of the full many body problem.

The aim of this work is to show that in a model which accounts for the orbital degeneracy, yet assumes that the spins are fully polarized, the broad incoherent spectral distribution of $\sigma(\omega)$, its increase with decreasing temperature, as well as the order of magnitude of $\sigma(\omega)$ at small ω can be explained. Our calculation also accounts for a small and narrow Drude peak as observed by Okimoto et al. in the range $\omega < 0.05$ eV. This is a clear indication of coherent motion of charge carriers with small spectral weight, i.e. in a model where due to the orbital degeneracy incoherent motion is dominant.

We start our discussion with a generic Hamiltonian for the manganite systems, i.e. the ferromagnetic Kondo lattice model with degenerate e_g-orbitals, and derive for the spin-polarized case an effective model which contains only the orbital degrees of freedom. This orbital model consists of a hopping term between the same and different orbitals α and β on neighbor sites and an orbital interaction. Renaming $\alpha = \sigma$ where $\sigma = \uparrow$ or $\downarrow$ the model maps on a generalized anisotropic t-J model. The usual t-J model known from the cuprates appears as a special case. Our derivation includes the 3-site hopping processes, which appear as a natural consequence of the strong coupling limit. Although such terms do not influence the orbital order for integer filling, they are important for the proper description of transport properties in the doped systems.[17-20]

Because of the complexity of the orbital model we present here a numerical study of the orbital correlations and of the frequency dependent conductivity.[21] Although there exist studies of the interplay of orbital and spin order at integer filling for LaMnO$_3$,[22,23] the effect of doping has not been considered so far. The finite temperature diagonalization[24,25] serves here as an unbiased tool to study the optical conductivity. The change of orbital correlations as function of doping and temperature is investigated by means of an *optimized orbital basis* (OOB).[21]

Furthermore we present results for the KLM *without* orbital degeneracy which clearly show that this model fails to explain the broad incoherent $\sigma(\omega)$ spectra observed in the ferromagnetic phase as $T \to 0$.[6,7] The KLM would instead lead to a sharp Drude absorption because of the ferromagnetic alignement,[15] i.e. to spinless fermion behavior in the low-temperature limit. This underlines the importance of the e_g orbital degeneracy, although there are alternative proposals invoking strong electron-phonon coupling and lattice polaron formation[4,3,16] to explain the incoherency of the absorption.

MODELS

We start from the generic ferromagnetic Kondo lattice model for the manganites, where the Mn e_g-electrons are coupled to core spins $\vec{S}_i$ (formed by the t_{2g} orbitals) and a local repulsion U between the electrons in the two e_g orbitals:

$$H = H_{band} + H_{int} + H_{Kondo}. \tag{1}$$

H_{band} describes the hopping of e_g-electrons between sites i and j, H_{int} the interaction between e_g electrons and H_{Kondo} the coupling of e_g-spins and t_{2g}-core spins. The e_g electrons to have two-fold orbital degeneracy labelled by a Roman index (a,b) and two fold spin degeneracy labelled by a Greek index (σ, σ'). Explicitly,

$$H_{band} = \sum_{ia\sigma} E_i^a d^\dagger_{ia\sigma} d_{ia\sigma} + \sum_{\langle ij \rangle ab\sigma} (t_{ij}^{ab} d^\dagger_{ia\sigma} d_{jb\sigma} + H.c.). \tag{2}$$

The hopping matrix elements t_{ij}^{ab} form a real symmetric matrix whose form depends on the choice of basis in orbital space and the direction of the $\mathbf{i-j}$ bond. For the present study we shall use $|x\rangle \sim x^2 - y^2$ and $|z\rangle \sim (3z^2 - r^2)/\sqrt{3}$ as basis for the e_g orbitals. From the Slater-Koster rules follows[26-28]:

$$t_{\mathbf{ij}\|x/y}^{ab} = -\frac{t}{4}\begin{pmatrix} 1 & \mp\sqrt{3} \\ \mp\sqrt{3} & 3 \end{pmatrix}, \quad t_{\mathbf{ij}\|z}^{ab} = -t\begin{pmatrix} 1 & 0 \\ 0 & 0 \end{pmatrix}, \tag{3}$$

which allows for inter-orbital hopping in the xy-plane, where the upper (lower) sign distinguishes hopping along x and y direction. The hopping matrix elements are defined in terms of the double exchange t of $d_{3z^2-r^2}$ orbitals along the c-axis. Here $t = V_{dp\sigma}^2/\Delta\epsilon_{dp}$ is determined by the Mn-d O-p hybridisation $V_{dp\sigma}$ and the corresponding level splitting $\Delta\epsilon_{dp}$. All matrix elements are negative except t^{xz} along x-direction, which is positive. The level splitting $\Delta E_i = E_i^z - E_i^x$ can be controlled by uniaxial pressure or in the case of layered compounds even with hydrostatic pressure.[29] We shall focus here on the orbital degenerate case, i.e. $\Delta E_i = 0$, however we will keep the orbital energy term for the derivation of the orbital model.

The large local electron-electron repulsion U is responsible for the insulating behavior in the case of integer band-filling

$$H_{int} = \sum_{ia} U_a n_{ia\uparrow} n_{ia\downarrow} + U_{ab} \sum_i n_{ia} n_{ib} + J_{ab} \sum_{i\sigma\sigma'} d^\dagger_{ia\sigma} d^\dagger_{ib\sigma'} d_{ia\sigma'} d_{ib\sigma}. \quad (4)$$

Here U_a, U_{ab} and J_{ab} are the intra- and inter-orbital Hubbard and exchange interactions, respectively. The relevant valences of Mn are Mn^{3+} (S=2) and Mn^{4+} (S=3/2) as was pointed out already by Zener.[2] The Hamiltonian $H_{band} + H_{int}$ was considered by Zaanen and Oleś who showed that in general rather complex effective Hamiltonians result for transition metal ions with partially filled d shell near orbital degeneracy.[30] Here a simplified approach is preferred, with the t_{2g} electrons of Mn ions forming core spins of $S = 3/2$, and thus we restrict the electron interactions in H_{int} to the e_g bands.

Even in the case with local orbital degeneracy there is one lower Hubbard band which is partially filled in the case of hole doping. With one e_g-electron per site these systems are Mott-insulators. Although the Jahn-Teller splitting can lead to a gap in the absence of U, it is not the primary reason for the insulating behavior for integer filling.[31]

The interaction of itinerant electrons with the (S=3/2) core spins is given by

$$H_{Kondo} = -K \sum_{ia\sigma\sigma'} \vec{S}_i \cdot d^\dagger_{ia\sigma} \vec{\sigma}_{\sigma\sigma'} d_{ia\sigma'} \quad (5)$$

leading to parallel alignment of the d-electron spin with the core-spin $\vec{S}_i$. Since the d-electron kinetic energy is favored by a parallel orientation of neighboring spins the ground state becomes ferromagnetic.[2]

In the low-temperature ferromagnetic phase we may introduce an effective Hamiltonian which contains only the orbital degrees of freedom assuming a fully spin-polarized ferromagnet. The spin degrees of freedom can also be eliminated by high magnetic fields. This opens the possibility to investigate the orbital order independent of the spin degrees of freedom, and to shed light on the nontrivial question how the orbital order is changed upon doping with e_g-electrons or holes.

The resulting model has similarities to the t-J model, where the spin-indices σ and σ' are now orbital indices. To stress the difference we use for the orbital indices the letters a, b and α, β. Unitary transformation $H_{orb} = e^{-S} H e^{S}$ [32] and the restriction to states without double occupancy leads to the orbital t-J model which has the following structure:

$$H_{orb} = \sum_{ia} E^i_a \tilde{d}^\dagger_{ia} \tilde{d}_{ia} + \sum_{\langle ij \rangle ab} (t^{ab}_{ij} \tilde{d}^\dagger_{ia} \tilde{d}_{jb} + H.c.) + H'_{orb}. \quad (6)$$

with the constraint that each site can be occupied by at most one electron, i.e. $\tilde{d}^\dagger_{ia} = d^\dagger_{ia}(1 - n_{i\bar{a}})$. Here and in the following the index $\bar{a}$ denotes the orthogonal e_g orbital with respect to orbital a. The orbital interaction H'_{orb} follows as a consequence of the elimination of doubly occupied sites with energy $U \sim U_{ab} - J_{ab}$.[21]

$$H'_{orb} = -\frac{1}{2} \sum_{\mathbf{j}\mathbf{u}\mathbf{u}'} \sum_{ab\alpha\beta} t^{\alpha\beta}_{\mathbf{j+u}\,\mathbf{j}} t^{ba}_{\mathbf{j}\,\mathbf{j+u'}} \left(\frac{1}{U + E^\beta_\mathbf{j} - E^\alpha_{\mathbf{j+u}}} + \frac{1}{U + E^b_\mathbf{j} - E^a_{\mathbf{j+u'}}} \right)$$
$$\left[\delta_{\beta,b}\, \tilde{d}^\dagger_{\mathbf{j+u}\alpha} \tilde{d}^\dagger_{\mathbf{j}\bar{\beta}} \tilde{d}_{\mathbf{j}\bar{b}} \tilde{d}_{\mathbf{j+u'}a} - \delta_{\beta,b}\, \tilde{d}^\dagger_{\mathbf{j+u}\alpha} \tilde{d}^\dagger_{\mathbf{j}b} \tilde{d}_{\mathbf{j}\beta} \tilde{d}_{\mathbf{j+u'}a} \right]. \quad (7)$$

Here $\mathbf{u}, \mathbf{u}' = (\pm a, 0)$ or $(0, \pm a)$ are lattice unit vectors. The orbital interaction $H'_{orb} = H'^{(2)}_{orb} + H'^{(3)}_{orb}$ consists of two types of contributions: (i) 2-site terms ($\mathbf{u} = \mathbf{u}'$), i.e. similar to the Heisenberg interaction in the standard t-J model, yet more complex because of the nonvanishing off-diagonal t^{ab}, and (ii) 3-site hopping terms ($\mathbf{u} \neq \mathbf{u}'$) between second nearest neighbors. In the half-filled case, i.e. one e_g-electron per site, only the 2-site interaction is operative and may induce some orbital order. In the presence of hole doping, i.e. for less than one e_g-electron per site, both the kinetic energy and the 3-site contributions in H'_{orb} lead to propagation of the holes and to a frustration of the orbital order.

The complexity of the model can be seen if we express the orbital interaction H'_{orb} in terms of pseudospin operators $T_i^z = \frac{1}{2}(n_{ia} - n_{ib})$, $T_i^+ = \tilde{d}_{ia}^\dagger \tilde{d}_{ib}$ and $T_i^- = \tilde{d}_{ib}^\dagger \tilde{d}_{ia}$. where $a = \uparrow$ and $b = \downarrow$ denote an orthogonal orbital basis in the e_g-space. To be specific we shall assume here $a(b) = z(x)$. Keeping only the two-site contributions, i.e. $\mathbf{u} = \mathbf{u}'$, one obtains an anisotropic Heisenberg Hamiltonian for the orbital interactions:

$$H'^{(2)}_{orb} = -\frac{2}{U} \sum_{\langle ij \rangle} \Big[(t_{ij}^{aa\,2} + t_{ij}^{bb\,2})(\frac{1}{4}n_i n_j - T_i^z T_j^z) - t_{ij}^{aa} t_{ij}^{bb}(T_i^+ T_j^- + T_i^- T_j^+)$$
$$+ (t_{ij}^{ab\,2} + t_{ij}^{ba\,2})(\frac{1}{4}n_i n_j + T_i^z T_j^z) - t_{ij}^{ab} t_{ij}^{ba}(T_i^+ T_j^+ + T_i^- T_j^-) \quad (8)$$
$$- (t_{ij}^{aa} t_{ij}^{ab} - t_{ij}^{bb} t_{ij}^{ba})(T_i^z T_j^+ + T_i^z T_j^-) - (t_{ij}^{aa} t_{ij}^{ba} - t_{ij}^{bb} t_{ij}^{ab})(T_i^+ T_j^z + T_i^- T_j^z) \Big].$$

The orbital interaction $H'^{(2)}_{orb}$ is equivalent to that studied by Ishihara et al..[22,29] Since we are interested here in the effect of holes on the orbital order we will generally base our study on H'_{orb} (7) which includes the 3-site hopping processes as well.

We note that in the special case $t_{ij}^{ab} = \delta_{ab} t$ and $E_i^a = 0$ equations (6) and (7) are identical to the usual t-J model (with $J = 4t^2/U$ and including 3-site processes). For the orbital model we shall adopt the same convention for the orbital exchange coupling, i.e. $J = 4t^2/U$.

The orbital degrees of freedom in combination with strong correlations (i.e. no doubly occupied sites are allowed) is expected to lead to incoherent motion of holes, although quasiparticle formation with reduced spectral weight is a plausible expectation on the basis of what is known about the usual t-J model.

In this paper we focus on the orbital degenerate case with $E_i^a = E_i^b = 0$. A detailed study of the influence of a finite level splitting will be presented elsewhere.

OPTICAL CONDUCTIVITY

We study the charge transport by calculating the optical conductivity

$$\sigma_0(\omega) = 2\pi e^2 D_c \delta(\omega) + \sigma(\omega). \quad (9)$$

The frequency dependent conductivity consists of two parts, the regular finite frequency absorption $\sigma(\omega)$ and the δ-function contribution which is proportional to the charge stiffness D_c[35,33]. The latter vanishes in insulators. This contribution is broadened into a usual Drude peak in the presence of other scattering processes like impurities which are not contained in the present model. The finite frequency absorption (or regular part) $\sigma(\omega)$ is determined by the current-current correlation function using the Kubo formula

$$\sigma(\omega) = \frac{1 - e^{-\omega/T}}{N\omega} Re \int_0^\infty dt e^{i\omega t} \langle j_x(t) j_x \rangle. \quad (10)$$

For the derivation of the current operator we introduce twisted boundary conditions via Peierls construction

$$t_{\mathbf{j+u}\,\mathbf{j}}^{ab}(\vec{A}) = t_{\mathbf{j+u}\,\mathbf{j}}^{ab} \exp\Big(-i\frac{e}{\hbar}\int_{\mathbf{j}}^{\mathbf{j+u}} \vec{A}(x)\vec{dx}\Big). \quad (11)$$

Assuming $\vec{A} = (A_x, A_y, 0)$ we obtain from the kinetic energy operator the x-component of the current operator $j_x = \partial H(A_x)/\partial A_x$:

$$j_x^{(1)} = -ie \sum_{\mathbf{j+u}ab} t^{ab}_{\mathbf{j+u\,j}} \, u_x \, \tilde{d}^\dagger_{\mathbf{j+u}a}\tilde{d}_{\mathbf{j}b}. \tag{12}$$

An additional contribution follows from the 3-site term in Eq.(6)

$$j_x^{(3)} = \frac{ie}{2U} \sum_{\mathbf{juu'}} \sum_{ab\alpha\beta} t^{\alpha\beta}_{\mathbf{j+u\,j}} t^{ba}_{\mathbf{j\,j+u'}} (u_x - u'_x) O^{\alpha\beta}_{ba}(\mathbf{j,u,u'}),$$

$$O^{\alpha\beta}_{ba} = \delta_{\beta,b} \, \tilde{d}^\dagger_{\mathbf{j+u}a}\tilde{d}^\dagger_{\mathbf{j}\bar{\beta}}\tilde{d}_{\mathbf{j}\bar{b}}\tilde{d}_{\mathbf{j+u'}a} - \delta_{\beta,\bar{b}} \, \tilde{d}^\dagger_{\mathbf{j+u}a}\tilde{d}^\dagger_{\mathbf{j}b}\tilde{d}_{\mathbf{j}\beta}\tilde{d}_{\mathbf{j+u'}a}, \tag{13}$$

where the expression $O^{\alpha\beta}_{ba}(\mathbf{j,u,u'})$ is an abbreviation for the operator within the last brackets of Eq.(6).

We stress here that the orbital order in the undoped case is determined exclusively by the (two-site) orbital interaction (7), while the 3-site processes in (6) only contribute in the doped case. Nevertheless the motion of holes and transport in general is influenced in a significant way by the 3-site current operator $j_x^{(3)}$. In fact from studies of the t-J model it is known that $\sigma(\omega)$ is not only quantitatively but even qualitatively changed by these terms. [18,19] Although the 3-site terms are usually ignored in studies of the t-J model they are an important part of the strong coupling model and should not be dropped in studies of the conductivity. One of the aims in our study of the orbital model is to analyse the effect of these 3-site terms in the model.

The charge stiffness D_c can be determined in two ways: (a) using Kohn's relation, [35] or (b) via the optical sum rule which relates the integrated spectral weight of the real part of the conductivity to the average kinetic energy:

$$\int_{-\infty}^{\infty} \sigma_0(\omega) d\omega = -\frac{\pi e^2}{N} \langle H^{kin}_{xx} \rangle. \tag{14}$$

Here H^{kin} contains two contributions: (a) the usual kinetic energy $\sim t$ (5) and (b) the 3-site hopping processes $\sim t^2/U$ in Eq. (6). The case with off-diagonal hopping considered here is a generalization of the sum rules for the t-J model[36] and for the t-J model including 3-site terms. [18,20] Together with (8) this implies

$$D_c = -\frac{1}{2N}\langle H^{kin}_{xx} \rangle - \frac{1}{\pi e^2} \int_{0+}^{\infty} \sigma(\omega) d\omega. \tag{15}$$

In the following we shall use $D_c = S - S_\omega$ as abbreviation for this equation, where S denotes the sum rule expression and S_ω the finite frequency absorption.

FINITE TEMPERATURE LANCZOS METHOD (FTLM)

For the calculation of $\sigma(\omega)$ we use a generalization of the exact diagonalization technique for finite temperature developed by Jaklič and Prelovšek. [24] In this approach the trace of the thermodynamic expectation value is performed by a Monte-Carlo sampling. The current-current correlation function in (9):

$$C(\omega) = Re \int_0^\infty dt\, e^{i\omega t} \langle j_x(t) j_x \rangle. \tag{16}$$

can be rewritten by introducing a complete set of basis functions $|r\rangle$ for the trace, and eigenfunctions $|\Psi_j^r\rangle$ and $|\tilde{\Psi}_j^r\rangle$ of H with eigenvalues E_i^r and $\tilde{E}_j^r$ (here r is an irrelevant label, which will get its meaning below):

$$C(\omega) \approx \frac{\pi}{Z} \sum_{r=1}^{R} \sum_{i,j=1}^{M} e^{-\beta E_i^r} \langle r | \Psi_i^r \rangle \langle \Psi_i^r | j_x | \tilde{\Psi}_j^r \rangle \langle \tilde{\Psi}_j^r | j_x | r \rangle \delta(\omega + E_i^r - \tilde{E}_j^r) \quad (17)$$

with

$$Z \approx \sum_{r=1}^{R} \sum_{i=1}^{M} e^{-\beta E_i^r} |\langle r | \Psi_i^r \rangle|^2 . \quad (18)$$

These expressions are exact if one uses complete sets. The FTLM is based on two approximations when evaluating Eqs.(17) for $C(\omega)$ and (18) for the partition function Z: (1) The trace is performed over a restricted number R of random states $|r\rangle$, and (2) the required set of eigenfunctions of H is generated by the Lanczos algorithm starting from the initial states $|\Phi_0^r\rangle = |r\rangle$ and $|\tilde{\Phi}_0^r\rangle = j_x |r\rangle / \sqrt{\langle r | j_x^2 | r \rangle}$, respectively. The latter procedure is truncated after M steps and yields the relevant intermediate states for the evaluation of $C(\omega)$.

Detailed tests[24] have shown that the results get very accurate already if $R, M \ll N_0$, where N_0 is the dimension of the Hilbert space. The computational effort is significantly reduced because only matrices of dimension $M \times M$ have to be diagonalized, i.e. much smaller than the dimension $N_0 \times N_0$ of the full Hamiltonian H. In practice typical dimensions are $R = 100$ and $M = 100$.

A complication in the case of the orbital t-J model is that T_{tot}^z does not commute with the Hamiltonian H_{orb} (6). Therefore the calculations for the orbital model are restricted to smaller clusters than in the t-J case, where different S_{tot}^z subspaces can be treated separately.

RESULTS

Kondo-Lattice Model without Orbital Degeneracy

We begin our discussion of the frequency dependent conductivity with results for the single orbital Kondo-lattice model excluding double occupancy

$$H = -\sum_{\mathbf{ij}\sigma} t_{\mathbf{ij}} \tilde{d}_{\mathbf{i}\sigma}^\dagger \tilde{d}_{\mathbf{j}\sigma} - J_H \sum_{\mathbf{i}\sigma\sigma'} \vec{S}_{\mathbf{i}} \cdot \tilde{d}_{\mathbf{i}\sigma}^\dagger \vec{\sigma}_{\sigma\sigma'} \tilde{d}_{\mathbf{i}\sigma'}. \quad (19)$$

Nearest neighbor hopping $t_{\mathbf{ij}} = t$ is assumed and the actual calculations were performed for $S = 1$ core spins for different doping concentrations x.

In Fig.2 the temperature dependence of the optical conductivity is shown for a chain at doping $x = 2/6$. The conductivity is given in dimensionless form $\sigma \rho_0$, where $\rho_0 = \hbar a/e^2$ sets the dimension of the resistivity in a 3D s.c. crystal structure with lattice constant a. At strong coupling $J_H \gg t$ the states of the model consist of a low-energy band corresponding to the $S + 1/2$ configurations, and of the exchange-splitted bands separated by an energy $\frac{2}{3}J_H$. The intraband transitions are shown in the large window. They give rise to an incoherent low-energy absorption extending up to the energy of the free carrier band-width $\sim 4t$. The spectrum of the high to low spin transitions is centered around $\omega \sim \frac{3}{2}J_H$ and is shown in the inset.

At low temperatures, however, when ferromagnetic correlations extend over the whole system, the low-energy spectrum is dominated by a narrow coherent peak at

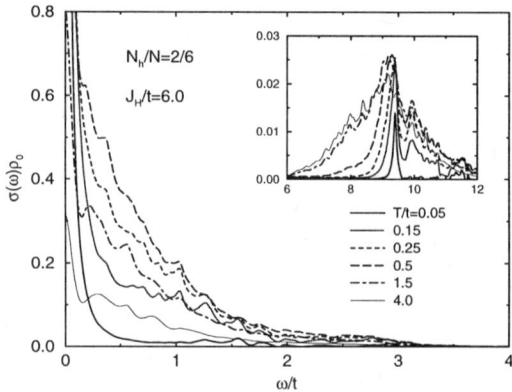

Figure 2: Frequency dependent conductivity $\sigma(\omega)$ of the *KLM without orbital degeneracy* for a six site chain with two electrons and $J_H = 6t$ for different temperatures. Inset shows the spectrum of excitations to the exchange-splitted band. Note that for $T \to 0$ all spectral weight is in the Drude peak (From Jaklič, Horsch and Mack[15]).

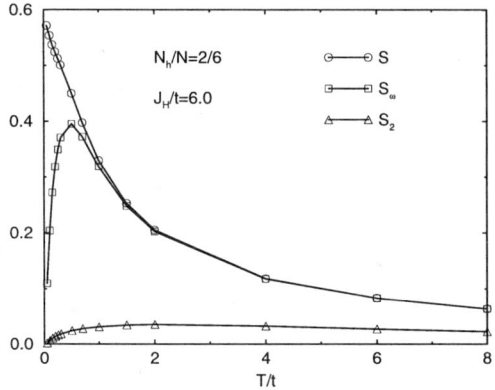

Figure 3: Temperature dependence of the kinetic energy and of the optical spectral weights of the ferromagnetic *Kondo-lattice model in the absence of orbital degeneracy* for a 6 site chain with two holes. Here $S = \langle H^{kin}_{xx}\rangle/2Nt$ and $S_\omega = \int \sigma(\omega)d\omega/\pi e^2 t$. The spectral weight of the interband optical transitions into low-spin states is measured by $S_2 = \int \sigma_2(\omega)d\omega/\pi e^2 t$.

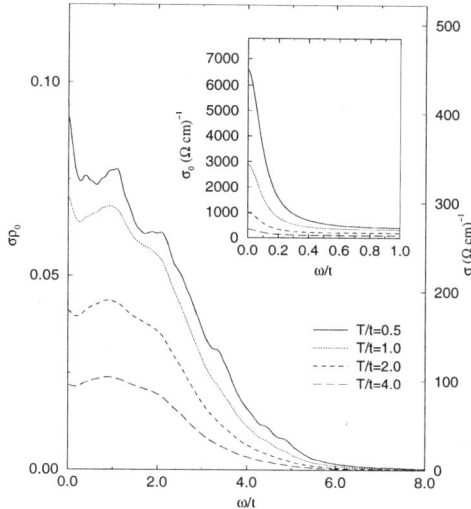

Figure 4: Optical conductivity $\sigma(\omega)$ calculated for the *orbital t-J model* for a N=10 site plane with N_e=8 electrons and J=0.25t (U=16t) at different temperatures. The calculation includes the 3-site terms. The inset shows the optical conductivity $\sigma_0(\omega)$ including the Drude-part. The spectra were broadened with Γ=0.1t.

$\omega = 0$, and the interband transitions vanish. At $T = 0$ all weight is in the Drude peak (see Fig.3). As the temperature is increased the spectral weight is transferred from the coherent part to the broad incoherent spectrum and to the exchange-split excitations. This clearly shows that the model without orbital degeneracy does not explain the anomalous absorption in the anomalous ferromagnetic state of manganites (Fig.1).

Optical Conductivity for the Orbital t-J Model

Typical results for the frequency and temperature dependence of $\sigma(\omega)$ for a two-dimensional 10-site cluster with 2 holes, i.e. corresponding to a doping concentration $x = 0.2$, are shown in Fig. 4 with and Fig. 5 without 3-site hopping terms, respectively. The $\sigma(\omega)$ spectra are bell-shaped and increase with decreasing temperature. The width of the distribution measured at half-maximum is $\omega_{1/2} \sim 2.5t$ and is essentially independent of temperature. Calculations for a 3D cluster yield a rather similar $\sigma(\omega)$ distribution with a slightly larger value for the width $\omega_{1/2} \sim 3.5t$.

If we compare this latter (3D) value with $\omega_{1/2}^{exp} \sim 0.7$ eV found from the data of Okimoto et al. (for $x = 0.175$ and $T = 9K$), we obtain an experimental estimate for the parameter t: $t^{opt} \sim 0.2$ eV. For comparison a rough theoretical estimate based on Harrison's solid state table yields $t = V_{dp\sigma}^2/\Delta\epsilon \sim 0.4$ eV. Hence the orbital t-J model explains in a natural way the ~ 1 eV energy scale of the incoherent absorption in the experiments. Although $\sigma(\omega)$ shows the anomalous increase with decreasing temperature, we do not claim here that the orbital model accounts for the full temperature dependence in the ferromagnetic phase $(T < T_c)$. The description of the complete T-dependence certainly requires the analysis of the degenerate-orbital KLM, i.e. the additional inclusion of the spin degrees of freedom. Yet we believe that the results can be compared with the data by Okimoto et al. in the low-T limit, i.e. in the regime

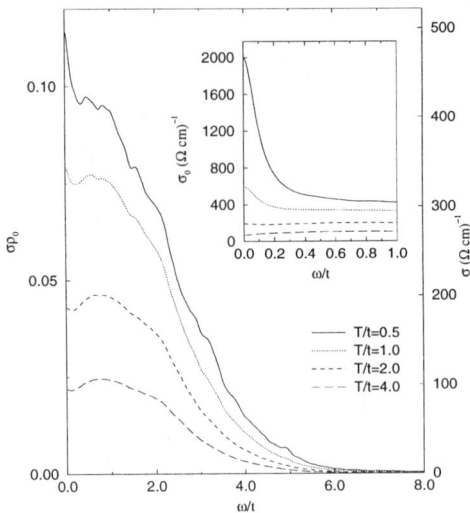

Figure 5: Optical conductivity as in Fig.4 but without 3-site contributions. Due to the stronger incoherence there is a smaller Drude peak (inset) and a smaller DC-conductivity. (From Horsch, Jaklič and Mack[21]).

where the magnetization is saturated.

To compare with experimental data we present the conductivity in dimensionless form and with the proper dimensions of a 3D conductivity, assuming that the 3D system is formed by noninteracting layers. For a cubic lattice $\rho_0 = \hbar a/e^2$. If we consider $La_{1-x}Sr_xMnO_3$ with lattice constant $a = 5.5$ Å and $\hbar/e^2 = 4.11 K\Omega$ we obtain $\rho_0 = 0.23 \cdot 10^{-3}$ Ωcm. Hence the low frequency limit of the regular part of the conductivity $\sigma(\omega \to 0) = \sigma_0^{inc} \sim 0.3$–$0.4 \cdot 10^3$ $(\Omega cm)^{-1}$. This is consistent with the order of magnitude for the low frequency limit of the incoherent part of the $\sigma(\omega)$ data of Okimoto et al. for $La_{1-x}Sr_xMnO_3$ $\sigma_0^{inc} \sim 0.4 \cdot 10^3$ $(0.3 \cdot 10^3)$ $(\Omega cm)^{-1}$ for the doping concentrations $x = 0.175$ (0.3), respectively, at $T = 10K$.

Hence we conclude that besides the energy scale also the absolute value of the incoherent part of the experimental $\sigma(\omega)$ spectrum is consistent with the orbital model. We stress that the value (order of magnitude) of σ_0^{inc} is essentially fixed by the conductivity sum rule and the scale $\omega_{1/2}$, as long as the conductivity is predominantly incoherent. The insets in Fig.4 and 5 show $\sigma_0(\omega)$, Eq.(9), with the Drude absorption at low frequency included, where we used an ad hoc chosen parameter $\Gamma = 0.1t$ to broaden the δ-function. This value corresponds to the experimental width $\Gamma \sim 0.02$ eV taken from the (small) Drude peak observed by Okimoto et al..[7] While our model yields the weight of the Drude peak D_c, it does not give Γ, which is due to extrinsic processes (e.g. scattering from impurities and grain boundaries). The charge stiffness D_c together with Γ determines the DC-conductivity.

Adopting the experimental estimate for Γ we find for the example in Fig.4 (inset) for the low temperature DC-conductivity $\sigma_{DC} \sim 6.5 \cdot 10^3$ $(\Omega cm)^{-1}$ and for the resistivity $\rho \sim 0.15 \cdot 10^{-3}$ Ωcm, respectively. For comparison, the experimental range of resistivities is e.g. $0.1 - 1.0 \cdot 10^{-3}$ Ωcm in the ferromagnetic metallic phase of $La_{1-x}Sr_xMnO_3$[34]. As we shall see below, the Drude weight and therefore the DC-conductivity depend considerably on the value of the exchange coupling J, and whether the 3-site hopping

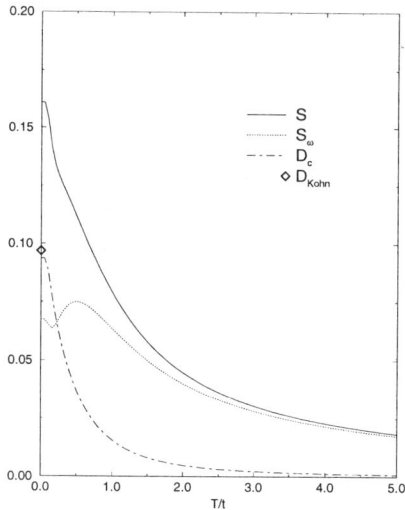

Figure 6: Temperature dependence of the kinetic energy S (solid line), the incoherent spectral weight S_ω (dotted) and the Drude weights D_c (dash-dotted line) and D_{Kohn} (diamond) for a N=10 site plane with N_e=8 electrons and J=0.25t (U=16t) in units of $t/\rho_0 e^2$.

processes in the model are taken into account or not. These terms have a strong effect on the coherent motion of charge carriers. After this discussion it should be clear, however, that the DC-conductivity at low temperatures is largely determined by extrinsic effects.

Figure 6 shows the temperature variation of the sum rule (kinetic energy), the finite frequency absorption S_ω and the Drude weight D_c for the model including the 3-site terms. Whereas at high temperature the sum rule is essentially exhausted by the finite frequency absorption $\sigma(\omega)$, we find at low temperatures a significant increase of the Drude weight. In this particular case D_c contributes about 40% to the sum rule at $T/t = 0.3$.

The conductivity data shown in Figs. 4 and 5 for $T/t > 0.5$ is characteristic for a 2D orbital liquid state, which is stabilized by thermal fluctuations. The significant increase of coherency in Fig. 6 below $T/t = 0.3$ is due to the onset of x^2-y^2 orbital order in the planar model (see discussion below). This ordering is characteristic for the 2D version of the orbital t-J model and does probably not occur in cubic systems for which an orbital liquid ground state was proposed.[13] Therefore we consider the $T/t = 0.3$ data for D_c to be more appropriate for a comparison with the low-temperature 3D data than the $T = 0$ values, which are enhanced due to orbital order.

We have also calculated D_c for $T = 0$ using Kohn's relation[35]

$$D_{Kohn} = \frac{1}{2N} \frac{\partial^2 E_0(\Phi)}{\partial \Phi^2}\bigg|_{\Phi=\Phi_0}, \quad (20)$$

which yields a consistent value (Fig.6). Here $\Phi = eaA_\alpha/\hbar$ is the Peierls phase induced by the applied vector potential with component A_α. For the evaluation of Eq.20 we followed Ref.[18] and first searched for the minimum of the ground state energy E_0 with respect to the vector potential,[37] and then calculated the curvature. Since this procedure is quite

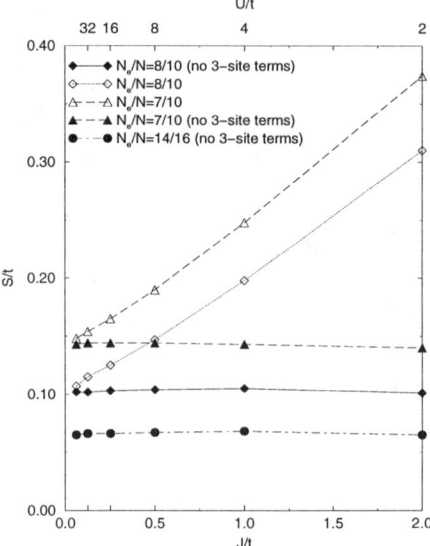

Figure 7: Kinetic energy S versus J for a N=10 site 2D cluster with N_e=7 and 8 electrons, with (full) and without (open symbols) the 3-site terms ($T/t = 0.3$), respectively. The dash-dotted line (circles) represents a 16 site cluster with N_e=14 and without 3-site terms (From Horsch, Jaklič and Mack[21]).

cumbersome, we have determined most of our D_c data via the sum rule. The resulting sets of data show systematic trends as function of J and doping.

From these results we expect in analogy with the t-J model that the single-particle electron Green's function is characterized by a pronounced quasiparticle peak to explain such a large fraction of coherent transport.

From Fig.6 we also see that the weight under $\sigma(\omega)$, i.e. S_ω, does not further increase for temperatures below $T = 0.5t$, although the temperature variation in Fig.4 seems to suggest a significant further increase towards lower temperatures. Conductivity data for $T < 0.5t$ is not shown in Figs. 4 and 5 because it shows pronounced discrete level structure, and probably requires larger clusters for a careful study. Nevertheless it appears that $\sigma(\omega)$ develops a pseudogap in the orbital ordered phase. [38] Integrated quantities on the other hand are much less influenced by such effects.

The Role of 3-Site Terms

In the following we wish to shed more light on the role of the 3-site processes in the orbital Hamiltonian and in the current operator. The importance of the 3-site term becomes particularly clear from the J-dependence of the sum rule S (Fig.7) and the charge stiffness D_c (Fig.8) taken at temperature $T/t = 0.3$. These low temperature values of S and D_c are approximatively independent of the exchange parameter J in the absence of 3-site terms. A similar observation was made for the t-J model. [39,18] Moreover one can see that the sum rule S is proportional to the doping concentration x for the 3 cases shown (x=0.125, 0.2, and 0.3). When 3-site terms are taken into account both S and D_c acquire a component which increases linearly with J. These general features are fully consistent with results for the t-J model. [18]

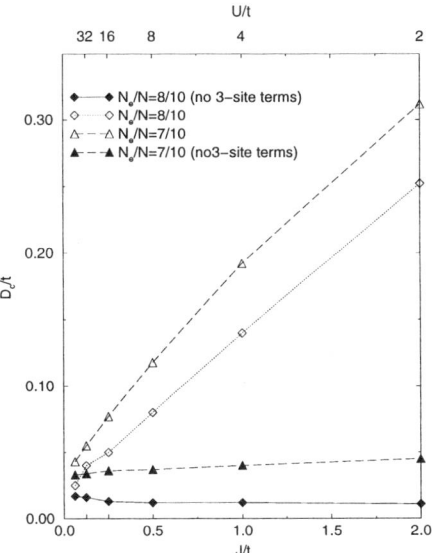

Figure 8: Drude weight D_c versus J calculated via the sum-rule (15) for a N=10 site planar cluster with N_e=7 and 8 e_g electrons, with and without the 3-site terms, respectively.

The change due to the 3-site terms is particularly large for the charge stiffness, which defines the weight of the low frequency Drude peak. The relative weight in the Drude peak D_c/S is shown in Fig. 8 as function of $J = 4t^2/U$. Small changes in J lead to considerable changes in the Drude weight and the coherent motion of carriers. For $J = 0.25$ (0.5) the 3-site terms lead to an increase of D_c by a factor of 3 (5), respectively, for the 8 electron case.

When we take the value $t \sim 0.2$ eV, which we determined from the comparison of $\sigma(\omega)$ with the experimental data, and $U \sim 3$ eV, [22] we obtain $U/t \sim 15$ and $J/t = 4t/U \sim 0.25$ ($J = 0.05$ eV). For such a value for J we expect a relative Drude weight $D_c/S \sim 0.4$, which is larger than the experimental ratio $D_c/N_{eff} \sim 0.2$ found by Okimoto et al. for $x = 0.175$, identifying here the effective number of carriers N_{eff}[7] with S. The small value for D_c/N_{eff} found in the experiments suggests that J is not larger than the value estimated, otherwise we would expect a too large relative Drude weight. A larger value for U would also have the effect to reduce D_c. However, we have to stress here that certainly calculations on larger clusters must be performed, before possible finite size effects in D_c can be quantified.

Comparison with the standard t-J Model

It is interesting to compare these results for the orbital t-J model with those obtained for the usual t-J model describing the carrier motion in the copper-oxygen planes in high-T_c superconductors. We note that the t-J model is a special case of the orbital model with $t_{aa} = t_{bb}$ and $t_{ab} = 0$. Figures 9 and 10 show results for a 4×4 cluster with $N_h = 3$ holes. The results for $\sigma(\omega)$ show for both models a large incoherent absorption, which is bell-shaped in the case of the orbital model, while in the t-J case there is a continuous increase of $\sigma(\omega)$ as ω approaches zero. The temperature dependence of the sum rule, the incoherent and the Drude weight shown in Fig. 10 are quite similar for

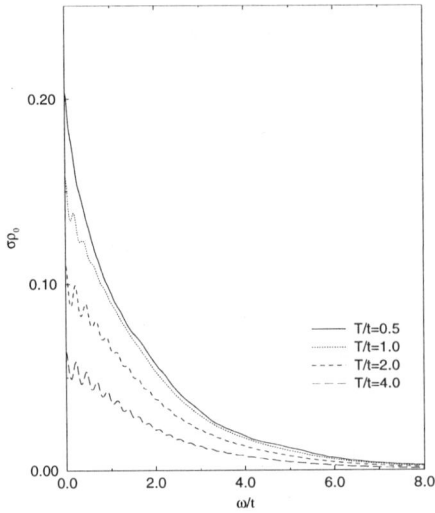

Figure 9: Optical conductivity for the *standard t-J model* ($t_{aa} = t_{bb} = t$ and $t_{ab} = 0$) at temperatures $T = 0.5, 1.0, 2.0$ and 4.0 which should be compared with the corresponding data for the orbital model. Here $N_h/N = 3/16$ and $J/t = 0.4$.

both models, with the important exception that the orbital model goes into the ordered phase at low temperatures ($T \leq 0.3$) which is accompanied by a marked increase of S and D_c (Fig.6). A more detailed study of the T-dependence may be found in the review by Jaklič and Prelovšek.[25]

Comparison with other Work

The optical conductivity for the orbital non-degenerate Kondo lattice model was studied by Furukawa using the dynamical mean-field approximation.[12] An important result of this calculation is, that the weight of intraband excitations within the lower exchange-split band should be proportional to the normalized ferromagnetic magnetization M/M_{sat}. Experimentally, however, $N_{eff}(\sim S)$ is still increasing when temperature is lowered even though M is already saturated. This contradiction to the prediction of the simple double-exchange model implies that some other large-energy-scale scattering mechanism survives at low temperature, where the spins are frozen.[7] The dynamical mean field theory yields one low energy scale, and not two, i.e. there is no Drude peak plus incoherent structure.

Shiba and coworkers *et al.*[14] approached the problem from the band picture. They analysed the noninteracting two-band model (2) (i.e. without constraint) and argued that the interband transitions within the e_g orbitals may explain the anomalous absorption in the ferromagnetic phase of $La_{1-x}Sr_xMnO_3$. While the frequency range of the interband transition is found consistent with the anomalous absorption, the structure of $\sigma(\omega)$ differs. In particular the noninteracting model leads to $\sigma(\omega) \sim \omega$ for small ω. Moreover they calculated the ratio $S_\omega/D_c \sim 0.85$, i.e. there is more weight in the Drude peak than in the regular part of the conductivity $\sigma(\omega)$. This ratio is found to be rather insensitive to doping in the range $0.175 < x < 0.3$. This result differs considerably from the experimental data of Okimoto *et al.* where this ratio is about 4 for $x = 0.175$.

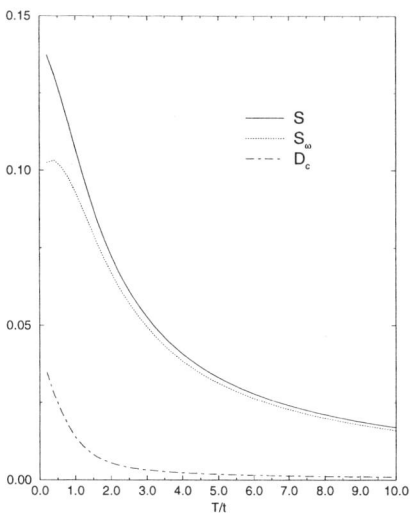

Figure 10: Temperature dependence of the kinetic energy S (solid line), the incoherent spectral weight S_ω (dotted) and the Drude weight D_c (dash-dotted line) for the t-J model. The data was calculated for a 2D system with 16 sites and 3 holes for J/t=0.4.

Orbital fluctuations in the ferromagnetic state and their effect on the optical conductivity were studied by Ishihara et al.[13] using slave fermions coupled to bosonic orbital excitations. Ishihara et al. deal with the concentration regime $xt \gg J$, where the kinetic energy is expected to dominate the orbital exchange energy, and arrive at the conclusion that the quasi two-dimensional nature of the orbital fluctuations leads to an *orbital liquid* in (3D) cubic systems. The orbital disorder is treated in a static approximation in their work. The $\sigma(\omega)$ spectrum obtained by this approach for $T = 0.1t$ has a quite similar shape as the experimental spectrum at low temperature (with $\omega_{1/2} \sim 3t$), however, because of the assumed static disorder, there is no Drude component. A further problem with the orbital liquid, as noted by the authors, is the large entropy expected for the orbital disordered state which seems to be in conflict with specific heat measurements. [40]

More recently a microscopic theory of the optical conductivity based on a slave boson parametrization which combines strong correlations and orbital degeneracy was developed by Kilian and Khaliullin[41] for the ferromagnetic state at zero temperature. Their approach yields a highly incoherent spectrum due to the scattering of charge carriers from dynamical orbital fluctuations, and a Drude peak with strongly reduced weight. However $\sigma(\omega)$ is depressed at low frequency in this calculation for the orbital model and an additional electron-phonon mechanism is invoked to obtain results similar to the experimental spectral distribution. This theory further accounts for the small values of the specific heat and therefore supports the orbital liquid scenario for the ferromagnetic cubic systems. [13]

We would like to stress that the proposed orbital liquid state[13] is a property of cubic systems, and not a property of the 2D or quasi-2D versions of the orbital model. In the planar model the cubic symmetry is broken from the outset and holes presumably cannot restore it. This is different from the physics of the t-J model, where holes restore the spin rotational symmetry (which is spontaneously broken in the antiferromagnetic ordered phase), because this symmetry is respected by the t-J Hamiltonian in any dimension.

For a deeper understanding of the results for the optical conductivity, we shall analyse below the structure of the orbital correlations in the doped and undoped two-dimensional orbital model in detail.

DOPING DEPENDENCE OF ORBITAL CORRELATIONS

In the absence of holes the interaction (9) between orbital pseudo-spins will lead to an orbital order below a certain temperature $\sim J$ due to the anisotropy of the Hamiltonian. Doping will destroy this order and lead either to a disordered orbital liquid, or may generate a new kind of ordered state which optimizes the kinetic energy of the holes. To investigate this question for an anisotropic model it is in general not sufficient to calculate simply e.g. the correlation function $<T_i^z T_j^z>$ defined with respect to the original orbital basis, because the relevant occupied and unoccupied orbitals may be different from the original orbital basis chosen. In the following we therefore introduce a local orthogonal transformation of the orbitals on different sublattices by angles ϕ and ψ, respectively. On the A-sublattice:

$$\begin{aligned}|\tilde{z}\rangle &= cos(\phi)|z\rangle + sin(\phi)|x\rangle \\ |\tilde{x}\rangle &= -sin(\phi)|z\rangle + cos(\phi)|x\rangle,\end{aligned} \quad (21)$$

and a similar rotation with angle ψ on the B-sublattice. This amounts to new operators, e.g.

$$\tilde{T}_i^z = cos(2\phi)T_i^z + sin(2\phi)T_i^x \quad (22)$$
$$\tilde{T}_{i+R}^z = cos(2\psi)T_{i+R}^z + sin(2\psi)T_{i+R}^x \quad (23)$$

and new correlation functions, e.g. $\langle \tilde{T}_i^z \tilde{T}_j^z \rangle$ is defined as

$$\begin{aligned}\langle \tilde{T}_i^z \tilde{T}_{i+R}^z \rangle = & cos(2\phi)cos(2\psi)\langle T_i^z T_{i+R}^z \rangle + sin(2\phi)sin(2\psi)\langle T_i^x T_{i+R}^x \rangle \\ & + cos(2\phi)sin(2\psi)\langle T_i^z T_{i+R}^x \rangle + sin(2\phi)cos(2\psi)\langle T_i^x T_{i+R}^z \rangle\end{aligned} \quad (24)$$

The angles ϕ and ψ are now chosen such that the nearest neighbor correlation function $\langle \tilde{T}_i^z \tilde{T}_j^z \rangle$ takes a maximal (ferromagnetic) value. With this convention the orbital order is specified by the corresponding local quantization axis, that is by the values of the rotation angles ϕ and ψ. We call this the *optimized orbital basis* (OOB). [21]

In the undoped 2D case (x-y plane) these angles are $\phi = 45°$ and $\psi = 135°$ corresponding to an $\frac{1}{\sqrt{2}}(|x\rangle + |z\rangle)$ and $\frac{1}{\sqrt{2}}(|x\rangle - |z\rangle)$ order at low temperatures (see Fig. 11 and 12). $\langle \tilde{T}_i^z \tilde{T}_{i+R}^z \rangle$ takes a value only slightly smaller than 0.25 at low temperature (Fig. 13), which implies that quantum fluctuations are small in this state. Hence the orbital correlation function $\langle \tilde{T}_i^z \tilde{T}_{i+R}^z \rangle$ of the undoped orbital model is *alternating* (i.e. antiferromagnetic in the pseudo-spin language for the orbital degrees of freedom). That the AF-order seems to be established at a finite temperature for the small clusters means that the correlation length ξ gets larger than the system size L. At small temperatures the correlation functions have about the same value independent of R, i.e. the data shows no spatial decay. Above $T/t = 0.1$ the orbital correlations are small and show a pronounced spatial decay, which can be considered as a signature of a thermally disordered orbital liquid state.

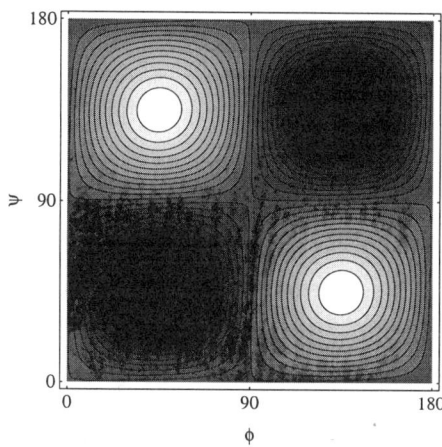

Figure 11: Contourplot of the *rotated* nearest neighbor orbital correlation function $\langle \tilde{T}_i^z \tilde{T}_{i+R}^z \rangle$, $R=(1,0)$, as function of the angles ϕ and ψ for a N=10 site planar cluster with N_e=10 electrons, J=0.25t (U=16t) and T=0.01t. White regions correspond to positive (ferromagnetic) and black areas to negative (antiferromagnetic) correlation functions, i.e. $\langle \tilde{T}_i^z \tilde{T}_{i+R}^z \rangle$>0.23 (<-0.23), respectively. The 25 contour lines are chosen equidistant in the interval [-1/4,+1/4]. (From Horsch, Jaklič and Mack[21])

Experimentally the coexistence of antiferromagnetic (or staggered) orbital order and ferromagnetic spin order has been established in the low-doping regime of LaMnO$_3$. [42,43] It should be recalled that LaMnO$_3$ has an A-type antiferromagnetic spin structure, [44] where ferromagnetic layers are coupled antiferromagnetically along the c-axis. In the orbital ordered case the MnO$_6$ octahedra are deformed, and Mn $3d_{3x^2-r^2}$ and $3d_{3y^2-r^2}$ orbitals with orientation along the x- and y-axis, respectively, have been considered as relevant occupied orbitals. [45,46] For the planar model we find here the alternate occupation of $\frac{1}{\sqrt{2}}(|x\rangle - |z\rangle)$ and $\frac{1}{\sqrt{2}}(|x\rangle + |z\rangle)$ orbitals (Fig. 12) which are also oriented along the x- and y-axis, however they differ with respect to their shape (and have pronounced lobes along z).

Doping of two holes in a ten site cluster is sufficient to remove the alternating orbital order and establishes a ferromagnetic orbital order ($\phi = \psi = 0$) as can be seen from Figs. 14 and 16. This corresponds to preferential occupation of orbitals with symmetry $x^2 - y^2$ as shown in Fig.15. The pseudospin alignment can be considered as a kind of *double exchange mechanism in the orbital sector*. An interesting feature in Fig.16 for the doped case is the fact that the correlations do not show significant spatial decay even at higher temperatures where the correlations are small. Moreover the correlations in the doped case appear to be more robust against thermal fluctuations than in the undoped case. The characteristic temperature (determined from the half-width in Figs.13 and 16) is $T^* \sim 0.2t$ for the 2-hole case while for the undoped system $T^* \sim 0.05t$. This trend is consistent with the fact that in the doped case order is induced via the kinetic energy and the corresponding scale t is larger than $J = 0.25t$. Although a more detailed analysis would be necessary to account properly for the doping dependence.

It has been shown recently by Ishihara *et al.*[29] that in the layered manganite compounds hydrostatic pressure leads to a stabilization of the x^2-y^2 orbitals as well.

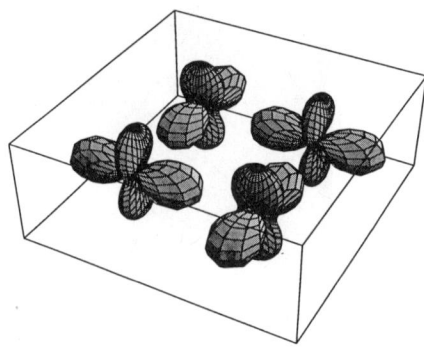

Figure 12: Alternating $\frac{1}{\sqrt{2}}(|x\rangle + |z\rangle)$ and $\frac{1}{\sqrt{2}}(|x\rangle - |z\rangle)$ orbital order in the undoped planar system.

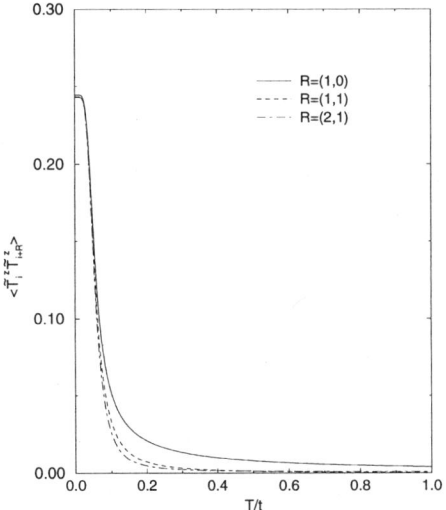

Figure 13: Temperature dependence of $\langle \tilde{T}_i^z \tilde{T}_{i+R}^z \rangle$ with $\phi = 45^0$ and $\psi = 135^0$ for nearest-neighbors and $\psi = 45^0$ for next-nearest neighbors, respectively. Results are shown for different distances R for an undoped $N = 10$ site planar cluster. Parameters as in Fig.11. The strong increase for $T/t \leq 0.1$ is due to the onset of alternating orbital order. (From Horsch, Jaklic and Mack[21])

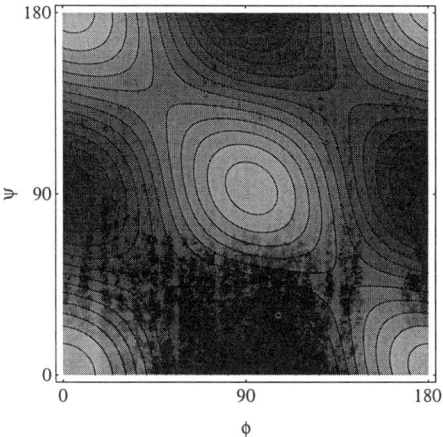

Figure 14: Contourplot of $\langle \tilde{T}_i^z \tilde{T}_{i+R}^z \rangle$ for $R = (1,0)$ as function of angles ϕ and ψ for a N=10 site cluster with two holes. Otherwise same parameters as for Fig.11.

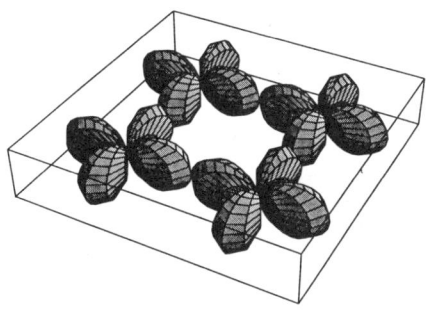

Figure 15: x^2-y^2 orbital order in the doped phase.

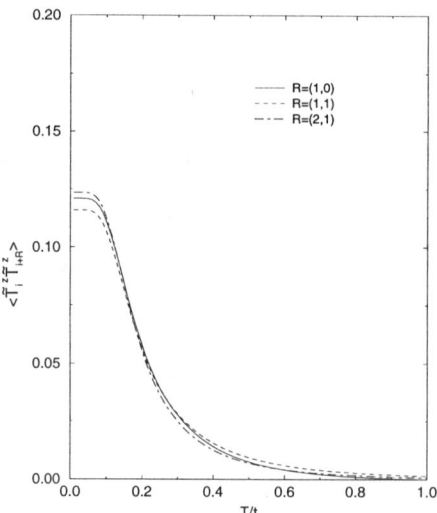

Figure 16: Temperature dependence of $\langle \tilde{T}_i^z \tilde{T}_{i+R}^z \rangle$ with $\phi = 0°$ and $\psi = 0°$ for a 10-site cluster with two e_g holes. Parameters as in Fig.11. The strong increase below $T/t = 0.3$ signals the onset of x^2-y^2 orbital order in the planar model.

A comparison of the doping dependence of spatial correlations with the t-J model is given in Fig.17. In the undoped case the orbital model shows almost classical (orbital) Néel order, whereas in the t-J model correlations are strongly reduced by quantum fluctuations. The $T = 0$ correlation function of the t-J model shows long-range order (LRO) consistent with a strongly reduced sublattice magnetization $m_z \sim 0.3$ (in the thermodynamic limit).

In the orbital model the effect of the two holes ($x = 0.125$ and 0.2) is strong enough to induce ferromagnetic (orbital) correlations, i.e. with a prefered occupation of x^2-y^2 orbitals. In the t-J model instead spin-correlations decay rapidly for two holes on a 20-site cluster ($x = 0.1$), which is consistent with the notion of an AF spin-liquid state (see inset Fig.17).

The preference of ferromagnetic orbital order in the doped case is due to the fact that $t^{\alpha\beta}$ depends on the orbital orientation, and $t^{aa} \gg t^{bb}$. One would expect that for sufficiently strong orbital exchange interaction $J = 4t^2/U$ one reaches a point where antiferromagnetic orbital interactions and ferromagnetic correlations due to the kinetic energy are in balance. This quantum critical point between FM and AF orbital order turns out to be at quite large J values. For the two-hole case ($x = 0.2$) we have found that this crossover happens for quite large orbital interaction $J \sim 2$, i.e. at a value about an order of magnitude larger than typical J values in manganite systems.

Before closing this section, we shall explain the different orbital occupancy in the doped and undoped case in more physical terms. The overlap of the two sets of orthogonal e_g orbitals on neighboring sites changes with the angles ϕ and ψ. (1) In the undoped state the direct hopping between the (predominantely) occupied orbitals $\frac{1}{\sqrt{2}}(|x\rangle + |z\rangle)$ and $\frac{1}{\sqrt{2}}(|x\rangle - |z\rangle)$ on neighboring sites is small (Fig.12). The hopping between these occupied orbitals is blocked anyhow because of Pauli's principle. However t^{ab} between an occupied orbital on one site and an unoccupied orbital on a neighbor site is extremal.

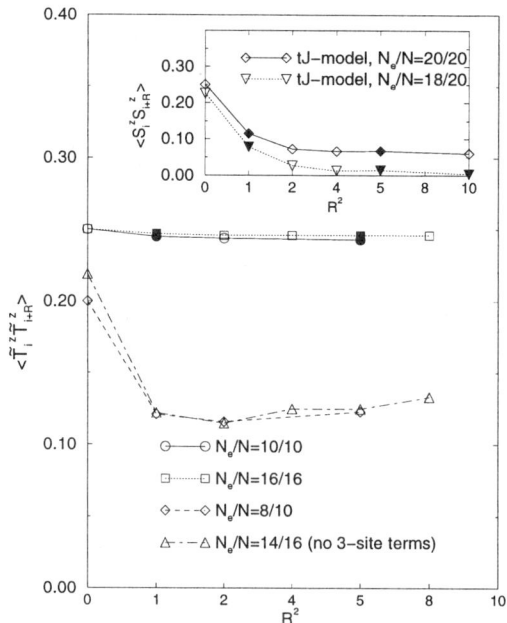

Figure 17: Orbital correlation functions for $T=0$ as function of distance (squared) for a 10-site and a 4×4 cluster with zero and two holes ($J/t=0.25$). In the undoped case the correlation function is antiferromagnetic, whereas in the doped case orbital correlations are ferromagnetic. Antiferromagnetic orbital order is indicated by full (negative) and open (positive) symbols, respectively. Inset: The spin correlations $\langle S_i^z S_{i+R}^z \rangle$ for a 20-site 2D Heisenberg model and the t-J model with two holes are shown for comparison (t-J data for $J=0.4$ taken from Horsch and Stephan[47]). The two-hole case shows a rapid decay of correlations characteristic for an antiferromagnetic spin liquid.

This leads to a maximal antiferromagnetic orbital exchange interaction, and thereby to a lowering of the energy. (2) In the doped case instead, the x^2-y^2 orbital occupancy is prefered on all sites (Fig.15), which implies a large hopping amplitude for holes in the partially occupied x^2-y^2 band, whereas the off-diagonal hopping matrix element t^{ab} and the orbital interactions are reduced for this choice of occupied orbitals. Hence the main energy gain in this case is due to the correlated motion of holes in the x^2-y^2-band. Nevertheless the off-diagonal hopping plays an important role in the x^2-y^2 ordered state, and leads to the strong incoherent features characteristic for the conductivity $\sigma(\omega)$.

Finally we note that recent experiments by Akimoto et al.[48] proved the existence of an A-type antiferromagnetic metallic ground state in a wide doping concentration regime of the 3D system $(La_{1-z}Nd_z)_{1-x}Sr_xMnO_3$ and in particular for $La_{1-x}Sr_xMn\,O_3$ at high doping $x = 0.52 - 0.58$. The latter compound was investigated by Okimoto et al. for smaller doping x where the ground state is the uniform ferromagnetic state. The A-type antiferromagnet consists of ferromagnetic layers which are coupled antiferromagnetically. Akimoto et al. propose that $x^2 - y^2$ orbitals should be occupied in this state and form a highly anisotropic 2D band. This seems consistent with our finding of $x^2 - y^2$ orbital order in the ferromagnetic planar system where the cubic symmetry is explicitely broken. An experimental study of the optical properties in this concentration regime at low temperature would be particularly interesting.

SUMMARY AND CONCLUSIONS

Our work was motivated by the experimental study of the frequency dependent conductivity of $La_{1-x}Sr_xMnO_3$ by Tokura's group. [6] In particular these experiments show in the saturated low-T ferromagnetic phase a broad incoherent spectrum and at small frequency in addition a narrow Drude peak, which contains only about 20% of the total spectral weight. This experiment by itself shows that *the ferromagnetic metallic state (although fully spin polarized) is highly anomalous* . These features cannot be explained by the standard double exchange model, but require an additional mechanism. The most natural mechanism, one can think of, is the twofold degeneracy of the e_g orbitals.

We have studied the optical conductivity, the charge stiffness and the optical sum rule for an effective Hamiltonian that contains only the e_g-orbital degrees of freedom. This model follows from the more general Kondo lattice model, when the spins are ferromagnetically aligned. The 'orbital t-J model' is expected to describe the low-temperature physics of manganese oxides in the ferromagnetic saturated phase.

Our results for $\sigma(\omega)$ in the 2D orbital disordered regime show that these characteristic features follow from the orbital model. We believe that the optical conductivity in the 2D orbital liquid regime ($T > T^*$) can be compared with the conductivity in the 3D-orbital liquid state. In particular there is (1) a broad continuum which decreases towards high frequency with a half-width $\omega_{1/2} \sim 2.5$ and $3.5t$ in two and three dimensions, respectively. (2) The absorption increases with decreasing temperature and (3) the absolute value of the incoherent part of $\sigma(\omega)$ at low frequency and temperature $\sigma_0^{inc} \sim 0.4 \cdot 10^3 (\Omega cm)^{-1}$ has the correct order of magnitude as in Tokura's experiments. (4) Despite this strong incoherence there is a finite Drude peak of about 20% of the total sum rule for small J. (5) The coherent motion and the Drude peak become more pronounced for larger values of J due to the 3-site hopping processes in the model. For the value $J/t = 0.25$ estimated here the relative Drude weight is 40 (20) % in the model with (without) 3-site hopping processes, respectively. (6) The low temperature values for the DC-conductivity and the resistivity can be estimated by assuming a width Γ for the Drude peak (taken from experiment). The additional scattering processes contributing to Γ are due to impurity or grain boundary scattering, i.e. these are extrinsic scattering processes which are not contained in the orbital model. The important information which follows from the model is the weight of the Drude peak, which together with Γ determines the DC-conductivity.

The relatively small Drude peak in comparison with the incoherent part of $\sigma(\omega)$ clearly indicates that the carrier motion is essentially incoherent. Nevertheless we expect a quasiparticle peak in the single particle Green's function, yet with small spectral weight. Our study of the frequency dependent conductivity shows that the (saturated) ferromagnetic state in the manganites has unconventional transport properties due to the scattering from orbital excitations in combination with the exclusion of double occupancy. A detailed study of the orbital dynamics in the doped system is necessary for a deeper understanding of these issues.

The orbital model resembles the t-J model, which describes the correlated motion of charge carriers in the cuprate superconductors. Yet an important difference is the cubic symmetry of the pseudospin representation of the orbital degrees of freedom. The hopping matrix elements depend on the orbital orientation, and are in general different in the two diagonal hopping channels. Moreover there is also an off-diagonal hopping matrix element, which does not exist at all in the t-J model. At low temperatures ($T < T^* \sim 0.2t$) doping induces (ferromagnetic) x^2-y^2 *orbital order* for realistic values for the exchange interaction J in the 2D orbital model. This is in striking contrast to t-J physics in cuprates, where the system changes upon doping from antiferromagnetic long-range order to an antiferromagnetic spin liquid. In both models coherent motion coexists with strong incoherent features.

Our calculations have shown that the orbital mechanism can explain the order of magnitude of the conductivity at low temperature. The orbital degrees of freedom are certainly also important for a quantitative calculation of the colossal magnetoresistance. This is obvious because in the full Kondo lattice model there is a close interplay between orbital and spin degrees of freedom. [22,23]

We acknowledge helpful discussions with F. Assaad, J. van den Brink, L. Hedin, G. Khaliullin, A. Muramatsu, A. M. Oleś and R. Zeyher.

REFERENCES

1. R. von Helmolt et al., *Phys. Rev. Lett.* **71**, 2331 (1993); K. Chahara et al., *Appl. Phys. Lett.* **63**, 1990 (1993); S. Jin et al., *Science* **264**, 413 (1994); Y. Tokura et al., *J. Phys. Soc. Jpn.* **63**, 3931 (1994).
2. C. Zener, *Phys. Rev.* **82**, 403 (1951); P. W. Anderson and H. Hasegawa, *Phys. Rev.* **100**, 675 (1955); K. Kubo and A. Ohata, *J. Phys. Soc Jpn* **33**, 21 (1972).
3. A. J. Millis, P. B. Littlewood and B. I. Shraiman, *Phys. Rev. Lett.* **74**, 5144 (1995); A. J. Millis, R. Mueller and B. I. Shraiman, *Phys. Rev. B* **54**, 5405 (1996).
4. A. J. Millis, *Nature* **392**, 147 (1998).
5. For a recent experimental review see: A. P. Ramirez, *J. Phys.: Condens. Matter* **9**, 8171 (1997).
6. Y. Okimoto, T. Katsufuji, T. Ishikawa, A. Urushibara, T. Arima and Y. Tokura, *Phys. Rev. Lett.* **75** 109 (1995).
7. Y. Okimoto, T. Katsufuji, T. Ishikawa, T. Arima, and Y. Tokura, *Phys. Rev. B* **55**, 4206 (1997).
8. S. G. Kaplan, U. M. Quijade, H. D. Drew, G. C. Xiong, R. Ramesh, C. Meuon, T. Venkatesan, and D. B. Tanner, *Phys. Rev. Lett.* **77**, 2081 (1996).

9. M. Quijada, J. Cerne, J. R. Simpson, H. D. Drew, K. H. Ahn, A. J. Millis, R. Shreekala, R. Ramesh, M. Rajeswari, and T. Venkatesan, cond-mat/9803201.
10. K. H. Kim, J. H. Jung, and T. W. Noh, cond-mat/9804167; K. H. Kim et al., cond-mat/9804284.
11. T. Ishikawa, T. Kimura, T. Katsufuji, and Y. Tokura, Phys. Rev. B **57**, R8079 (1998) and references therein.
12. N. Furukawa, J. Phys. Soc. Jpn. **64**, 2734 (1995), ibid. **64**, 3164 (1995).
13. S. Ishihara, M. Yamanaka, and N. Nagaosa, Phys. Rev. B **56**, 686 (1997).
14. H. Shiba, R. Shiina, and A. Takahashi, J. Phys. Soc. Jpn. **66**, 941 (1997).
15. J. Jaklič, P. Horsch, and F. Mack (to be published).
16. H. Röder, J. Zang, and A. R. Bishop, Phys. Rev. Lett. **76**, 1356 (1996).
17. K. J. von Szczepanski, P. Horsch, W. Stephan, and M. Ziegler, Phys. Rev. B **41**, 2017 (1990).
18. W. Stephan and P. Horsch, Int. J. Mod. Phys. B **6**, 589 (1992); P. Horsch and W. Stephan, in *The Hubbard Model*, ed. by Dionys Baeriswyl et al. (Plenum, New York, 1995), p. 193.
19. P. Horsch and W. Stephan, Phys. Rev. B **48**, R 10595 (1993).
20. H. Eskes, A. M. Oleś, M. Meinders, and W. Stephan, Phys. Rev. B **50**, 17980 (1994).
21. P. Horsch, J. Jaklič, and F. Mack, cond-mat/9807255 .
22. S. Ishihara, J. Inoue, and S. Maekawa, Phys. Rev. B **55**, 8280 (1997).
23. L. F. Feiner and A. M. Oleś, (preprint).
24. J. Jaklič and P. Prelovšek, Phys. Rev. B **49**, 5065 (1994); ibid **50**, 7129 (1994); ibid **52**, 6903 (1995).
25. For a review see J. Jaklič and P. Prelovšek, cond-mat/9803331.
26. J. C. Slater and G. F. Koster, Phys. Rev. **94**, 1498 (1954).
27. K. I. Kugel and D. I. Khomskii, Sov. Phys. JETP **37**, 725 (1973).
28. W. A. Harrison, *Electronic Structure and Properties of Solids*, (Freeman, San Francisco, 1980).
29. S. Ishihara, S. Okamoto, and S. Maekawa, cond-mat/9712162.
30. J. Zaanen and A. M. Oleś, Phys. Rev. B **48**, 7197 (1993).
31. C. M. Varma, Phys. Rev. B **54**, 7328 (1996).
32. A. H. MacDonald, S. M. Girvin, and D. Yoshioka, Phys. Rev. B **37**, 9753 (1988) and references therein.
33. B. S. Shastry and B. Sutherland, Phys. Rev. Lett. **65**, 243 (1990).
34. A. Urushibara et al., Phys. Rev. B **51**, 14103 (1995).
35. W. Kohn, Phys. Rev. **133**, A171 (1964).
36. D. Baeriswyl, J. Carmelo, and A. Luther, Phys. Rev. B **16**, 7247 (1986).
37. The same twisted boundary conditions were used for the finite temperature studies in this work. These boundary conditions guarantee a nondegenerate closed-shell ground state.
38. F. Mack and P. Horsch, unpublished.
39. D. Poilblanc and E. Dagotto, Phys. Rev. B **44**, 466 (1991).
40. B. F. Woodfield, M. L. Wilson, and J. M. Byers, Phys. Rev. Lett. **78**, 3201 (1997).
41. R. Kilian and G. Khaliullin, preprint.
42. J. B. Goodenough, Phys. Rev. **100**, 564 (1955).
43. J. B. Goodenough, in *Progress in Solid State Chemistry*, Vol.5, ed. by H. Reiss (Pergamon, London, 1971).
44. E. D. Wollan and W. C. Koehler, Phys. Rev. **100**, 545 (1955).
45. G. Matsumoto, J. Phys. Soc. Jpn. **29**, 606 (1970).
46. J. B. A. A. Elemans, B. van Laar, K. R. van der Veen, and B. O. Loopstra, J. Solid State Chem. **3**, 238 (1971).
47. P. Horsch and W. Stephan, Physica C **185-189**, 1585 (1991).
48. T. Akimoto, Y. Maruyama, Y. Moritomo, A. Nakamuram K. Hirota, K. Ohoyama, and M. Ohashi, Phys. Rev. B **57**, R5594 (1998).

ELECTRON-LATTICE INTERACTIONS IN MANGANESE-OXIDE PEROVSKITES

J. B. Goodenough
Texas Materials Institute
University of Texas at Austin,
Austin, TX 78712-1063

The virial theorem is invoked to argue for a first-order transition from localized to itinerant electronic behavior. A competition between localized and itinerant electronic behavior leads to phase segregation via cooperative oxygen displacements that may be manifest either as a stationary charge-density wave or as a condensation of two-manganese polarons into ferromagnetic, conductive clusters in a paramagnetic, polaronic matrix. Growth in a magnetic field of the ferromagnetic clusters to beyond their percolation threshold gives the intrinsic colossal magnetoresistance (CMR) found above T_c and the transition to a "bad metal" below T_c. A rapid rise in T_c with increasing $(180° - \phi)$ Mn - O - Mn bond angle at cross-over where the CMR is found is attributed to an angle-dependent oxygen-vibration frequency $\omega_o(\phi)$, a prediction that can be checked experimentally.

INTRODUCTION

Two features of the colossal-magnetoresistance (CMR) data of Hwang et al [1] on the system $La_{0.7-x}Pr_xCa_{0.3}MnO_3$ with fixed ratio Mn(IV)/Mn = 0.3 are noted: (1) a remarkable change of the ferromagnetic Curie temperature T_c from 250 K to 80 K in the range $0 \leq x \leq 0.6$ and (2) an increase in log $[(\rho_{0T} - \rho_{5T})/\rho_{5T}]$ from less than 0.5 to over 5 with decreasing T_c, where ρ_H is the resistivity in an applied field H. To explain this phenomenon, a review is first given of relevant structural properties of perovskites, of the ligand-field description of localized-electron configurations, of magnetic-exchange interactions, and of the character of the transition from localized to itinerant electronic behavior.

STRUCTURAL CONSIDERATIONS

The deviation from unity of the geometric tolerance factor of an AMO_3 perovskite

$$t \equiv <A\text{-}O> / \sqrt{2} <M\text{-}O>. \qquad (1)$$

is a measure of the mismatch between the mean equilibrium bond lengths <A - O> and <M - O>. In the $AMnO_3$ perovskites, a t < 1 places the <Mn-O> bond under compression and the <A - O> bond under tension. These internal stresses are relieved by a cooperative rotation of the corner-shared $MnO_{6/2}$ octahedra. In the manganese oxides, rotations about

[110] axes reduce the symmetry from cubic to orthorhombic (Pbnm or Pnma) symmetry, rotations about [111] axes give rhombohedral ($R\bar{3}c$) symmetry. These rotations bend the Mn - O - Mn bond angles from $180°$ to $(180° - \phi)$, and ϕ is greater in the orthorhombic than in the rhombohedral structure [2]. In the $La_{0.7-x}Pr_xCa_{0.3}MnO_3$ system, the oxidation state of the MnO_3 array is fixed at a ratio Mn(IV)/Mn = 0.3, and the relevant variable is the tolerance factor t; the $4f^2$ configuration on the Pr^{3+} ion plays no significant role.

Normally, a larger thermal expansion and compressibility of the <A - O> bond makes [3,4]

$$dt/dT > 0 \text{ and } dt/dP < 0 \qquad (2)$$

but a $dt/dP > 0$ may occur where the <M - O> bond has an anomalously large compressibility indicative of a double-well potential [5].

Cooperative oxygen displacements away from one M atom of an M - O - M bond toward the other may be superimposed on the cooperative $MO_{6/2}$ rotations. These displacements can be detected by a diffraction experiment where they are stationary below a transition temperature T_t; but where the displacements are dynamic, some other experimental probe is required. Static displacements are, for example, responsible for the ferroelectric transitions of $BaTiO_3$ that create shorter and longer Ti - O bonds in the $TiO_{6/2}$ octahedra [6]. Contraction and dilation of alternate $FeO_{6/2}$ octahedra in $CaFeO_3$ create a "negative - U" charge-density wave (CDW) in which $Fe(v)O_6$ complexes alternate with high-spin Fe^{3+} ions [7,8]. In the system $La_{1-x}Ca_xMnO_3$, orbital ordering at high-spin Mn^{3+} ions creates local Jahn-Teller deformations that become long-range ordered below an ordering temperature T_t in a manner that minimizes the elastic energy. In $LaMnO_3$, for example, orbital ordering creates alternating long and short O - Mn - O bonds in the (001) planes of the orthorhombic phase, which changes the axial ratio from $c/a > \sqrt{2}$ to $c/a < \sqrt{2}$. We distinguish the O' - orthorhombic ($c/a < \sqrt{2}$) from the O-orthorhombic ($c/a > \sqrt{2}$) phase so as to indicate where static orbital ordering occurs [9,10]. Static cooperative distortions also create a series of CDW phases at $1/2 \leq x = n/(n + 2) \leq 7/8$, where n is an integer in the range $2 \leq n \leq 14$, in which slabs containing orbitally ordered localized configurations alternate with empty delocalized σ - bonding orbitals in all-Mn(IV) slabs [11]. On the other hand, cooperative oxygen displacements that are dynamic give rise to a phase segregation characterized by the condensation of mobile polarons into a Mn(IV) - rich ferromagnetic phase that is conductive within a Mn(IV) - poor paramagnetic polaronic matrix. The Mn(IV) - rich clusters grow by trapping mobile polarons from the paramagnetic matrix, and the trapping energy is increased in the presence of a magnetic field to give the CMR phenomenon; the clusters grow in a magnetic field to beyond their percolation threshold. Above T_c, the mobile polarons that are trapped are two-manganese polarons created by cooperative, dynamic oxygen displacements; below T_c, the Weiss molecular field grows the clusters to well beyond percolation, which allows study of the conductive, ferromagnetic phase.

The lattice instabilities responsible for phase segregation occur at a transition from localized to itinerant electronic behavior. At this transition, a double-well potential gives two equilibrium <M - O> bond lengths; and the signature for the double well potential is a $dt/dP > 0$. The double-well potential at cross-over is a result of the virial theorem of mechanics [5], which states that for central-force fields

$$2<T> + <V> = 0 \qquad (3)$$

where the mean electronic kinetic energy <T> decreases discontinuously if the volume the electrons occupy increases discontinuously at the transition from localized to itinerant (or molecular-orbital) behavior. The magnitude |<V>| of the mean negative potential energy |<V>| = - <V> must also decrease discontinuously, and for antibonding electrons a decrease in |<V>| is accomplished by shortening the <M - O> bond, which makes the transition first-order. Moreover, a discontinuous shortening of the <M - O> bonds means there is a double-well potential and a lattice instability that leads to phase segregation at cross-over. The phase segregation may be long-range ordered in a static CDW or short-range dynamic.

ELECTRONIC CONSIDERATIONS

The cubic crystalline field at an octahedral-site transition-metal ion splits the fivefold-degenerate d orbitals into three degenerate t orbitals (xy, yz ± izx) that π-bond to the oxygen $2p_\pi$ orbitals and two degenerate e orbitals ($x^2 - y^2$, $3z^2 - r^2$) that σ-bond to the nearest-neighbor oxygen 2s and $2p_\sigma$ orbitals. Back transfer of electrons from the oxygen to the empty d orbitals is treated in second-order perturbation theory to give wave functions of the form [12]

$$\psi_t = N_\pi(f_t - \lambda_\pi \phi_\pi) \tag{4}$$

$$\psi_e = N_\sigma(f_e - \lambda_\sigma \phi_\sigma - \lambda_s \phi_s) \tag{5}$$

where the ϕ are properly symmetrized oxygen wave functions, f_t and f_e are atomic t and e orbitals, and the covalent-mixing parameters

$$\lambda_i \equiv (f_i, H'\phi_i)/\Delta E_i \tag{6}$$

are proportional to the (f_i, ϕ_i) overlap integrals; ΔE_i is the energy required to transfer an electron from the oxygen to an empty orbital on the M cation in an ionic model. These parameters stand in the relation

$$\lambda_s \ll \lambda_\pi < \lambda_\sigma \tag{7}$$

to give a cubic-field splitting

$$\Delta_e = \Delta_M + (\lambda_\sigma^2 - \lambda_\pi^2) \Delta E_p + \lambda_s^2 \Delta E_s \tag{8}$$

in which the electrostatic term Δ_M is relatively small. In the manganese-oxide perovskites, an intraatomic Hund direct exchange lifts the spin degeneracy by an energy $\Delta_{ex} > \Delta_c$, so the localized-electron configuration at a Mn^{3+} ion is high-spin $t_\alpha^3 e_\alpha^1$; we drop the spin subscript α hereinafter.

The twofold e-orbital degeneracy may be removed by a deformation of the octahedral site to tetragonal or orthorhombic symmetry. The elastic energy associated with such a Jahn-Teller deformation is minimized if the local deformations are cooperative, and long-range orbital ordering normally gives a static global deformation of the structure as in the case of the transition from O to O' orthorhombic symmetry on cooling $LaMnO_3$ from above 750 K.

The spin-spin Mn - O - Mn interactions involve an electron transfer from one manganese atom to the other. Where the electron transfer requires an energy, *i.e.* is virtual, the interaction is called superexchange; where the electron transfer requires no energy, *i.e.* is real, the interaction is called double exchange. In either case, the spin angular momentum is conserved in an electron transfer. At both Mn^{3+} and Mn^{4+} ions, the t^3 configurations are localized with a spin S = 3/2 and the electron transfer of a t^3 - O - t^3 interaction not only requires an energy $U = U_\pi + \Delta_{ex}$; it is also constrained by the Pauli exclusion principle to have an antiparallel spin component, which makes the transfer integral $t_{ij}^{\uparrow\downarrow} = b_\pi \sin(\theta_{ij}/2)$ dependent on the angle θ_{ij} between spins on neighboring atoms. The superexchange interaction is therefore antiferromagnetic and described by second-order perturbation theory to give a stabilization energy of the Heisenberg form [13]:

$$\Delta \varepsilon_{ex}^s \approx -J_{ij} \mathbf{S}_i \cdot \mathbf{S}_j \quad \text{with} \quad J_{ij} \sim -2b_\pi^2/(4S^2 U) \tag{9}$$

where b_π is the nearest-neighbor spin-independent transfer integral. Orthogonality in a 180° Mn - O - Mn bond eliminates a t^3 - O - e^0 electron transfer. Accordingly, $CaMnO_3$ is a Type G antiferromagnet in which every Mn - O - Mn interaction is antiferromagnetic.

The interactions associated with e-electron transfer are made complex not only by the orbital degeneracy of a $Mn^{3+}:t^3e^1$ configuration, but also by the competition between a virtual and a real charge transfer in a $Mn^{3+}:t^3e^1$ - O - $Mn^{4+}:t^3e^0$ interaction. In $LaMnO_3$, where the $Mn^{3+}:t^3e^1$ configurations are localized, the e-orbital ordering introduces anisotropic spin-spin interactions. In the (001) planes, alternating long and short O - Mn - O bonds create e^1 - O - e^0 interactions, and electron transfer is not constrained by the Pauli exclusion principle. However, a strong Hund exchange energy Δ_{ex} favors transfer of an electron with spin parallel to the t^3 spin on the receptor ion, so the superexchange interaction is ferromagnetic and described by third-order perturbation theory:

$$\Delta\varepsilon^s_{ex} \approx -J_{ij}\mathbf{S}_i \bullet \mathbf{S}_j \text{ with } J_{ij} \sim -2b_\pi^2 \Delta_{ex}/(4S^2U^2) \qquad (10)$$

These σ-bond interactions are stronger than the π-bond interactions, so the (001) planes order ferromagnetically. However, the c-axis oxygen is not displaced from the middle of the Mn - O - Mn c-axis bond, and any c-axis interaction between e-orbitals of like occupancy is, like the t^3 - O - t^3 interactions, antiferromagnetic to give a Type - A antiferromagnetic order consisting of ferromagnetic (001) planes coupled antiparallel to one another [9,14].

Both $CaMnO_3$ and $LaMnO_3$ are orthorhombic, which introduces a Dzialoshinskii vector [15] $\mathbf{D}_{ij}$ parallel to the b-axis and an antisymmetric spin-spin coupling $\mathbf{D}_{ij} \bullet \mathbf{S}_i \times \mathbf{S}_j$ that cants the spins to give a weak ferromagnetic component along the c-axis; the antisymmetric exchange forces the spins into the a - c plane and magnetostatic energy constrains the antiferromagnetic component along the a-axis.

Substitution of Ga for Mn in $LaMn_{1-x}Ga_xO_3$ dilutes the Jahn-Teller Mn^{3+} ions and therefore reduces the elastic energy to be gained by long-range cooperativity of the local octahedral-site distortions. Accordingly, the O' - O transition temperature decreases with x. By x = 0.5, all static orbital ordering is suppressed and every Mn - O - Mn interaction becomes ferromagnetic. The observation of a ferromagnetic, isotropic coupling indicates that the interactions are dominated by the e^1 - O - e^0 electron transfer as a result of cooperative short-range dynamic Jahn-Teller deformations [16,17]. Therefore, at the transition from O' to O orthorhombic symmetry, the Mn^{3+} - O - Mn^{3+} superexchange interactions become isotropic and ferromagnetic, so the Weiss constant of the high-temperature Curie-Weiss plot is more positive than that found for the low-temperature O' phase [18].

The e-orbital degeneracy is not removed by a rhombohedral distortion; and in the R phase, the e electrons occupy a narrow σ^* band of e-orbital parentage. A strong Hund exchange field couples the σ electrons parallel to the local t^3 spins, so it is necessary to introduce a spin-dependent electron-transfer integral $t_{ij} = b_\sigma \cos(\theta_{ij}/2)$ where b_σ is the conventional spin-independent nearest-neighbor electron-transfer for the σ-bonding e electrons [19]. From tight-binding theory, the de Gennes [20] ferromagnetic double-exchange interaction becomes

$$\Delta\varepsilon^s_{ex} \approx -2z \times b_\sigma <\cos(\theta_{ij}/2)> \qquad (11)$$

where z = b is the number of nearest manganese neighbors.

The de Gennes double-exchange interaction is relatively strong, so the ferromagnetic R phase has a much higher Curie temperature T_c than the superexchange-coupled ferromagnetic phase appearing at the transition from the O' to the O phase. The CMR phenomenon occurs in a transitional O-orthorhombic phase appearing in a narrow compositional range in the O phase near the O' - O phase boundary; and for a fixed ratio Mn(IV)/Mn = 0.3, T_c increases sharply with increasing tolerance factor t on going from the O' - O to the O - R phase boundary. I turn to the original double-exchange mechanism proposed by Zener in 1951 and modify it to account for the evolution with t of physical properties found in the transitional O phase.

Zener [21] proposed a fast real charge transfer between a Mn^{3+} and a Mn(IV) in a two-manganese cluster, the e electron occupying, in this case, a two-manganese molecular orbital and, by the strong intraatomic Hund exchange, coupling the two localized t^3 configurations ferromagnetically. In order to account not only for long-range ferromagnetic order by this double-exchange coupling but also for a metallic temperature dependence of the resistivity below T_c, Zener postulated that his two-manganese polarons hopped without any activation energy even though they moved diffusively. Implicit in Zener's model is a polaron mobility $\mu_p \sim \tau_p^{-1}$ that has a transfer time $\tau_p < \tau_s$, the spin-lattice relaxation time. Moreover, Zener did not consider the e-orbital degeneracy at a Mn^{3+} ion.

Modification of the Zener model begins with a recognition that the e-orbital degeneracy constrains real electron transfer to be from an occupied Mn^{3+}-ion e orbital that is directed toward the Mn(IV) near neighbor. This requirement stabilizes the formation of a two-manganese polaron as a transitional state between small-polaron and itinerant-electron behavior. If W_σ is the itinerant-electron bandwidth and Δ_{JT} is the Jahn-Teller splitting of the e-orbital degeneracy, we should expect a Zener-type double-exchange coupling to be proportional to

$$x\mu_p \sim x\tau_p^{-1} \sim x(W_\sigma/\Delta_{JT})\omega_o(\phi) \sim x\,(W_\sigma/\Delta_{JT})\,M_o^{-1/2} \qquad (12)$$

where $\omega_o(\phi)$ is the frequency of the dynamic Jahn-Teller displacement made by an oxygen atom of mass M_o in the O-orthorhombic phase. In this model, T_c should increase from a ferromagnetic superexchange value to the de Gennes double-exchange model as τ_p falls from $\tau_p > \tau_s$ to $\tau_p < \tau_s$. This cross-over of time scales would give rise to a sharp increase in T_c with tolerance factor t since $W_\sigma = W_o \cos\phi$ and we can anticipate that $\omega_o(\phi)$ will increase as ϕ decreases.

The final model that emerges may be summarized as follows:

(1) Zener polarons form above T_c in the O-orthorhombic phase. At high temperatures, the t^3 configurations are paramagnetic, but a strong ferromagnetic exchange within a polaron makes the polarons superparamagnetic below a $T_c^* > T_c$. (2) Since a transition from small-polaron, localized-electron behavior in the O' phase to itinerant-electron behavior is occurring in the transitional O phase, cooperative oxygen displacements trap the Zener polarons into larger superparamagnetic clusters below a critical temperature near T_c^*. The trapping energy increases as $\tau_p < \tau_s$ increases to $\tau_p > \tau_s$ with decreasing t in the O phase. In addition, the mobile two-manganese Zener polarons remaining in the paramagnetic matrix may tend to transform back to small polarons as the angle ϕ of the matrix increases both with decreasing temperature and increasing volume of the Mn(IV)-rich phase with smaller ϕ. This latter phenomenon is manifest in the thermoelectric power as an additional configurational entropy in the polaronic matrix majority phase as t decreases to the O' - O phase boundary. (3) Since the double-exchange contribution to the global ferromagnetic coupling is proportional to the number of mobile charge carriers in the majority phase, which is the paramagnetic matrix, the $T_c < T_c^*$ decreases with t, i.e. with increasing trapping energy of the Zener polarons into ferromagnetic clusters. (4) From equation (12), the exchange of ^{18}O for ^{16}O would decrease $\omega_o(\phi) \sim M_o^{-1/2}$ and increase τ_p, thereby enhancing the trapping energy of the mobile Zener polarons and lowering T_c. (5) Above T_c in the O-phase, an applied magnetic field increases the trapping energy, which increases the volume of the Mn(IV)-rich ferromagnetic phase and lowers T_c for the matrix. The CMR phenomenon occurs where an applied magnetic field increases the volume of the Mn(IV)-rich phase to beyond its percolation limit. (6) Below T_c in the O-phase, the Weiss molecular field increases the volume of the Mn(IV)-rich phase to make it the majority phase with a higher T_c than the Mn(IV)-poor majority phase above T_c. The transition at T_c is second-order in the O' and R phases where there is no change in the character of the electrons in the majority phase at T_c; the transition is first-order in the O phase where the majority phase changes from the polaronic, paramagnetic Mn(IV)-poor matrix above T_c to the Mn(IV)-rich phase below T_c.

The volume change at T_c is greatest at the O' - O phase boundary where the change in the density of mobile charge carriers at T_c is greatest. (7) Near the O' - O phase boundary, strong electron-lattice coupling in the majority phase below T_c as well as the persistence of two magnetic phases makes difficult clarification of the nature of the conduction electron. A vibronic state that stabilizes a traveling charge-density wave consisting of Mn(IV)-rich and Mn(IV)-poor regions is probable.

EXPERIMENTAL

The reader is referred to a more extensive review [22] for a summary of a series of systematic experiments on the $(La_{1-x}Nd_x)_{0.7}Ca_{0.3}MnO_3$ system in which x, pressure, and $^{18}O/^{16}O$ isotope exchange are used to pass from the O' to the O phase without changing the ratio Mn(IV)/Mn = 0.3. Pressure transforms the O' to the O phase, demonstrating a dt/dP > 0 at the transition from small-polaron to itinerant-electron behavior. These experiments [23-26] support the model outlined. Moreover, an additional study [27] of the transport properties of single-crystal $La_{1-x}Sr_xMnO_3$, x = 0.12 and 0.15, have provided evidence for a transition between 5 and 6 kbar from vibronic to itinerant electronic behavior in the O - orthorhombic x = 0.15 sample in the temperature range $T_{co} < T < T_c$, where T_{co} is a charge-ordering temperature below which Mn(IV)-rich (001) planes alternate with orbitally ordered all Mn^{3+} (001) planes in a static CDW [28].

The author thanks the NSF and the Robert A. Welch Foundation for financial support.

REFERENCES

1. H.Y. Hwang, S.-W. Cheong, P.G. Radaelli, M. Marezio, and B. Batlogg, Phys. Rev. Lett. **75**, 914 (1995).

2. J.B. Goodenough and J.M. Longo, "Crystallographic and Magnetic Properties of Perovskite and Perovskite-Related Compounds", *Landolt-Bornstein Tabellen*, New Series III/4a K.H. Hellwege, ed. (Springer-Verlag, Berlin, 1970) p. 126.

3. J.B. Goodenough, J.A. Kafalas, and J.M. Longo, in *Preparative Methods in Solid State Chemistry*, P. Hagenmuller, ed. (Academic Press, New York, 1972), Chap. 1.

4. A. Manthiram and J.B. Goodenough, *J. Solid State Chem.* **92**, 231 (1991).

5. J.B. Goodenough, *Ferroelectrics* **130**, 77 (1992).

6. H.D. Megow, Nature **155**, 484 (1945); P. Vousden, Acta Crystallogr. **4**, 68 (1991).

7. M. Takano, N. Nakamishi, Y. Takeda, S. Naka, and T. Takada, Mat. Res. Bull. **12**, 923 (1977).

8. D. Woodward, (in press)

9. J.B. Goodenough, Phys. Rev. **100**, 564 (1955).

10. J.B. Goodenough, Magnetic Properties of Perovskites, *Landolt-Börnstein Tabellen*, Vol. **II**, Pt. 9 (Springer-Verlag, Berlin, 1962), p. 187.

11. S. Mori, C.H. Chen, and S.-W. Cheong, Nature **392**, 438 (1998).

12. J.B. Goodenough, Progress in Solid State Chemistry **5**, 141 (1971).

13. P.W. Anderson, Phys. Rev. **115**, 2 (1959).

14. E.O. Wollan and W.C. Koehler, Phys. Rev. **100**, 545 (1955).

15. I.E. Dzialoshinskii, J. Phys. Chem. Solids **4**, 241 (1958).

16. J.B. Goodenough, A. Wold, R.J. Arnott and N. Menyuk, Phys. Rev. **124**, 373 (1961).

17. J. Töpfer and J.B. Goodenough, Eu. J. Solid State and Inorg. Chem. **34**, 481 (1997).

18. G.H. Jonker, J. Appl. Phys. **37**, 1424 (1966)

19. P.W. Anderson and H. Hasegawa, Phys. Rev. **100**, 675 (1955).

20. P.-G. de Gennes, Phys. Rev. **118**, 141 (1960)

21. C. Zener, Phys. Rev. **82**, 403 (1951).

22. J.B. Goodenough, Australian J. Physics (in press).

23. W. Archibald, J.-S. Zhou, and J.B. Goodenough, *Phys. Rev.* B **53**, 14445 (1996).

24. J.-S. Zhou, W. Archibald, and J.B. Goodenough, *Nature,* **381**, 770 (1996).

25. J.B. Goodenough, J. Appl. Phys. **81**, 5330 (1997).

26. J.-S. Zhou and J.B. Goodenough, Phys. Rev. Lett. **80**, 2665 (1998).

27. J.-S. Zhou, J.B. Goodenough, A. Asamitsu, and Y. Tokura, Phys. Rev. Lett. **79**, 3234 (1997).

28. Y. Yamada, O. Hino, S. Nolido, R. Kanao, T. Inami, and S. Katano, Phys. Rev. Lett. **77**, 904 (1996).

SPIN WAVES IN DOPED MANGANITES

T. A. Kaplan and S. D. Mahanti
Physics-Astronomy Dept. and Center for Fundamental Materials Research,
Michigan State University
East Lansing, MI 48824

I. INTRODUCTION

We focus here on the spin waves in the low-temperature ferromagnetic-metal phase of doped manganites. While the experimental discovery of essential remarkable characteristics of these materials occurred as long ago as 1950[1], experiments on the fundamental low-lying excitations called magnons or spin waves only appeared within the last 2 years.[2-8] The first theoretical consideration of these excitations was in the work of Kubo and Ohata (1972)[9], who presented a semi-classical approximation within the double exchange (DE) model[10]. This approximation was emphasized recently by Furukawa (1996).[11] The first exact calculations, Zang et al [12] and the present authors[13], didn't appear until 1997. An important result of these exact calculations is that the shape of the spin wave dispersion (SWD) curve, energy vs. wave vector, differs in general from that of the famous Heisenberg spin model (with nearest neighbor interactions). In fact such a difference was not at all unexpected in view of fundamental differences found earlier[14,15] between the magnetic behavior of the two models. We note though that the semiclassical DE model happens to give precisely Heisenberg shape for the SWD.[9, 11]

In our study[13] of the DE model in one dimension, we found that while the SWD differs drastically from Heisenberg shape (HS) for some carrier concentrations, x, there is a range where the shapes are very close. The example $La_{1-x}Ca_xMnO_3$ defines x. Since this range included the value x = 0.3, we offered our result as an explanation for the surprising experimental finding[2] that the SWD shape for $La_{.7}Pb_{.3}MnO_3$ at low temperature is HS throughout the Brillouin zone. (We define HS as that of the nearest-neighbor Heisenberg model.) But quite recently there appeared an experimental study[6] of $Pr_{.63}Sr_{.37}MnO_3$ that found SWD of drastically different shape from Heisenberg, in contrast to the situation for the La-Pb compound. And since our calculations within the DE model also showed essentially Heisenberg shape for x=.37, we are forced to ask what is the source of the observed shapes.

The change in shape of the SWD accompanies changes in dopant and rare earth ions. There is evidence that magnetic properties are correlated with the size of these ions, resulting from a size-dependence of the bandwidth or hopping integral t.[16] There are three presently pursued corrections to the usual DE model, all of which might possibly be involved in the effects of interest here. Namely, there is orbital degeneracy[17-21], there are polarons with strong electron-lattice interactions[20-25], and there are terms of order t^2 which give rise to Heisenberg-like interactions between the Mn spins[17-19]. As a beginning, we have chosen to explore the last possibility first. Some motivation for this choice is that it is relatively simple. There are no extra degrees of freedom beyond those of the DE model (namely a single e_g band with filling $\leq$ ½, and a 3-fold degenerate t_{2g} band which is ½ filled). And the addition of t^2 terms is a necessary correction to the DE model in order that the end-points x=0 and x=1 be described by (different) Heisenberg models. Also, there are indications that in the ferromagnetic metals polarons are not present.[22] Finally, a change in t will of course result in changing the ratio of the O(t) and O(t^2) terms.

To this end we calculate the effective Hamiltonian H_{eff} through O(t^2), derived from an appropriate and commonly-considered electronic model; the 1st-order form of H_{eff} is precisely the DE Hamiltonian. We follow recent work[17] in the effective Hamiltonian approach, but as we said, we take just one e_g orbital per site (ignoring the so-called orbital degeneracy). There are also 3 t_{2g} orbitals, populated essentially with 3 parallel-spin electrons (often called the localized spins, each with S=3/2). At the limiting concentrations, x=0 and 1, the 1st-order terms vanish, and the 2nd-order terms constitute a Heisenberg Hamiltonian (different at the two x-values), as expected.

We find that for general x, the form of the new 2nd order terms in H_{eff} appears as a *modified Heisenberg Hamiltonian*, where the "exchange coupling constants" are *dynamic variables*, through their dependence on the e_g-occupation at the nearest-neighbor pair of interest. Somewhat surprising is that this is also true even for the coupling between the localized spins, contrary to common assumption[15,17-19] Also surprising is the fact that we find terms of O(t^2) which are missing in the earlier work[17] and which are larger than terms kept there.

Another remarkable experimental observation[6,8] is that the spin waves are very broad (short lifetime) at very low temperatures in some of the ferromagnetic-metal manganites. We speculate that an important ingredient in understanding the widths of the magnon excitations at low temperature might originate in H_{eff} to 1st order, i.e., in the DE model. We show that this model gives, in the single-magnon space, a continuum that might lie above and adjoin or overlap the spin wave dispersion curve.

In Section II we review and extend our earlier work[13] on spin waves in the DE model. The development of the effective Hamiltonian to O(t^2) is given in Section III. Section IV, which is a summary and look to the future, contains the speculation about the low-temperature magnon widths.

II. SPIN WAVES IN THE DOUBLE EXCHANGE MODEL

Here we review and extend our earlier results[13]. We begin with the following Hamiltonian

$$H_1 = -t\sum_{i,\sigma}(c_{i\sigma}^+ c_{i+1,\sigma} + h.c.) - j\sum \vec{s}_i \cdot \vec{S}_i + U\sum n_{i\uparrow}n_{i\downarrow} \qquad (1)$$

Here $c_{i\sigma}^+$ (the adjoint of $c_{i\sigma}$) creates an e_g- or e-electron at site i with spin σ, $n_{i\sigma} = c_{i\sigma}^+ c_{i\sigma}$, $s_i^+ = c_{i\uparrow}^+ c_{i\downarrow}$, and $s_i^z = (1/2)(n_{i\uparrow}-n_{i\downarrow})$; $\vec{S}_i$, with quantum number S, is the spin operator for the localized or t-electrons at site i, which runs from 1 to N. The number of e-electrons $N_e = \sum n_{i\sigma}$. The three terms represent, in order, the hopping, the intra-atomic exchange (the "Hund's rule" term), and the on-site Coulomb repulsion. H commutes with the square $\vec{S}_T^2$ of the total spin and any Cartesian component of $\vec{S}_T = \sum(\vec{S}_i + \vec{s}_i)$.

The double exchange model occurs in the limit $j \to \infty$, in which case U becomes irrelevant. In this limit the finite part of the spectrum is invariant under $t \to -t$ for a bipartite lattice. Our calculations were actually carried out using (1) with $U = \infty$, and $j = 40$ (the results are very insensitive to further increase of j).

We have considered only the cases where S_T^z = its maximum value S_M and $S_M - 1$. The latter contains the 1-magnon space. The cases $S_T^z = S_M$, with each conduction or e-electron in a Bloch function (ferromagnetic states), are eigenstates of (1), the lowest of which we call $|F\rangle$, with energy E_F. We have studied the following: (i) $(N, N_e) = (N, 1)$ and $(N, N-1)$ for $N = 4, 6, 8, 10, 12, 16, 20, 40, 60, 80$; (ii) $(N, N_e) = (N, 3)$ and $(N, N-3)$ for even $N = 4 - 18$; (iii). $(N, N_e) = (6,3)$ and $(10,5)$. Odd N_e, for which $|F\rangle$ is the ground state, is chosen because for even N_e the ground state is not ferromagnetic, as found in ref. 24; non-ferromagnetic behavior was also found[12] for 2- and 3-dimensional clusters with periodic boundary conditions and certain numbers of electrons ("open-shell cases"). Arguments in ref. 12 suggest that this deviation from ferromagnetism is a "1 in N" effect, a suggestion supported by Kubo's finding[24] that the ferromagnetic state is the ground state for *open chains*. We note that the very-long-wavelength spiral found for the non-ferromagnetic state[24,12] would probably be shifted to the ferromagnetic state by a small realistic anisotropy (via spin-orbit interactions). For further recent discussion of the ground state problem, see ref. 25.

To illustrate the calculations, consider 6 sites and 5 e-electrons. Periodic boundary conditions are assumed. We first need to set up the matrix of H in the space $S_T^z = S_M - 1$. We denote the possible states of the localized or t-electrons at site i as $|S\rangle_i$ and $|S-1\rangle_i$, which will be described as ↑ and ↓ (or up and down) respectively; similarly the possibilities for an e-electron spin at a site are $|1/2\rangle$ and $|-1/2\rangle$, also denoted by up and down arrows. A picture of a state of the system will thus have one down arrow and 10 up arrows. Let Φ_{ij} have a down-spin t-electron at site i and no e-electron (a "hole") at j; and let Φ'_{ij} have a down-spin e-electron at site i and a hole at j; $U = \infty$ excludes the other possibilities (a doubly-occupied e at site i). These states are illustrated by some examples, where the sites are labeled $1, 2, \ldots, 6$ from left to right:

Φ_{14}: ↑↑↑ ↑↑ Φ'_{14}: ↓↑↑ ↑↑   (2)
↓↑↑↑↑↑   ↑↑↑↑↑↑

More precisely, $\Phi_{nm} = (-)^{m-1} c_{m\uparrow}^+ c_{1\uparrow}^+ c_{2\uparrow}^+ c_{3\uparrow}^+ c_{4\uparrow}^+ c_{5\uparrow}^+ c_{6\uparrow}^+ |0\rangle \chi_n$, where $|0\rangle$ is the e-electron vacuum and χ_n is the t-electron spin state with site n down, the others up, and similarly for Φ'_{nm}. Note that the state Φ_{ii} exists, but there is no Φ'_{ii}. The number of Φ_{ij} states is 36, or in general for one hole, N^2, and there are 30 Φ'_{ij}'s, or $N(N-1)$ in general. One straightforwardly can calculate the matrix elements of H in this basis. Transforming to Bloch functions,

$$\Psi_k^{(p)} = N^{-1/2} \Sigma \exp(ikn) \Phi_{n,n+p}, \quad (3)$$

and similarly for $\Psi'^{(p)}_k$, one obtains for each allowed k, an 11x11 Hamiltonian matrix (or 2N-1x2N-1 for general N and one hole). The hopping term in H mixes $\Psi_k^{(p)}$ for different p and the same k, and similarly for $\Psi'^{(p)}_k$. The spin flip part of the Hund's rule term connects $\Psi_k^{(p)}$ and $\Psi'^{(p)}_k$. For $j = \infty$, one replaces, e.g. for S=1/2, the up-down pair at a site by the triplet with $S_{Ti}^z = 0$ (the spins at a site must be parallel); thus in the example above the primed and unprimed states become identical. (This amounts to applying the appropriate projection operator.[9,13]) In this case, the 11x11 reduces to a 6x6, the other 5 energies having $\to \infty$. Since we are interested here in large $j / |t|$, in the remainder of this paper we will exclude those states with energies that diverge as $j \to \infty$.

The energy eigenvalues minus the ground state energy $E_F = -2t$ are shown in Figure 1, sorted according to the wave vector k. They are invariant under k→-k, so negative k-values are omitted. The qualitatively similar graph for 6 sites, 1 electron, with S=1/2 was shown in ref. 13. A good check: Because of the commutation of H and the total spin lower operator S_T^-, one sees that for each k a (particular) free-electron eigenvalue must appear at least once, namely 0 at k=0, 1 at k=π/3, 3 at k=2π/3 and 4 at π. We note the low-lying branch, identified as the spin wave branch, whose highest energy is a small fraction (about 1/30) of the total width of the spectrum. We also note the bunching of the eigenvalues

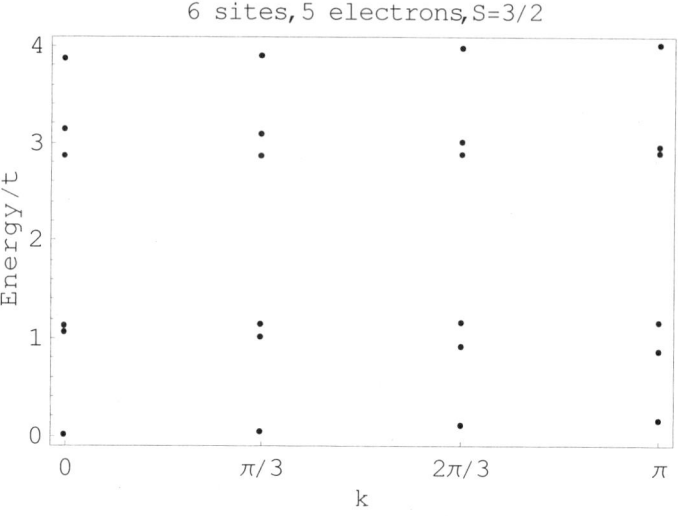

Figure 1. Energy eigenvalues measured from E_F in units of t, vs. wave vector k. $N, N_e = 6, 5$.

around the free-electron values 0,1,3,4. A similar property of the spin wave branch is found for other cases, e.g., for (N,N_e) = (10,7), (8,5), (6,5), (6,1), all with S=3/2, the ratios (spin-wave width)/(total free-electron width) = 0.025, 0.030, 0.036, 0.036, respectively (the corresponding hole concentrations are x = 1-N_e/N = 0.3, 0.375, 0.17, 0.83). Thus there is a natural second 'energy scale', despite having only one parameter t in the Hamiltonian. This is probably the same second energy scale found by Sarkar (in a mean field approximation).[26] However the bunching at the intermediate energies tends to smear out for larger systems, since in the latter the energies as well as the k-values become denser. This tendency is seen in Figure 2, where the spin wave branch remains separated from what appears to be approaching a continuum in the mid-range of energies. Whether this isolation continues as N increases, holding x fixed, and its significance vis a vis experiment will be discussed in Section IV.

Some justification for identifying the lowest branch as the spin wave dispersion is its qualitative similarity to the spin wave dispersion for the nearest neighbor (nn) ferromagnetic Heisenberg-model. We found further support for this by showing (for 1 electron and 6 sites, S=1/2) that the spin-spin correlation functions $\langle S_i^z S_j^z \rangle$ and $\langle S_i^+ S_j^- \rangle$ are very similar to those in the single magnon states or Bloch spin waves for the nn. Heisenberg ferromagnet: The z-z or longitudinal correlation is identical for the two models; the transverse correlation functions are close for the case considered.[13] In the remainder of

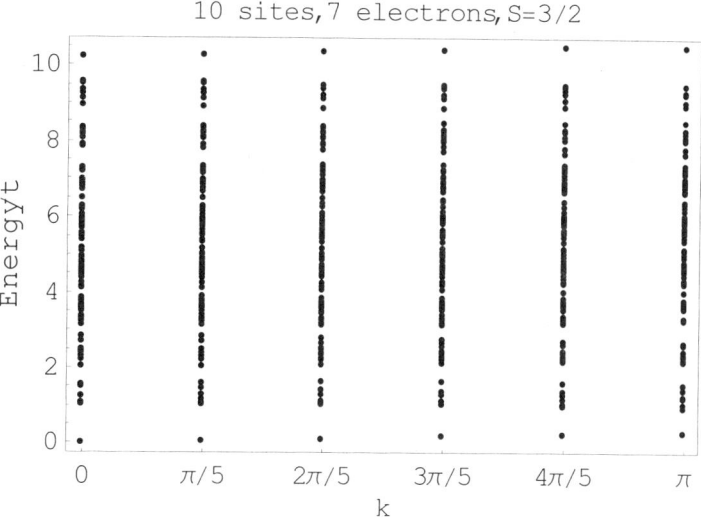

Figure 2. Energy eigenvalues measured from E_F in units of t, vs. wave vector k. N,N_e = =10,7.

this paper we will refer to the nn ferromagnetic Heisenberg model simply as the Heisenberg model.

We also studied a different kind of correlation function, for the case N,N_e=6,1, namely the probability, in spin-wave state k, of the conduction electron being a distance r from the down spin. This showed an interesting spin polaron type effect. Namely, there is an effective attraction for k=0, this gradually changing as k increases to repulsion at k = π. This effect was also found independently by Zang et al[12].

Figure 3 shows the spin wave energies for S=3/2, j = 40t, and for the Heisenberg model (solid curve), normalized to unity at the maximum (at k = π) so as to compare the *shapes* of the energy vs k relations. (In Fig 3 the cases (N, 1) and (N, N-1) are shown for a subset of the N-values calculated, selected for clarity of presentation.) We notice the large deviation from Heisenberg shape (HS) for small electron- and small hole-concentrations. We also see asymmetry between electrons and holes (as is consistent with the analùsis of ref. 9): the large-N limits are similar, but not identical, and both differ radically from HS. Most importantly, while even the largest electron concentration cases shown for (N,1), (N = 4, 12) deviate appreciably from HS, the largest *hole* concentrations in the (N,N-1) set (again, N = 4, 12) actually *straddle* the Heisenberg shape, and are quite close to it. To study this behavior in more detail, we calculated the deviation of each (normalized) spin wave spectrum, $\varepsilon(k)$, from the HS, as

$$\text{Dev} = (2/N)\sum_{k\geq 0}[\varepsilon(k)-\varepsilon_{\text{Heis}}(k)] \qquad (4)$$

This is a useful definition of the overall difference from HS because in every case (each (N, N_e), the bracketed quantity is of one sign; also it approaches the difference in areas as N $\to\infty$. Dev is plotted in Figure 4 as a function of x, for all the cases we studied, described above. One sees a range of x for which the DE and Heisenberg spectra have about the same shape, while for other x, near 0 and near 1, the deviation from HS is large. In the terminology of others, e.g. refs. 6 and 12, where the spin waves are normalized to the Heisenberg dispersion at small k, "softening" at large k corresponds to Dev>0.

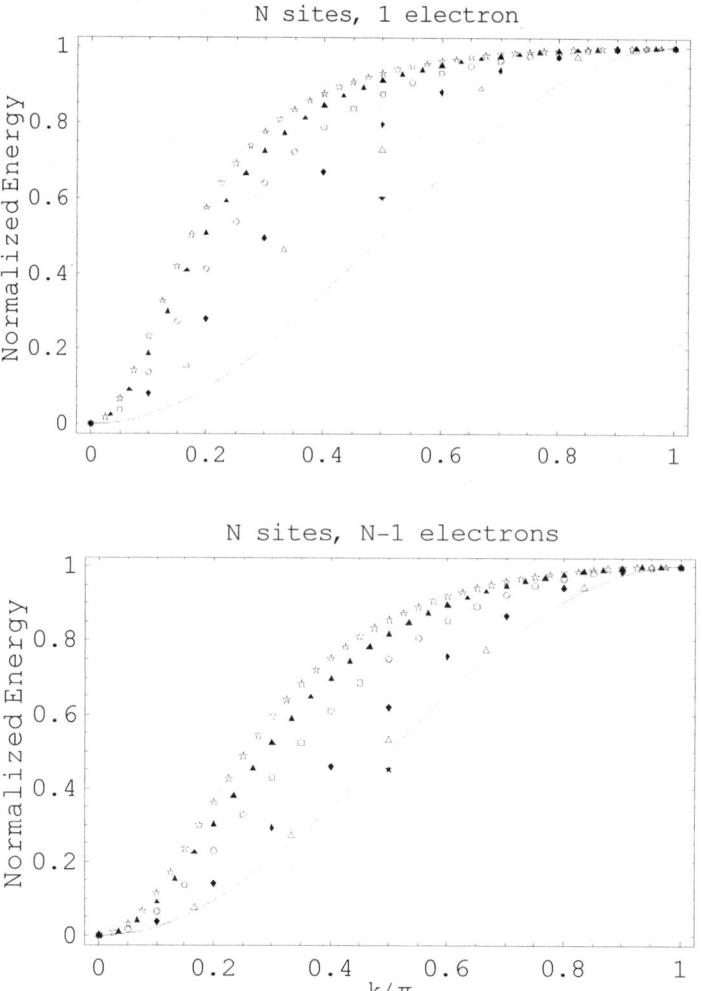

Figure 3. Spin wave energies for S=3/2, J=40t, and for the Heisenberg model (solid line), normalized to unity at k=π, for N=4,12,20,40,60,80. At k=π/2, all the N-values appear; in both cases, (N,1) and (N,N-1), the energy increases with N.

A figure similar to Figure 4, published earlier,[13] contained fewer points; in particular $(N, N_e) = (18,15)$, $(16,13)$, and $(14,11)$ were missing, as well as similar points at larger x.[27] The following possibility was raised on the basis of the previous figure: The small- and large-x values, where Dev is large, originated in the (N,N-1), and (N,1) cases, where x→0 or 1 as N→∞; but one is really only interested in the infinite-N limit with x fixed and not 0 or 1; perhaps if x is fixed at a small value > 0, Dev will become small as N increases, so that in this proper limit Dev will be small for *all* x > 0 (and < 1).[28] However, the added points, particularly (18,15) added to (6, 5) at the same x, suggest the opposite, namely that larger N will lead to larger values of Dev at small x. At the x where we have calculated more than one point, we find for Dev(N,N_e): Dev(6,5) = -.0237, Dev(18,15) = .0074; Dev(4,3) = -.0247, Dev(12,9) = -.0258; Dev(6,3) = -.0020, Dev(10,5) = -.0066; Dev(4,1) = .0494, Dev(12,3) = .0506. These data suggest that as N increases, the Dev vs. x "curve" will deepen in the central part and heighten in the edge regions. Of course one

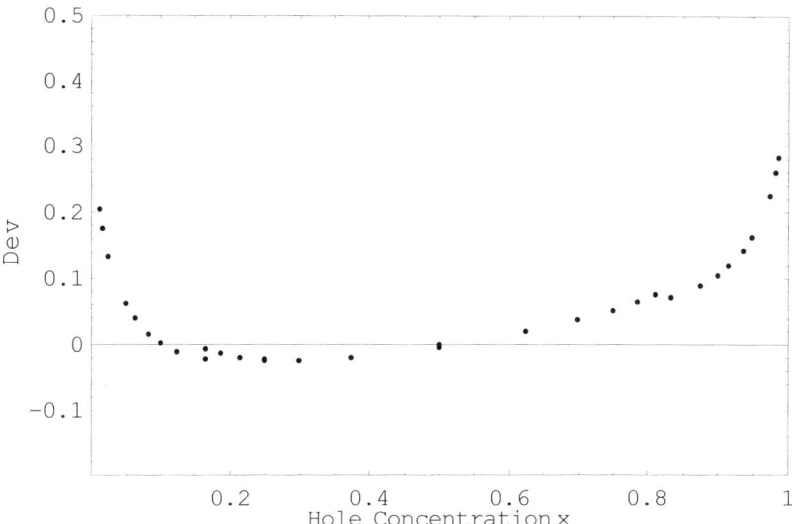

Figure 4. Deviation of the spin-wave spectrum shape from that of the Heisenberg model vs. concentration x.

would like to have more than 2 points (at given x) for extrapolation; but it is very difficult to attain since the diagonalization problem grows rapidly with N, and the difficulty is exacerbated by the requirement that N_e be even.

To give a feeling for the significance of the measure Dev of the deviation from HS,

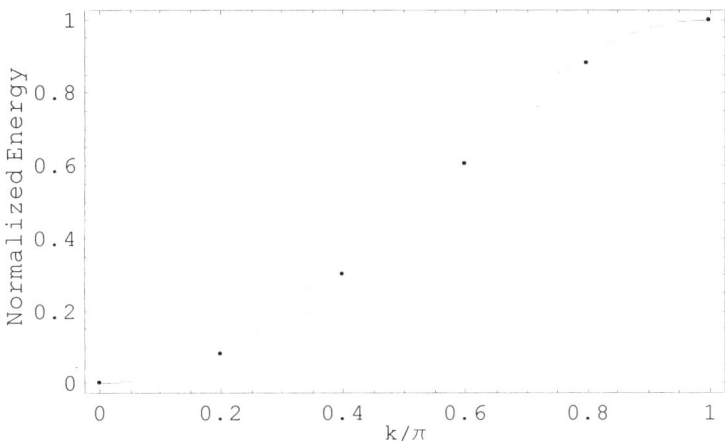

Figure 5. Spin wave energies. Points: DE, $(N,N_e) = (10,7)$. Line: Heisenberg. Dev = -0.027

we show in Figure 5 the two normalized SWD's for the case (10,7), where x=0.3 and Dev = -0.027.

We now consider the absolute values of the spin wave excitation energies. In any sequence (N, n) or (N, N-n), with fixed n, the width of the spectrum vanishes as $N \to \infty$ (x $\to$ 1 or 0). To address values of x like the experimental values, we consider $(N,N_e) = (8,5)$ and (10,7) (x=.375 and .3). The widths of the spin wave spectra in these cases are .294t and .246t, respectively. The estimate $t \approx 0.2$ eV (Millis et al[29]) then gives about .06 and 0.05eV for these cases. We compare, of course, with the experiment where HS is found, namely the Pb-doped sample with $x = 0.3$.[2] There the measured width[2] is about 0.11eV; probably a more appropriate comparison with our 1-dimensional theory is the width in the 100 direction, namely about .04 eV. In view of the uncertainty in the t-value, and possible finite-size corrections to our (10,7) result, this semi-quantitative agreement between theory and experiment suggests that the DE model is sufficient to describe SWD in those materials where HS is observed.

We looked numerically at large S to compare with Furukawa's results.[11] Indeed, we found in several cases that as S increases the spin wave spectrum approaches the Heisenberg shape, in agreement with his result (which is to leading order in 1/S). It is interesting to check the actual spin wave bandwidth. In 1 dimension, his[11] formula (8), which takes $J \to \infty$, gives

$$\omega_k = A \frac{1 - \cos k}{2}, \tag{5}$$

with $A = \frac{2t}{NS} \sum_q^{occ} \cos q$. For $(N, N_e) = (6,1)$ and $(8,5)$, this gives $A/t = 0.222$ and 0.402, respectively, to be compared with our exact results, 0.127 and 0.301, all for $S=3/2$. Thus the result to leading order in 1/S is of the correct order of magnitude, although not very accurate.

When presented with the results of Figure 3, we have been asked, why can the SWD shape differ so dramatically from Heisenberg shape. In our opinion, the right answer is another question, namely, why should the SWD in the DE model be close to HS in *any* case? In other words, we ask what does the Heisenberg model have to do with the DE model? Clearly both are isotropic; i.e. both are invariant under uniform rotation of all the spins. But *any* function of the various scalar products $\vec{S}_i \cdot \vec{S}_j$ is isotropic. The only legitimate reason for comparing DE with the Heisenberg model that we can see is the latter's familiarity to the community. In a similar vein, Zang et al[12] tried to explain the "softening" found at large k: Since the electrons spend less time near the down spin as k increases, "the effective spin-spin coupling J decreases". It is clear that if the electrons spend no time near the down spin, the excitation energy would be zero. But we don't see why for finite excitation energy, it should be less than the *Heisenberg* excitation energy (why not the Heisenberg energy plus terms like $(\vec{S}_i \cdot \vec{S}_j)^2$, normalized of course at small k).

III. EFFECTIVE HAMILTONIAN TO SECOND ORDER IN HOPPING (ADDS HEISENBERG-TYPE EXCHANGE PLUS OTHER TERMS).

That Heisenberg-type interactions, usually referred to as "superexchange" interactions, exist and are important has been recognized by a number of workers.[15,17-19] They introduce them by simply adding the familiar Heisenberg interactions between the t_{2g} or localized-electron spins. The exchange parameters J_{ij} are constants, as usual. It is well known that these Heisenberg terms come about from perturbation processes that are 2nd order in the hopping t, whereas DE is 1st order. Although one would ordinarily expect therefore that the new terms would be much smaller (roughly by t/U), the fact that the SWD-width is << hopping, as discussed in Sec. II, suggests that the added terms might be important to the nature of the spin waves even for intermediate x. However, instead of simply adding a Heisenberg type term to the DE model, we step back and consider where it might come from, in a deductive way. We show this rather straightforward calculation in some detail because we obtain results which haven't appeared in the literature, and which differ conceptually from previous considerations of Heisenberg exchange.

Following the usual assumption of DE, we consider a minimum-basis-set model, where there is for each Mn just one "e_g" orbital, fractionally occupied, plus the 3 t_{2g} electrons. (Thus the "orbital degeneracy" is excluded.) I'll describe this for the simpler case of one t_{2g} orbital per site. The model Hamiltonian includes what are usually considered the most important terms (we drop the g and 2g subscripts in e_g and t_{2g} and replace t for the hopping parameter by τ):

$$H = -\tau_e \sum_{<i,j>,\sigma}(e_{i\sigma}^+ e_{j\sigma} + h.c.) - \tau_t \sum_{<i,j>,\sigma}(t_{i\sigma}^+ t_{j\sigma} + h.c.)$$
$$-2j\sum_i \vec{s}_i \cdot \vec{S}_i + U\sum_i (n_{i\uparrow}^e n_{i\downarrow}^e + n_{i\uparrow}^t n_{i\downarrow}^t + n_i^e n_i^t), \quad (6)$$

where $e_{i\sigma}$ destroys an e electron at site i with spin σ, $t_{i\sigma}$ is the same for a t electron. τ_e and τ_t are the e-e and t-t hopping parameters (we neglect e-t hopping, which is zero for the cubic perovskite), j > 0 is the Hund's rule coupling parameter between the e-spin $\vec{s}_i$ and the t-spin $\vec{S}_i$. E.g., $s_i^z = \frac{1}{2}(e_{i\uparrow}^+ e_{i\uparrow} - e_{i\downarrow}^+ e_{i\downarrow})$, $s_i^+ = e_{i\uparrow}^+ e_{i\downarrow}$. U is the intra-atomic Coulomb interaction strength. We have simplified the "U" terms by taking the intra-orbital and inter-orbital on-site terms of equal strength. The difference, while small,[30] might not be negligible; it can be taken into account without serious modification. H is seen to be similar to H_1; it has added t-t hopping and contributions to the on-site repulsion from the t-electrons.

H would be just the usual Kondo lattice model with omission of the t-t hopping and the U term; H gives the DE model for $j \to \infty$. We are interested however in the case of large but finite j and U. So we're interested in treating this model in perturbation theory, where the unperturbed Hamiltonian H_0 = H – hopping terms.

There is a large manifold of ground states of H_0, which we call the P-manifold. In this manifold the states have the properties that at each site there can be either a hole or an electron with spin parallel to the t-spin. For a site with parallel spins, the e- and t-electrons must be in a triplet state, having total z-component either 1, 0 or –1. Letting P project onto this space and Q=1-P, the effective Hamiltonian familiar in perturbation theory is[31]

$$H_{eff} = PHP - PV[Q(H-E)Q]^{-1}VP, \quad (7)$$

where V=H-H_0. Here E is an exact eigenvalue of H and the eigenstates of H_{eff} are related to those of H. So to 2nd order,

$$H_{eff} = E_0 - \tau_e P \sum_{<i,j>,\sigma}(e_{i\sigma}^+ e_{j\sigma} + h.c.)P$$
$$- PV_e Q \frac{1}{H_0 - E_0} QV_e P - PV_t Q \frac{1}{H_0 - E_0} QV_t P \quad . \quad (8)$$

The 1st-order contribution of the τ_t terms vanishes because every t-orbital is occupied in the P-manifold. Here V_e and V_t are the e-e and t-t hopping terms, V_e appearing in the first-order term. (E_0 is the ground state energy of H_0.) The first order term is precisely the DE Hamiltonian.

For the 2nd order terms consider 5 electrons (we refer here to the e-electrons as electrons) on 6 sites (one hole). Some sample states in the P-manifold:

e: ↑↑↑↑↑   ↑[↑] ↑↑↑   ↑↑↓↑↑
t: ↑↑↑↑↑↑   ↑[↓]_t ↑↑↑↑   ↑↑↓↑↑↑ .

$\begin{bmatrix}↑\\↓\end{bmatrix}_u$ with u = t stands for the triplet with total S_z for the site = 0, and for u = s, the singlet.

    The first state does not contribute to the 2nd order term. The second and third states do. For example, in the 2nd state, electron 1 can hop to site 2 creating doublets on sites 1 and 2, and then hop back, with either spin. The following shows possible initial and intermediate states, and the energy changes (two sites are sufficient):

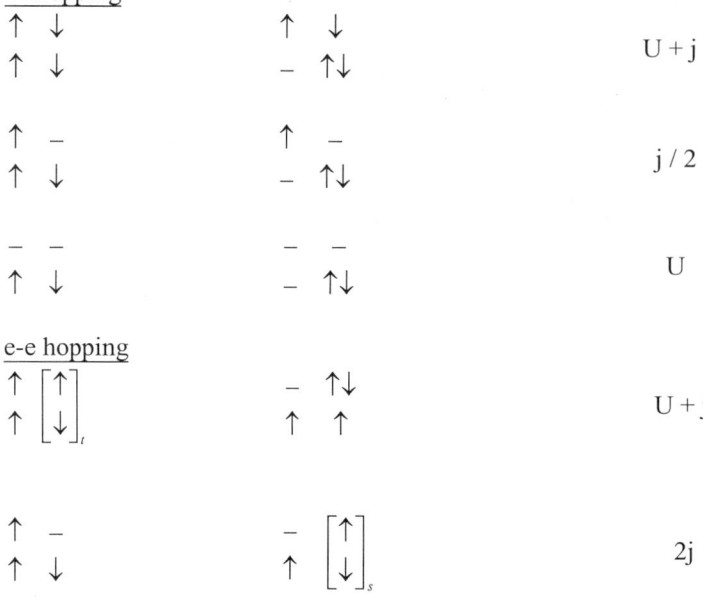

The Q projection forces the second site in the last intermediate state to be a singlet.

    There are two possibilities for the final state: (i) after one electron hops to site j, an electron from j hops back (2-site terms), or (ii) if there's a hole nearby, after the 1st hop to j, an electron can hop from j to a third site, (3-site terms), e.g.

↑ – –   – [↑] –   – – ↑
↑ ↓ ↑   ↑ [↓]_s ↑   ↑ ↓ ↑

A similar term occurs in perturbation theory for the Hubbard model. It is argued there that one is interested in such small hole concentrations, at least for high-T superconductivity, that the 3-site terms added to the t-J model are negligible. For the manganites, the concentrations of interest are not necessarily so small. The final result obtained is

$$H_{eff} = H_{eff}^{DE} + H_{eff\,2\text{-site}} + H_{eff\,3\text{-site}}. \qquad (9)$$

The spin-dependent part of the 2-site terms is

$$H_{eff2\text{-site}} = \sum_{<i,j>} (J_2 \delta_{n_i+n_j,2} + J_1 \delta_{n_i+n_j,1} + J_0 \delta_{n_i+n_j,0}) \vec{S}_{Ti} \cdot \vec{S}_{Tj}, \tag{10}$$

where

$$J_2 = \frac{\tau_e^2 + \tau_t^2}{U+j}, \tag{11}$$

$$J_1 = \frac{\tau_e^2/4 + 8\tau_t^2}{j}, \tag{12}$$

$$J_0 = 4\tau_t^2/U \tag{13}.$$

$n_I = e_{i\uparrow}{}^+ e_{i\uparrow} + e_{i\downarrow}{}^+ e_{i\downarrow}$, and $\vec{S}_{Ti} = \vec{S}_i + \vec{s}_i$. (14)

We can understand these terms: J_2 corresponds to the undoped case, J_0 holds for the completely doped situation. Intermediate doping requires all 3 terms. Thus we obtain a Heisenberg Hamiltonian with the "exchange parameter" a dynamic variable, depending on the e-orbital occupation at the interacting sites. As far as we know, such an effect has not been pointed out (although it is similar to spin-phonon coupling via the Heisenberg interaction). Note: even the Heisenberg interactions between the localized or t-spins depend on the e-occupancies. Also, the $8\tau_e^2/j$ term is new; the other similar term $\tau_e^2/(4j)$, was noted by Sarkar [26].

The 3-site terms: Again this depends on the e-occupancy of the "middle" site. We write this for a chain of sites, although the generalization to higher dimensions is clear.

$$H_{eff3\text{-site}} = -\frac{\tau_e^2}{U+J_{hd}} h_{mid\ n=1} - \frac{\tau_e^2}{4J_{hd}} h_{mid\ n=0}, \tag{15}$$

where

$$h_{mid\ n=1} = \sum_i \{\sum_\sigma (\frac{1}{2} - \sigma s_{i+1}^z) e_{i\sigma}{}^+ e_{i+2,\sigma} - s_{i+1}^- e_{i\uparrow}{}^+ e_{i+2\downarrow} - s_{i+1}^+ e_{i\downarrow}{}^+ e_{i+2\uparrow} + h.c.\}, \tag{16}$$

and $h_{mid\ n=0}$ is the same as (16) but with $s_{i+1}{}^\nu \to S_{i+1}{}^\nu$. Note the "middle"-spin mediation of these 2nd-neighbor hopping terms. For example, $e_{i\downarrow}{}^+ e_{i+2,\uparrow} S_{i+1}{}^+$ hops an up e-electron from site i+2 to i, flips its spin and raises the t-spin at site i, thus conserving spin. The process doesn't occur if the t-spin at site i is up (that's included in 1st-order).

IV. SUMMARY AND FUTURE WORK

We reviewed and extended our earlier work on exact calculations of spin waves in the DE model for small 1D systems. We found that our original conclusion that, for some concentrations x the shape of the spin wave dispersion differs dramatically from Heisenberg shape, while there is a region of x for which the shape of the SWD is close to HS,[13] is supported by our extension to larger systems. We also added a picture (Figure 2) which indicates how a continuum of energy eigenvalues develops as the size of the system increases. We derived the extension of the double exchange model to second order in hopping strength; the effective Hamiltonian contains Heisenberg-like spin-spin interactions, but the exchange couplings are dynamical variables, in contrast to the usual

Heisenberg model where they are constants. This is true even for the interactions between the localized spins (the t_{2g}-electron spins). These results differ conceptually from the way Heisenberg interactions have been introduced previously[15,17-19].

Faced with the fact that the effective Hamiltonian has a quite complex structure, and the parameters are not known precisely, our calculations on the consequences of the new terms are very preliminary, sufficiently so that we will only mention the following results on the stability of the ferromagnetic ground state. We considered $(N,N_e)=(6,5)$, and discuss here two cases (see eq.(10)): (i) $J_2 = 0$, and (ii) $J_1 = 0$, with the 3-site terms neglected in both cases. (J_0 does not contribute when there is one hole.) We find, after generalizing to t-spin S=3/2, that stability of the ferromagnetic state requires $J_1 < .047t$ for case (i), and $J_2 < .022t$ for (ii). These results make sense qualitatively: Recall that for a non-zero J_1-term there must be one hole for the pair and for a non-zero J_2-term, there must be no holes for the pair. Thus the J_1-terms act only on 2 of the 6 nearest-neighbor pairs, while the J_2-terms act on 4 such pairs. Hence it is reasonable that a larger J_1 is required to destabilize the ferromagnetic state than is required for J_2. Also, the destabilizing values are in the ballpark of the width of the spin wave dispersion ($\approx 0.1t$, which follows from data given on page 4).

Clearly an extensive study of the effect on the spin-wave spectrum of the new terms in H_{eff} must be carried out. This will be done by calculating exactly the spin wave energies for the same cases (N, N_e) considered for the DE model. We will continue a study of each term alone (the J_n-terms, n = 2,1,0 within $H_{eff\ 2\text{-site}}$, plus $H_{eff\ 3\text{-site}}$) to get an idea of the physical meaning of each of them; and then we must consider them together for reasonable values of the basic parameters in H. Hopefully, they might explain the observed spin wave spectra, at least qualitatively (extension from 1 to higher dimensions probably will be needed to get a quantitative understanding). In any case, the $O(t^2)$ terms we have calculated must be present in addition to the DE Hamiltonian, and even in the presence of orbital degeneracy and electron-lattice interactions (Jahn-Teller effect), terms like those obtained here must occur.

Finally, we speculate on the possible relevance of the DE model to the magnon lifetimes observed[6,8] at low temperature ($<< T_c$, the Curie temperature) in some of the doped manganites. As we have noted, and illustrated by Figure 2, a continuum is expected to occur in the DE model, within the "single-magnon subspace", i.e., within the space of states with total z-component of spin equal to its maximum value minus 1. Recall that the spin wave branch is isolated from what appears to be the developing continuum for the cases discussed (Figures 1 and 2). The central question is, will the gap between the spin wave branch and the continuum vanish for large systems. To try to extrapolate from our small systems is made especially difficult by the clear requirement that the extrapolation be made at constant concentration, $x = x_o$, or at least by a path $x(N)$ where $x(N) \to x_o$ as $N \to \infty$. Having obtained the solution for the case $(N,N_e) = (18,15)$ does give us two cases with $x = 1/6$, $(6,5)$ being the other. As limited as this is, we plot the gap vs. $1/N$ in Figure 6 for wave vector $k = 0$ and π; the upper points are for $k = 0$. It is seen that a straight line through the k=0 points essentially goes through 0 at $1/N = 0$, and through the lower points hits the $1/N = 0$ axis at a slightly negative value. Despite the crudeness of this analysis, the behavior does suggest that the continuum will, in the limit $N \to \infty$, approach or overlap the spin wave branch.

It will be of considerable interest to calculate various static correlation functions in these higher states to understand their physical nature. Also it is important to obtain the

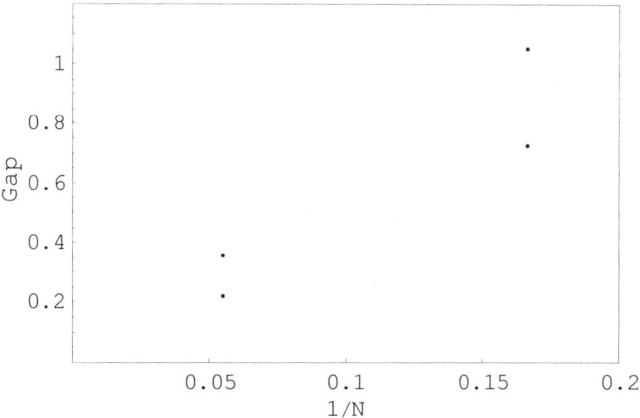

Figure 6. The gap between the spin wave energy and the next highest energy for k=0 (upper points) and for k=π (lower points). The numbers are for x=1/6, N=6 and 18.

neutron scattering function $S(q, \omega)$ within this manifold of states to see the intensity of the inelastic scattering at very low temperatures. The latter, perhaps in conjunction with the various corrections ($O(t^2)$ terms, orbital degeneracy and/or electron-lattice interactions) might explain the observed[6,8] large magnon lifetimes at low temperatures.

ACKNOWLEDGMENTS

We thank K. Kubo and J. Lynn for valuable discussions.

REFERENCES

1. G. H. Jonker and J. H. Van Santen, Physica **16**, 337 (1950); J. H. Van Santen and G. H. Jonker, *ibid*. page 599.
2. T. G. Perring, G. Aeppli, S. M. Hayden, S. A. Carter, J. P. Remeika, and S-W. Cheong, Phys. Rev. Lett. **77**, 711 (1996).
3. J. W. Lynn, R. W. Erwin, J. A. Borchers, Q. Huang, and A. Santoro, Phys. Rev. Lett. **76**, 4046 (1996).
4. L. Vasiliu-Doloc, J. W. Lynn, A. H. Moudden, A. M. de Leon-Guevara, and A. Revcolevschi, J. Appl. Phys. **81**(8), 5491 (1997)
5. J. W. Lynn, R. W. Erwin, J. A. Borchers, A. Santoro, Q. Huang, J.-L. Peng, and R. L. Greene, J. Appl. Phys. **81**(8), 5488 (1997).
6. H. Y. Hwang, P. Dai, S-W. Cheong, G. Aeppli, D. A. Tennant, and H. A. Mook, Phys. Rev. Lett. **80,** 1316 (1998).
7. J. Fernandez-Baca, P. Dai, H. Y. Hwang, C. Cloc, and S-W. Cheong, Phys. Rev. Lett. **80**, 4012 (1998)
8. J. W. Lynn, this volume.
9. K. Kubo and N. Ohata, J. Phys. Soc. Japan **33**, 21 (1972).
10. C. Zener, Phys. Rev. **82**, 403 (1951)
11. N. Furukawa, J. Phys. Soc. Japan **65**, 1174 (1996).
12. J. Zang, H. Röder, A. R. Bishop and S. Trugman, J. Phys.: Condens. Matter **9**, L157 (1997)

13. T. A. Kaplan and S. D. Mahanti, J. Phys.: Condens. Matter **9**, L291 (1997)
14. P. W. Anderson and H. Hasagawa, Phys. Rev. 100, 675 (1955)
15. P. G. DeGennes, Phys. Rev. **118**, 141 (1960)
16. Sang-Wook Cheong and Harold Y. Hwang, in *Collosal Magnetoresistance Oxides*, Y. Tokura, ed., *Monographs in Condensed Matter Science* (Gordon & Breach, Amsterdam, to be published)
17. S. Ishihara, J. Inoue and S. Maekawa, Phys. Rev. B **55**, 8280 (1997)
18. S. Maekawa, this volume.
19. R. Maezono, S. Ishihara, and N. Nagaosa, Phys. Rev. B **57**, R13993 (1998)
20. A. Millis, B. I. Shraiman, and R. Mueller, Phys. Rev. Lett. **77**, 175 (1996); A. Millis, R. Mueller, and B. Shraiman, Phys. Rev. B **54**, 5389 and 5405 (1996).
21. H. Röder, Jun Zang, and A. R. Bishop, Phys. Rev. Lett. **76**, 1356 (1996).
22. S. Billinge, this volume.
23. M. Jaime, this volume.
24. K. Kubo, J. Phys. Soc. Japan **51**, 782 (1982)
25. J. Riera, K. Hallberg and E. Dagotto, cond-mat/9609111 12 Sep. 1996
26. S. K. Sarkar, J. Phys.: Condens. Matter **8**, L515 (1996).
27. In ref. 13, we claimed the figure contained all our calculated points; however that was incorrect: we inadvertently sent the wrong figure to the journal.
28. K. Kubo, private communication.
29. A. J. Millis, P. B. Littlewood, and B. I. Shraiman Phys. Rev. Lett. **74**,5144 (1995).
30. See A. M. Oleś, Phys. Rev. B **28**, 327 (1983); T. Momoi and K. Kubo, Phys. Rev. B **58**, R567 (1998)
31. J. Lindgren and J. Morrison, *Atomic Many-Body Theory*, J. Peter Toennies, ed., *Springer Series in Chemical Physics*, Vol. 13 (Springer-Verlag, New York, 1982), p. 207 ff.

SPIN DYNAMICS OF MAGNETORESISTIVE OXIDES

J. W. Lynn

NIST Center for Neutron Research
National Institute of Standards and Technology
Gaithersburg, MD 20899
and
Center for Superconductivity Research
Department of Physics
University of Maryland
College Park, MD 20742

INTRODUCTION

The magnetic properties of the doped $LaMnO_3$ class of materials have attracted renewed interest because of the dramatic increase in conductivity observed when the spins order ferromagnetically.[1] This enormous increase in the carrier mobility originates from a metal-insulator transition that is closely associated with the magnetic ordering. Here we briefly review the ongoing neutron scattering measurements to investigate the magnetic order and spin dynamics of the magnetoresistive oxides, such as the perovskite (La, Pr, or Nd)$_{1-x}$A$_x$MnO$_3$ (A=Ca, Sr, and Ba) system. A number of anomalous features have been identified in the manganites, including very strong damping of the spin waves in the ground state and as a function of temperature, anomalous spin wave dispersion, and the development of a strong spin-diffusion component to the fluctuation spectrum well below T_C in the region of optimal doping. For the colossal magnetoresistive $Tl_2Mn_2O_7$ pyrochlore, on the other hand, the behavior of the spin dynamics is conventional, which argues that the mechanism responsible for the magnetoresistive effect has a different origin in these two classes of materials.

PEROVSKITES

Starting with the undoped material, pure $LaMnO_3$ is an antiferromagnetic insulator in which the $Mn^{3+}O_3$ octahedra exhibit a Jahn-Teller distortion that strongly couples the magnetic and lattice system[2,3]. The material can be converted to a ferromagnet (with a much smaller distortion) with oxygen annealing.[4] Doping with divalent ions such as Ca, Sr, or Ba introduces Mn^{4+}, and with sufficient doping ($x > \sim 0.1$) the holes become mobile and the system transforms into a metal that exhibits ferromagnetic

Physics of Manganites. Edited by Kaplan and Mahanti
Kluwer Academic/Plenum Publishers, 1999

ordering.[5,6] In this metallic regime the double-exchange mechanism only allows holes to move if adjacent spins are parallel, which results in a "colossal" decrease in the resistivity (CMR) when the spins order ferromagnetically, either by lowering the temperature or applying a magnetic field. The carrier mobility is thus intimately tied to both the lattice and magnetism, and considerable effort has been devoted to identifying the basic interactions that dominate the energetics and control the magnetoresistive properties. One avenue to unraveling these interactions is by measuring the spin dynamics.

Spin Wave Dispersion and Linewidths

In the regime where the materials are ferromagnetic metals (0.15 <x <0.5), the magnetic system at long wavelengths (small wave vectors q) behaves as an isotropic ferromagnet at low T, as is the case for all the metallic ferromagnetic manganites[7-18]. The magnetic excitations are conventional spin waves, with a dispersion relation given by

$$E = \Delta + D(T)q^2, \quad (1)$$

where Δ represents the spin wave energy gap, and the spin stiffness coefficient $D(T)$ is directly related to the exchange interactions. There are two features that make a ferromagnet isotropic. One is that the spin wave gap Δ, which represents the energy to uniformly rotate the entire spin system away from the easy direction of magnetization, is much smaller than the exchange energies in the problem. This is indeed the case, as Δ is too small ($\Delta < 0.02$ meV) to be measured directly with neutrons anywhere in the metallic regime of these materials. In particular, for the $x=\frac{1}{3}$ doping where the magnetoresistance anomalies are largest, the perovskites are ideal isotropic ferromagnets, with an essentially gapless spin wave dispersion relation.[19]

The second question concerns the isotropy of the exchange interactions along different directions of the crystal in these distorted (non-cubic) perovskites. In general, the dispersion relation can be expanded in a power series in q^2, with the first two terms as given in Eq. (1). In the undoped and lightly doped antiferromagnetic materials the dispersion relation is of course qualitatively different for the in-plane ferromagnetic sheets compared with the direction along which these sheets are stacked antiferromagnetically. However, as soon as the materials become ferromagnetic, they exhibit a remarkably isotropic dispersion relation.[9] This isotropy is characteristic of ferromagnetic metals, even though the dc conductivity near the compositional phase boundary can have semiconducting behavior in some temperature regimes.

The dispersion relations have been measured to the zone boundary in quite a number of materials now.[9-15,17] The early work suggested that a nearest-neighbor Heisenberg model was adequate to explain the basic shape of the dispersion curve,[12,15] but more recent work has found large variations of the dispersion relation from such a model, including splitting of the dispersion relations, a stiffening of the dispersion relation at intermediate q, and a softening or flattening of the spin wave dispersion relation near the zone boundary.[9-11,13,14,17] Hence it is clear that there is considerable structure in the dispersion relations for some of these systems.

Results on single crystals at larger q also reveal that the low temperature spin waves exhibit large intrinsic linewidths, which are likely associated with a strong magnon-electron interaction. These observations demonstrate that spin waves are not true eigenstates of the system, even in the ground state.[10,13]. In addition, there is a dramatic increase in the damping of the spin waves at elevated temperatures[12,10], along with the development of a strong spin-diffusion component to the fluctuation spectrum well below T_C,[7-9,11,14,16,17] as we now discuss.

Quasielastic Central Component

As the temperature is increased towards T_C, a quasielastic component develops in the small wave vector fluctuation spectrum for all the perovskite systems. For the Ca-doped CMR system in particular, this central diffusive component completely dominates the spectral weight as $T \to T_C$ and drives the phase transition, while the spectral weight for the spin wave excitations decreases in strength.[7] The application of magnetic fields of a few tesla strongly reduces this quasielastic component in favor of the spin waves,[8] while for Ca concentrations away from $x=\frac{1}{3}$ the intensity of the quasielastic scattering is greatly reduced compared to the spin wave intensities.[7,16] The spin waves, on the other hand, do not appear to renormalize to zero as $T \to T_C$, in contrast to the behavior expected for a second-order ferromagnetic phase transition. Therefore the ferromagnetic transition cannot be driven by the usual route of the thermal population of conventional spin waves. Rather, the ferromagnetic phase transition appears to be driven by the development of the quasielastic component of the fluctuation spectrum, which has been identified as the spin component of the polaron in this system[8]. A central component to the fluctuation spectrum has also been observed in the Sr doped[9,11] and Ba doped materials[17], as well as for $Nd_{0.7}Sr_{0.3}MnO_3$ and $Pr_{0.7}Sr_{0.3}MnO_3$[14]. In this latter work the authors found that the spin wave dispersion relation is quantitatively the same for several of the $x=\frac{1}{3}$ materials, while the ordering temperature varies by a factor of two. They found that the anomalous properties of the transition, such as the strength of the central peak and the failure of the spin waves to renormalize to zero at the transition, were inversely related to the value of T_C. Hence the lower the value of T_C, the more anomalous (compared to a conventional ferromagnet) are the spin dynamical properties.

One point that is worth noting concerns the (weak) quasielastic scattering found in ordinary ferromagnets as the ferromagnetic transition is approached. This component originates from the $S^z S^z$ (longitudinal) fluctuation spectrum rather than from the (transverse) scattering associated with the spin wave operators S^+, S^-. Generally this component is not observed in conventional unpolarized beam data, but can only be distinguished with polarized beam techniques.[20] In the present perovskite systems, however, the central component *dominates* the magnetic fluctuation spectrum. It was also found from our field-dependent inelastic scattering results that the quasielastic scattering below T_C in $La_{0.67}Ca_{0.33}MnO_3$ has both transverse and longitudinal character. Thus the central component observed in the manganites is not related to the ordinary longitudinal component in a ferromagnet.

Paramagnetic Spin Correlations

At very high temperatures the spins in a material are uncorrelated, but they are expected to develop correlations as the transition temperature is approached. The scattering typically follows the Ornstein-Zernike form

$$I \propto \frac{1}{\kappa^2 + q^2} \quad (2)$$

where $\xi = 1/\kappa$ is the correlation range in real space. For a ferromagnetic system these correlations can be measured by small angle neutron scattering, and the first experiments carried out on the CMR material $La_{0.67}Ca_{0.33}MnO_3$[7] showed that the scattering obeyed Eq. (2) to a good approximation as expected. Surprisingly, however, the correlation range was only weakly temperature dependent, and remained quite limited in range ($\sim 12 \text{Å}$) rather than diverging at T_C. These correlations are also affected by a magnetic field, and appear to be related to the (spin) polarons in the system.[21,8]

The (overall) size of the magnetic correlation range also appears to be related to the other anomalous properties of the spin dynamics. In particular, the anomalously small and weakly temperature-dependent correlation range is observed in the lower T_C systems, which also exhibit the strong quasielastic component of the fluctuation spectrum below T_C, and where the spin wave excitation spectrum does not collapse at T_C. The correlation range in the higher T_C systems, on the other hand, rapidly increases as the Curie temperature is approached, and appears to exhibit more conventional behavior.

Bilayer Manganite

A related material that has been investigated with neutrons more recently is the $(La_{1-x}Sr_x)_xMn_2O_7$ Ruddlesdon-Popper phase.[22] This material consists of two layers of the perovskite (bilayer) that are separated by La(Sr)-O layers which tend to magnetically isolate the Mn-O bilayers. This lowers the magnetic ordering temperature, and allows the bilayers to acquire a two-dimensional character. In the ferromagnetic regime, the composition $La_{1.2}Sr_{1.8}Mn_2O_7$ has been studied in some detail, but the studies are at a less advanced stage than the cubic perovskites. The bilayers have a basic ferromagnetic behavior,[23,24] but a substantial portion of the magnetic fluctuation spectrum was identified as long-lived antiferromagnetic clusters.[23] However, at the same nominal composition quantitative studies show that these antiferromagnetic fluctuations are only 2% of the ferromagnetic ones.[24] It appears unlikely then that the antiferromagnetic fluctuations play an important role in the underlying physics of these materials. We also have very recently observed a diffuse component of the scattering in both x-ray and neutron measurements, which may be related to polarons in the system. Further work is ongoing.

PYROCHLORE

Recently a new CMR compound has been discovered, namely the pyrochlore $Tl_2Mn_2O_7$[25,26], and an important question concerns whether this new class of CMR materials contains the same underlying physics, or represents a completely new and different CMR mechanism. We have investigated the magnetic correlations, phase transition (T_C of 123K), and long wavelength spin dynamics using neutron scattering techniques. The pyrochlore system also behaves as an ideal isotropic ferromagnet at low T, but the anomalous quasielastic component is not apparent in the spectrum and the transition is driven by the usual thermal population of spin waves.[27] This suggests that the mechanism for exchange and the magnetoresistance effects in the pyrochlore is different than for the perovskites.

ACKNOWLEDGMENTS

It is a pleasure to thank my colleagues who have worked with me on various aspects of these problems; L. Vasiliu-Doloc, M. A. Subramanian, S. Skanthakumar, S. K. Sinha, D. A. Shulyatev, O. Seeck, A. Santoro, S. Rosenkranz, A. Revcolevschi, J. L. Peng, R.

Osborn, Y. M. Mukovskii, A. H. Moudden, J. F. Mitchell, J. Mesot, Z. Y. Li, L. A. Kurnevitch, Q. Huang, R. L. Greene, K. E. Gray, K. Ghosh, R. W. Erwin, A. M. de Leon-Guevara, G. L. Bychkov, J. A. Borchers, S. N. Barilo, S. D. Bader, A. A. Arsenov, and D. N. Argyriou. Research at the University of Maryland is supported by the NSF, DMR 97-01339 and NSF-MRSEC, DMR 96-32521.

References

1. S. Jin, T. H. Tiefel, M. McCormack, R. A. Fastnacht, R. Ramesh, and L. H. Chen, Science **264**:413 (1994); S. Jin, M. McCormack, T. H. Tiefel and R. Ramesh, J. Appl. Phys. **76**:6929 (1994).
2. For references to some of the earlier literature see, for example, G. H. Jonker and J. H. Van Santen, Physica **16**:337 (1950); *ibid* **19**:120 (1950); E. O. Wollan and W. C. Koehler, Phys. Rev. **100**:545 (1955); C. Zener, Phys. Rev. **81**:440 (1951), *ibid* **82**:403 (1951); J. B. Goodenough, Phys. Rev. **100**:564 (1955); P. W. Anderson and H. Hasegawa, Phys. Rev. **100**:675 (1955).
3. A.J. Millis, P.B. Littlewood, and B.I. Shraiman, Phys. Rev. Lett. **74**:5144 (1995); A.J. Millis, Phys. Rev. B **55**:6405 (1997).
4. Q. Huang, A. Santoro, J. W. Lynn, R. W. Erwin, J. A. Borchers, J. L. Peng, and R. L. Greene, Phys. Rev. B**55**:14987 (1997).
5. S.-W. Cheong, C. M. Lopez, and H. Y Hwang (preprint).
6. Q. Huang, A. Santoro, J. W. Lynn, R. W. Erwin, J. A. Borchers, J. L. Peng, K. Ghosh, and R. L. Greene, Phys. Rev. B**58** (in press).
7. J. W. Lynn, R. W. Erwin, J. A. Borchers, Q. Huang, A. Santoro, J. L. Peng, and Z. Y. Li, Phys. Rev. Lett. **76**:4046 (1996).
8. J. W. Lynn, R. W. Erwin, J. A. Borchers, Q. Huang, A. Santoro, J. L. Peng, and R. L. Greene, J. Appl. Phys. **81**:5488 (1997).
9. L. Vasiliu-Doloc, J. W. Lynn, A. H. Moudden, A. M. de Leon-Guevara, and A. Revcolevschi, J. Appl. Phys. **81**:5491 (1997).
10. L. Vasiliu-Doloc, J. W. Lynn, A. H. Moudden, A. M. de Leon-Guevara, and A. Revcolevschi (preprint).
11. L. Vasiliu-Doloc, J. W. Lynn, Y. M. Mukovskii, A. A. Arsenov, and D. A. Shulyatev, J. Appl. Phys. **83**:7342 (1998).
12. T. G. Perring, G. Aeppli, S. M. Hayden, S. A. Carter, J. P. Remeika, and S-W. Cheong, Phys. Rev. Lett. **77**:711 (1996).
13. H. Y. Hwang, P. Dai, S-W. Cheong, G. Aeppli, D. A. Tennant, and H. A. Mook, Phys. Rev. Lett. **80**:1316 (1998).
14. J. A. Fernandez-Baca, P. Dai, H. Y. Hwang, S-W. Cheong, and C. Kloc, Phys. Rev. Lett. **80**:4012 (1998)
15. M. C. Martin, G. Shirane, Y. Endoh, K. Hirota, Y. Moritomo, and Y. Tokura, Phys. Rev. B**53**:14285 (1996).
16. J. J. Rhyne, H. Kaiser, H. Luo, G. Xiao, and M. L. Gardel, J. Appl. Phys. **83**:7339 (1998).
17. J. W. Lynn, L. Vasiliu-Doloc, K. Ghosh, S. Skanthakumar, S. N. Barilo, G. L. Bychkov, and L. A. Kurnevitch (preprint).
18. For a review of the data at low doping, see A. H. Moudden, L. Vasiliu-Doloc, L. Pinsard, and A. Revcolevschi, Physica B **241-243**:276 (1998).
19. In some of the early work a substantial gap was reported,[12,15] but this was a result of an extrapolation of the spin wave dispersion relation from large q.
20. For a recent discussion of this point see J. W. Lynn, N. Rosov, M. Acet and H. Bach, J. Appl. Phys. **75**:6069 (1994).

21. J. M. De Teresa, M. R. Ibarra, P. A. Algarabel, C. Ritter, C. Marquina, J. Blasco, J. Garcia, A. del Moral, and Z. Arnold, Nature **386**:256 (1997).
22. Y. Moritomo, A. Asamitsu, H. Kuwahara, and Y. Tokura, Nature **380**:141 (1996). J. F. Mitchell, D. N. Argyriou, J. D. Jorgensen, D. G. Hinks, C. D. Potter, and S. D. Bader, Phys. Rev. **B55**:63 (1997).
23. T. G. Perring, G. Aeppli, Y. Moritomo, and Y. Tokura, Phys. Rev. Lett. **78**:3197 (1997).
24. S. Rosenkranz, R. Osborn, J. F. Mitchell, L. Vasiliu-Doloc, J. W. Lynn, S. K. Sinha, and D. N. Argyriou, J. Appl. Phys. **83**:7348 (1998). R. Osborn, S. Rosenkranz, D. N. Argyriou, L. Vasiliu-Doloc, J. W. Lynn, S. K. Sinha, J. F. Mitchell, K. E. Gray, and S. D. Bader (preprint).
25. Y. Shimakawa, Y. Kubo, and T. Manako, Nature **379**:53 (1996).
26. For a review of the pyrochlores, see M. A. Subramanian and A. W. Sleight, Chapter 107 in *Handbook on the Physics and Chemistry of Rare Earths*, Vol. 16, Ed. by K. A. Geschneider, Jr. and L. Eyring (Elsevier, New York 1993). See also A. P. Ramirez and M. A. Subramanian, Science **277**:546 (1997).
27. J. W. Lynn, L. Vasiliu-Doloc, and M. A. Subramanian, Phys. Rev. Lett. **80**:4582 (1998).

METAL-INSULATOR PHENOMENA RELEVANT TO CHARGE/ORBITAL ORDERING IN PEROVSKITE MANGANITES

Y. Tomioka[1], A. Asamitsu[1], H. Kuwahara[1*] and Y. Tokura[2]

[1]Joint Research Center for Atom Technology (JRCAT), 1-1-4 Higashi, Tsukuba 305-8562, Japan
[2]Dept. of Applied Physics, University of Tokyo, 7-3-1 Hongo Bunkyo-ku, Tokyo 113-8656, Japan

I. INTRODUCTION

In the reduced bandwidth systems, such as $3d$ transition metal oxides, correlation among charge carriers plays an important role for their magnetic and electronic properties. The charge/orbital ordering is a representative phenomenon releavant to the long range Coulomb interaction among charge carriers. Such examples are frequently seen in $3d$ transition metal oxides, for instance, in Fe_3O_4 [1], $La_{1-x}Sr_xNiO_4$ ($x = 1/3$, $1/2$) [2], $La_{1-x}Sr_xFeO_3$ ($x = 2/3$) [3,4], $(La_{1-y}Nd_y)_{2-x}Sr_xCuO_4$ ($x = 1/8$) [5] and so forth. We may further pick up examples of the orbital ordering in RVO_3 (R = La, Y) [6] and $RNiO_3$ (R = Pr, Nd, Sm,...) [7,8]. In the case of perovskite-type manganese oxides, as recent extensive studies have revealed, the close interplay among spin, charge and orbital (crystal lattice) is quite important to understand many unconventional physical properties of these oxides. The perovskite manganites are in general formulated as $RE_{1-x}AE_xMnO_3$, where RE and AE are a trivalent rare earth and an alkaline earth element, respectively. In $REMnO_3$ ($x = 0$), the electron configuration of ($3d^4$; $t_{2g}^3 e_g^1$) is realized for the Mn^{3+} site, and due to ordering of e_g-orbital (or cooperative Jahn-Teller distortion [9-12]) the substance is a layered (or A-type) antiferromagnetic insulator. In $RE_{1-x}AE_xMnO_3$, Mn^{4+} ($3d^3$; $t_{2g}^3 e_g^0$) is created by the cation substitution and holes are introduced in the e_g-band. In $La_{1-x}Sr_xMnO_3$ and $La_{1-x}Ca_xMnO_3$, which have long been known as conducting ferromagnets [13-15], the mobile e_g-electrons mediate the ferromagnetic interaction between the neighboring Mn^{3+} and Mn^{4+} sites, which is the so-called double exchange interaction [16-18].

In $RE_{1-x}AE_xMnO_3$, the substitution of RE^{3+} with AE^{2+} at the perovskite A-site controls the mean Mn valence, while the average radius of the cations at the A-site, $r_A = (1-x)r_{RE3+} + x\, r_{AE2+}$, affects a distortion of the perovskite lattice. The transfer integral t of an e_g-electron between the neighboring Mn sites is mediated by the oxygen $2p$ state and hence dependent on the degree of the hybridization between the Mn $3d$-orbital and O $2p$-orbital. Therefore, t is sensitive to the lattice distortion or the tilting of MnO$_6$ octahedra ($t \approx |\cos\phi|^2$ where ϕ is the Mn-O-Mn bond angle) [19]. As the tolerance factor, $\Gamma = (r_A+r_O)/\sqrt{2}(r_B+r_O)$ (r_A, r_B and r_O being ionic radii of the cations at perovskite A-site, B-site and oxygen in ABO_3, respectively), or the r_A reduces, a lattice distortion increases. In perovskite manganites, $RE_{1-x}AE_xMnO_3$, the smaller the r_A the larger the tilting of MnO$_6$ octahedra, which causes narrowing of the effective one-electron bandwidth W of the e_g-band [19]. Such a narrowing of W tends to destabilize the double exchange mediated ferromagnetic state due to the presence of competing instabilities, such as antiferromagnetic superexchange interaction between the localized t_{2g} spins, charge/orbital ordering, and Jahn-Teller type electron-lattice coupling [20-22]. In particular, a charge/orbital ordering, namely, an ordering of Mn^{3+}/Mn^{4+} with 1/1 accompanied by a simultaneous ordering of e_g-orbital of Mn^{3+} occasionally occurs in $RE_{1-x}AE_xMnO_3$ at $x \approx 1/2$ as the average radius of the cations at the perovskite A-site reduces and accordingly W decreases. Such a charge/orbital ordering of La$_{1-x}$Ca$_x$MnO$_3$ ($x \approx 1/2$) has also been well known [23], as an earlier neutron diffraction study reported [24]. In Sec. II, the prototypical charge/orbital ordering transition and associated change of electronic properties in perovskite manganites are described by taking an example of the case of Nd$_{1/2}$Sr$_{1/2}$MnO$_3$ [25]. As the W is further reduced, such as in Pr$_{1-x}$Ca$_x$MnO$_3$, a charge/orbital-ordered state similar to that at $x = 1/2$ exists in much broader range of x ($0.3 \leq x < 0.75$) [26,27] (Sec. III).

The variation of the mean Mn valence and the effective one-electron bandwidth with the substitution at the perovskite A-site are indicated in Fig. 1 (a) in which the radii of various r_{RE3+} and r_{AE2+} [28] are plotted at the upper and lower abscissa, respectively. The substitution of La with Sr, for instance, corresponds to connecting the line from La to Sr. It is understood from Fig. 1 (a) that if x is fixed constant the W of $RE_{1-x}AE_xMnO_3$ increases with increase in r_A. Correspondingly, Fig. 1 (b) indicates critical temperatures of ferromagnetic transition (T_C) and charge/orbital ordering transition (T_{CO}) for various $RE_{1-x}AE_xMnO_3$ ($x = 1/2$) as a function of r_A. The ferromagnetic double exchange interaction is destablized while the charge/orbital ordering stabilized with decrease in W. For instance, T_C and T_{CO} of Nd$_{1/2}$Sr$_{1/2}$MnO$_3$ are 255 K and 160 K, respectively, while those of Sm$_{1/2}$Ca$_{1/2}$MnO$_3$ are 0 K (no ferromagnetic state) and 270 K.

In these oxides the coupling among spin, charge and lattice (orbital) is thus changed by the chemical composition on the A-site, which also affects the feature of the colossal magnetoresistance (CMR) phenomena. In the CMR phenomena, an application of an external magnetic field tends to align the local t_{2g} spins ferromagnetically, which increases the transfer integral of an e_g-electron due to the strong Hund's-rule coupling between the t_{2g} and e_g

electrons. Similarly to such CMR phenomena, an application of an external magnetic field by several Tesla in narrow-W systems energetically favors the ferromagnetic metallic state, and promotes a transition from the antiferromagnetic charge/orbital-ordered state accompanying the resistivity change by more than several orders of magnitude (Secs. II and III). This field-induced insulator-metal transition is of first order in nature due to the competition between the charge/orbital ordering and double exchange interaction. In such a first order transition, a supercooled or superheated metastable state is allowed for a restricted temperature region and the metastable state can change to a stable state by some stimuli, not only a magnetic field but also photo-excitation [29-32] and current injection [33]. The phase control in terms of the first order transition has recently been reported for $Pr_{1-x}Ca_xMnO_3$ with a metastable charge/orbital-ordered state [29-33].

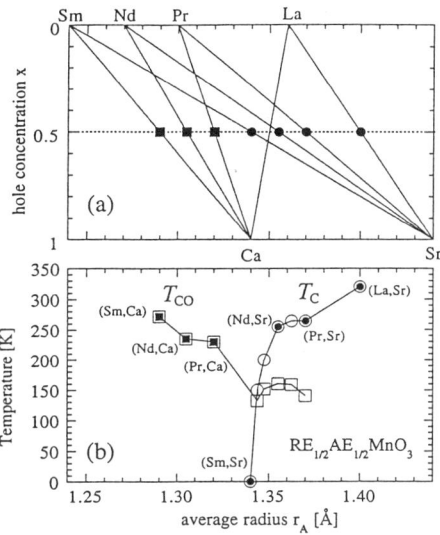

Figure 1. (a). Variations of the mean Mn valence (1-x; on the ordinate) and the effective one-electron bandwidth (on the abssisa) with the substitution at the perovskite A-site. The various r_{RE3+} and r_{AE2+} [28] are plotted at the upper and lower abssisa, respectively. The substitution of La with Sr, for instance, corresponds to the line connecting from La to Sr. (b). Critical temperatures of ferromagnetic transition T_C (circles) and charge/orbital ordering transition T_{CO} (squares) for various $RE_{1-x}AE_xMnO_3$ ($x = 1/2$) as a function of r_A.

II. CHARGE/ORBITAL ORDERING VS. DOUBLE EXCHANGE INTERACTION AT $x = 1/2$

II-1. Manifestation of the Charge/Orbital Ordering in Perovskite Manganites

In Figure 2 are shown typical examples of the charge/orbital ordering transitions observed for $Nd_{1-x}Sr_xMnO_3$ ($x = 0.5$) [25] and $Sm_{1-x}Ca_xMnO_3$ ($x = 0.5$). As shown in the temperature dependencies of resistivity and magnetization, the charge/orbital ordering transitions manifest themselves as decreases in magnetization and increases in resistivity, which locate at T_{CO} (= T_N) ≈ 160 K in $Nd_{1/2}Sr_{1/2}MnO_3$ (left) and T_{CO} ≈ 270K in $Sm_{1/2}Ca_{1/2}MnO_3$ (right), respectively. As another noteworthy aspect, the changes in lattice parameters are also observed at T_{CO}; elongation in the orthorhombic a and b axes and contraction in the c-axis are observed, which suggests that an ordering of e_g-orbital occurs simultaneously. Such changes in lattice parameters upon the charge/orbital ordering transition as shown in Fig. 2 has also been reported on polycrystalline $La_{1/2}Ca_{1/2}MnO_3$ (T_{CO} (= T_N) ≈ 160 K) [34,35]. In $Nd_{1/2}Sr_{1/2}MnO_3$, the ferromagnetic and metallic state due to the double exchange interaction is seen below T_C ≈ 255 K, and subsequently the transition to the antiferromagnetic charge/orbital-ordered state occurs at T_{CO} (= T_N) ≈ 160 K [25]. In $Sm_{1/2}Ca_{1/2}MnO_3$, similar changes in resistivity and magnetization are seen at T_{CO} ≈ 270K, where the charge/orbital ordering takes place. As is clear from the temperature dependence of magnetization, however, the antiferromagnetic spin-ordering occurs not concurrently but at a lower temperature, T_N ≈ 170 K and no ferromagnetic state is present at zero field. Namely, in $Sm_{1/2}Ca_{1/2}MnO_3$, the temperature of spin-ordering is lowered than that of charge/orbital ordering. In the reduced-bandwidth systems, such as $RE_{1/2}Ca_{1/2}MnO_3$ (RE = Pr, Nd, Sm,..), in which no ferromagnetic and metallic state is realized, a split of T_{CO} and T_N is generally observed. A similar feature is also observed for the related material with the layered-perovskite structure, $La_{1-x}Sr_{1+x}MnO_4$ (x = 1/2), where T_{CO} ≈ 217 K and T_N ≈ 110 K [36,37].

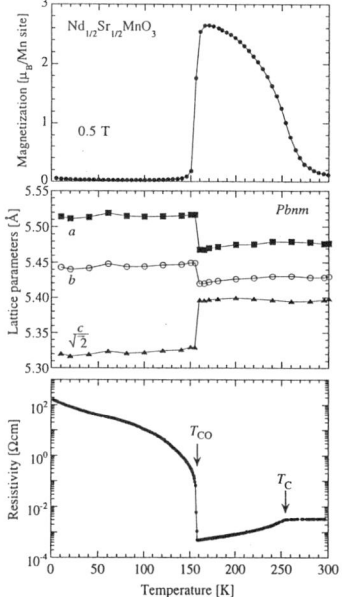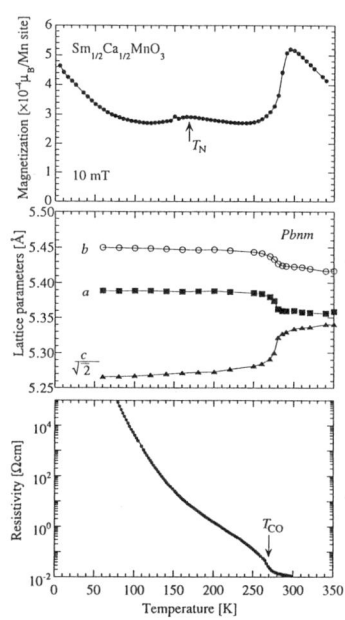

Figure 2. The temperature dependencies of magnetization (top), resistivity (middle) and lattice parameters (bottom) observed for $Nd_{1/2}Sr_{1/2}MnO_3$ [25] (left) and $Sm_{1/2}Ca_{1/2}MnO_3$ (right). In $Nd_{1/2}Sr_{1/2}MnO_3$ the critical temperature of the charge/orbital ordering transition coincides with the antiferromagnetic Neel temperature (T_{CO} = T_N ≈ 160 K), while in $Sm_{1/2}Ca_{1/2}MnO_3$ the former is higher than the latter (T_{CO} ≈ 270 K and T_N ≈ 170 K).

In $Nd_{1/2}Sr_{1/2}MnO_3$, an antiferromagnetic spin-ordering of the CE-type structure has been confirmed below T_{CO} by the neutron diffraction study [38]. Figure 3 shows a schematic picture of the CE-type spin-ordering [24] including the charge/orbital ordering. As a result of the antiferromagnetic CE-type ordering, the magnetic unit cell expands to $2 \times 2 \times 1$ of the original orthorhombic crystal lattice. As shown in this figure, the cations of Mn^{3+} and Mn^{4+} alternatively order within the orthorhombic ab-plane, and the same cations do along the c-axis. Furthermore, the $d3x^2-r^2$ and $d3y^2-r^2$ orbitals on Mn^{3+} sites show an ordering as depicted in Fig. 3, which causes the doubling of the orthorhombic unit cell along the b-axis. In $Sm_{1/2}Ca_{1/2}MnO_3$ the charge/orbital- and spin-ordering as shown in Fig. 3 are realized below T_N. The charge/orbital ordering at $x = 1/2$ is thus characterized by the antiferromagnetic CE-type structure which has originally been revealed for $La_{1-x}Ca_xMnO_3$ ($x \approx 1/2$) [24]. The ordering of e_g-orbital of Mn^{3+} sites related with the antiferromagnetic CE-type structure has been postulated by Goodenough [39].

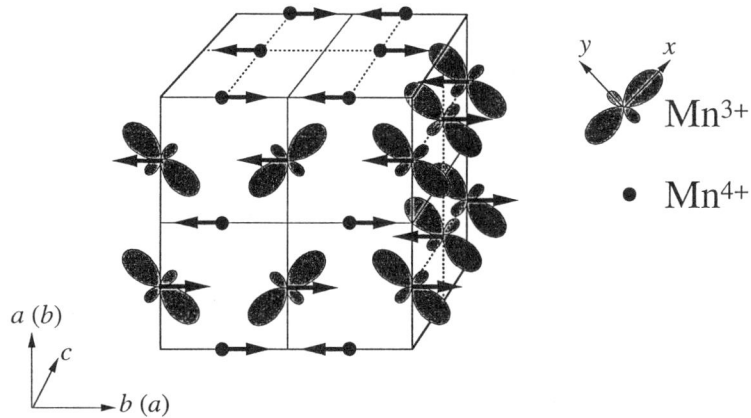

Figure 3. A schematic picture of antiferromagnetic CE-type structure [24] including the pattern of ordering of charge and e_g-orbital. The magnetic unit cell expands to $2 \times 2 \times 1$ of the original orthorhombic crystal lattice.

II-2. Magnetic Field induced Insulator-Metal Transition

In perovskite manganites, the very large magnetoresistance is frequently observed around the ferromagnetic Curie temperature, typically observed for $La_{1-x}Sr_xMnO_3$ ($x = 0.15 \sim 0.3$) [40,41] and $La_{1-x}Ca_xMnO_3$ [23]. Such a large MR is intuitively understood according to the double exchange model [16-18], although other important factors must be taken into account for the quantitative understanding. The transfer integral, t_{DE} of the e_g-state electron between the neighboring Mn^{3+} and Mn^{4+} increases in such a manner that $t_{DE} = t_0 \cos(\theta/2)$, where t_0 is the transfer integral in a fully spin-polarized state and θ the relative angle between their local t_{2g} spins of the neighboring Mn^{3+} and Mn^{4+}. Since an application of an external magnetic field aligns the local t_{2g} spins ferromagnetically, t_{DE} increases and hence the large MR is observed.

How is the effect of an external magnetic field on the charg/orbitale-ordered state? We show in Fig. 4 the magnetization and resistivity as a function of magnetic field for the $Nd_{1/2}Sr_{1/2}MnO_3$ crystal, which is taken at $T = 141K$ [25]. In the magnetization curve, we can see the metamagnetic transitions at around 2.2 T and 0.9 T in the field increasing and decreasing runs, respectively. In accord with the behavior in magnetization, the resistivity also shows steep decrease and increase in the respective runs. Namely, the transition from the antiferromagnetic charge/orbital-ordered insulating state to the ferromagnetic metallic one is caused by application of an external magnetic field. From the thermodynamic point of view, the both states are energetically almost degenerated, but the free energy of the ferromagnetic (FM) state decreases by the Zeeman energy $-M_sH$ (M_s; the spontaneous magnetization) so that the magnetic field induced transition takes place by applying an external magnetic field of a few Tesla. This relatively destabilizes the antiferromagnetic charge/orbital-ordered state and drives the phase transition to the FM metallic state.

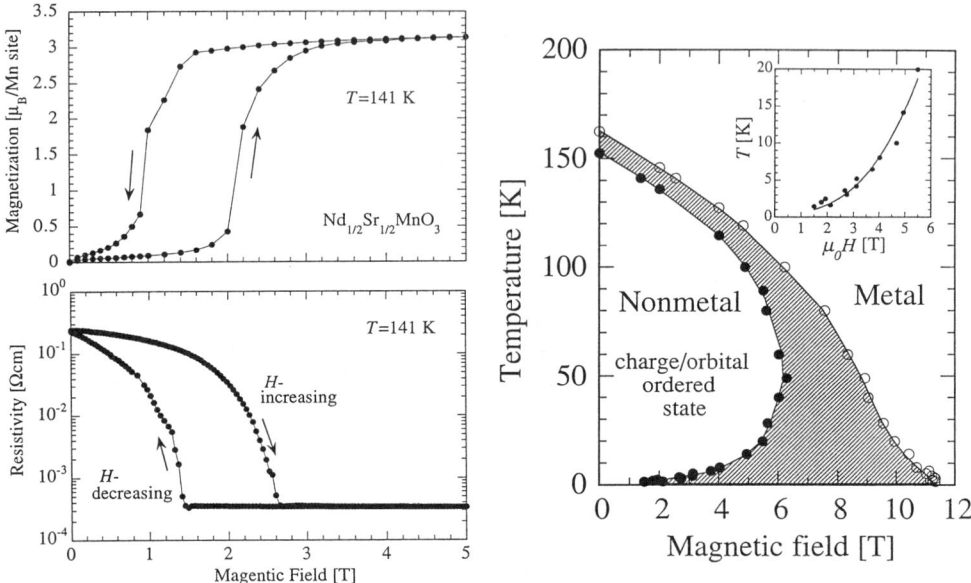

Figure 4. The magnetic field dependence of magnetization (upper) and resistivity (lower) at 141 K for a $Nd_{1/2}Sr_{1/2}MnO_3$ crystal [25].

Figure 5. The charge/orbital ordering phase diagram for a $Nd_{1/2}Sr_{1/2}MnO_3$ crystal, which is obtained by measurements of magnetic field dependence of resistivity [25]. Open and closed circles represent the phase boundaries from and to the charge/orbital-ordered state, respectively. The hatched area indicates the hysteresis. In the inset is shown the fitting of the critical fields from the metallic to the charge/orbital ordered state below 20 K with the theoretical formulas (see text) [25].

Another point to be noted in Fig. 4 is the large hysteresis upon the transition being characteristic of the first order transition coupled with the change of a crystal lattice. The change in lattice parameters originates from the field-destruction of charge/orbital ordering. We show in Fig. 5 the phase diagram obtained by the field dependence of resistivity at several temperatures for a $Nd_{1/2}Sr_{1/2}MnO_3$ crystal [25]. In Fig. 5, the stability of the AF charge/orbital-ordered state against magnetic field is well indicated. A similar phase diagram have been reported on the AF charge/orbital-ordered state of a polycrystalline $La_{1/2}Ca_{1/2}MnO_3$ sample [42] of which the one-electron bandwidth is comparable to that of $Nd_{1/2}Sr_{1/2}MnO_3$. It is noted in Fig. 5 that the hysteresis region (hatched area) expands with decreasing temperature especially below about 20 K. In the case of the first order phase transition, the transition from the metastable to the stable state occurs by overcoming a potential barrier. At a finite temperature, the thermal energy can assist to overcome the barrier to form the nucleus for the transition. Since the thermal energy reduces with decrease in temperature, a larger (smaller) field is needed to induce the transition from (to) the AF charge/orbital-ordered to (from) the FM metallic state. Thus the hysteresis between the field increasing and decreasing runs increases with decrease in temperature. The inset of Fig. 5 shows a magnification of the transition lines in a low temperature region, where empirical relation that $H_c - H_{c0} = k\,T^\alpha$ with $1/4 < \alpha < 1/3$ (H_{c0} being the magnetic field at which the potential barrier vanishes) appears to hold.

If the average radius of the perovskite-A site, r_A reduces, namely, the one-electron bandwidth W reduces, the charge/orbital ordering phase should relatively be stabilized as a result of that the double exchange interaction is weakened. In Fig. 1, the critical temperature of ferromagnetic transition decreases from $\approx 255K$ for $Nd_{1/2}Sr_{1/2}MnO_3$ to 0 K for $Sm_{1/2}Sr_{1/2}MnO_3$. In $(Nd_{1-y}Sm_y)_{1/2}Sr_{1/2}MnO_3$, the reduction in T_C accompanied by an appearance of an antiferromagnetic spin correlations above T_C is pronounced with increase in y, which implies that the charge/orbital ordering instability is enhanced and comparable to the ferromagnetic double exchange interaction due to narrowing in W [43-45]. For these cases with controlled-W, low-field CMR phenomena accompanied by a large change in striction [46] are observed as a result of the destruction of the antiferromagnetic correlation by an external magnetic field [47,48]. As the r_A (or W) further reduces, the charge/orbital ordering dominates the ferromagnetic double exchange interaction, as shown in Fig. 1.

Figure 6 shows the charge/orbital ordering phase diagrams for various $RE_{1/2}AE_{1/2}MnO_3$ crystals which are presented on the magnetic field-temperature (H - T) plane. The phase boundaries have been determined by the measurements of the magnetic field dependence of resistivity (ρ - H) and magnetization (M - H) at fixed temperatures, and those for $R_{1/2}Ca_{1/2}MnO_3$ (R = Pr, Nd and Sm) have been obtained by measurements utilizing pulsed high magnetic fields up to 40 T [49]. In this figure, the critical field to destruct the charge/orbital-ordered state in $Nd_{1/2}Sr_{1/2}MnO_3$ is about 11 T at 4.2 K, while that in $Pr_{1/2}Ca_{1/2}MnO_3$ increases to about 27 T. In the case of $Sm_{1/2}Ca_{1/2}MnO_3$, the charge/orbital ordering is so strong that the critical field becomes as large as about 50 T at 4.2 K. Fig. 6 thus demonstrates that the robustness of the charge/orbital ordering at $x = 1/2$ changes depending on the W, which is understood as a competition between the double exchange interaction and an ordering of Mn^{3+}/Mn^{4+} with 1/1 accompanied by a simultaneous ordering of e_g-orbital of Mn^{3+}.

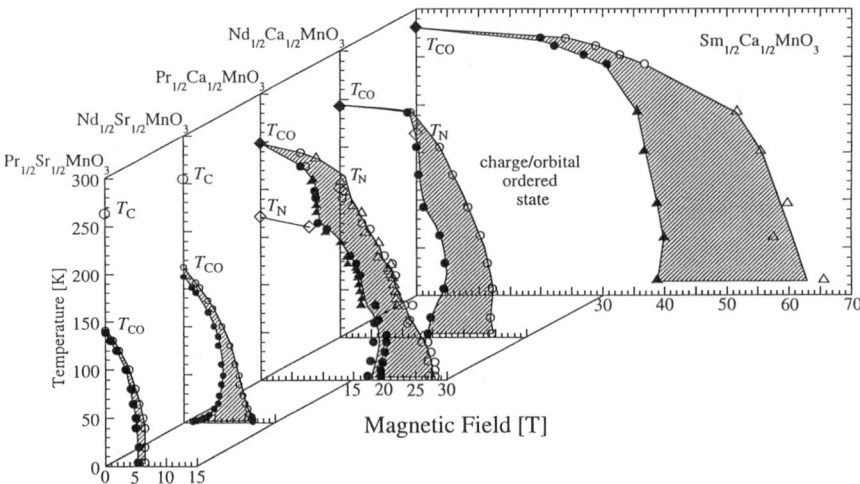

Figure 6. The charge/orbital-ordered phase of various $RE_{1/2}AE_{1/2}MnO_3$ plotted on the magnetic field-temperature plane. The phase boundaries have been determined by measurements of the magnetic field dependence of resistivity (ρ - H) and magnetization (M - H) at fixed temperatures, and those for $R_{1/2}Ca_{1/2}MnO_3$ (R = Pr, Nd and Sm) have been obtained by measurements utilizing pulsed high magnetic fields up to 40 T [49].

III. EFFECT OF DISCOMMENSURATION OF CARRIER CONCENTRATION ON CHARGE/ORBITAL ORDERING (THE CASE STUDY ON $Pr_{1-x}Ca_xMnO_3$)

III-1. Variation of Charge/Orbital Ordering with x

The charge/orbital-ordering phenomena should tend to be most stabilized when the carrier concentration coincides with a rational number for the periodicity of the crystal lattice. In fact, the charge/orbital-ordered state in $Nd_{1-x}Sr_xMnO_3$, as shown in Fig. 3, emerges only around $x = 0.5$, and disappears for $x < 0.48$ and $x > 0.52$ [50]. In perovskite manganites of which W is much further reduced, such as $Pr_{1-x}Ca_xMnO_3$, however, the charge/orbital ordering of the similar CE type appears in a much broader range of x [26,27]. We show in Fig. 7 the electronic phase diagram of $Pr_{1-x}Ca_xMnO_3$ ($0 \leq x \leq 0.5$) [51]. In Fig. 7, the ferromagnetic and metallic state is not realized under zero magnetic field and at ambient pressure due to the reduced W, while the ferromagnetic but insulating phase appears for $0.15 < x < 0.3$. With further increase in x, the charge/orbital-ordered state with an 1:1 ordering of Mn^{3+}/Mn^{4+} appears for $x \geq 0.3$. As an earlier neutron diffraction study [27] reported, the charge/orbital ordering exists in a broad range of x ($0.3 \leq x < 0.75$), where the pattern of spin, charge and orbital ordering is basically described by that of $x = 0.5$, namely the AF CE type structure shown in Fig. 3. The coupling of spins along the c-axis is antiferromagnetic at $x = 0.5$, while it is not antiferromagnetic but caning for $x < 0.5$. Such a *pseudo*-CE-type AF structure at $x = 0.4$ is schematically shown in Fig. 8 [27]. As the carrier concentration deviates from the commensurate value of 0.5, the extra electrons are doped in the Mn^{4+} sites in a naive sense. To explain the modification of the arrangement of spins along the c-axis, Jirak et al. [27] postulated that the extra electrons hop along the c-axis mediating the ferromagnetic double exchange interaction. (This is analogous to the lightly hole-doped

LaMnO$_3$, where the canted antiferromagnetic spin structure is realized due to the hole motion along the c-axis [18].) Such an effect of extra electrons on the magnetic structure is expected to be enhanced with decrease in x from 0.5. In fact, neutron diffraction studies [27,52] have revealed that the coupling of spins along the c-axis becomes almost ferromagnetic at $x = 0.3$ in spite of that the CE-type ordering is maintained within the orthorhombic ab-plane. The observed modification of the magnetic structure is quite important to understand the metal-insulator phenomena induced by an external magnetic field in this substance, which is described in following sections.

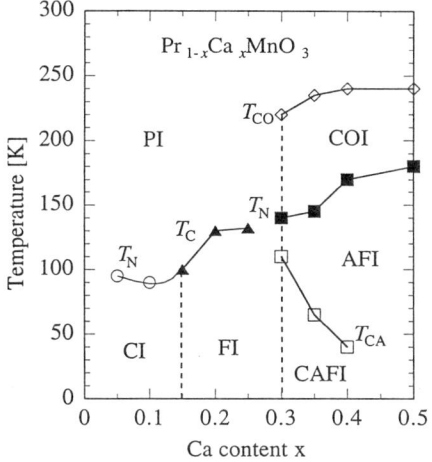
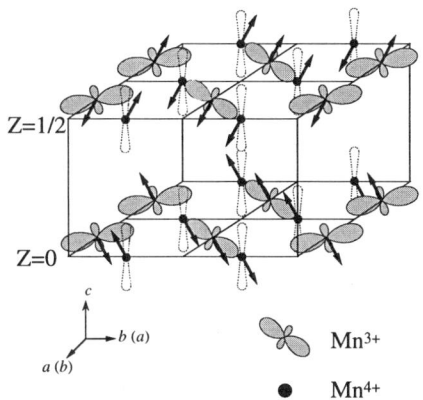

Figure 7. The magnetic as well as electronic phase diagram of Pr$_{1-x}$Ca$_x$MnO$_3$ [51]. The PI and FI denote the the paramagnetic insulating and ferromagnetic insulating states, respectively. For $0.3 \leq x \leq 0.5$, the antiferromagnetic insulating (AFI) state exists in the charge/orbital-ordered insulating (COI) phase. The canted antiferromagnetic insulating (CAFI) state also shows up below the AFI state in the COI phase for $0.3 \leq x \leq 0.4$.

Figure 8. A schematic picture of the antiferromagnetic *pseudo*-CE-type structure in the case of Pr$_{1-x}$Ca$_x$MnO$_3$ ($x = 0.4$) [27]. The coupling of spins along the c-axis becomes canted while the CE-type ordering is kept within the orthorhombic ab-plane.

III-2. Magnetic Field Induced Insulator-Metal Transition

Figure 9 shows the temperature profile of resistivity under several magnetic fields for the Pr$_{1-x}$Ca$_x$MnO$_3$ ($x = 0.3$) crystal [53]. Although the crystal of $x = 0.3$ undergoes the charge/orbital ordering transition at $T_{CO} \approx 200$K [52], it changes to the metallic state under a magnetic field. The resistivity under zero field shows an insulating behavior with $d\rho/dT < 0$, while the other curves under fields (> 2 T) show a change from an insulating to metallic behavior below some critical temperatures. The critical temperature, which is tentatively defined as that with maximum resistivity, increases as the magnetic field is intensified. As exemplified in the data under 7 T, the hysteresis is observed between the cooling and warming runs. The obtained phase diagram is shown in the inset.

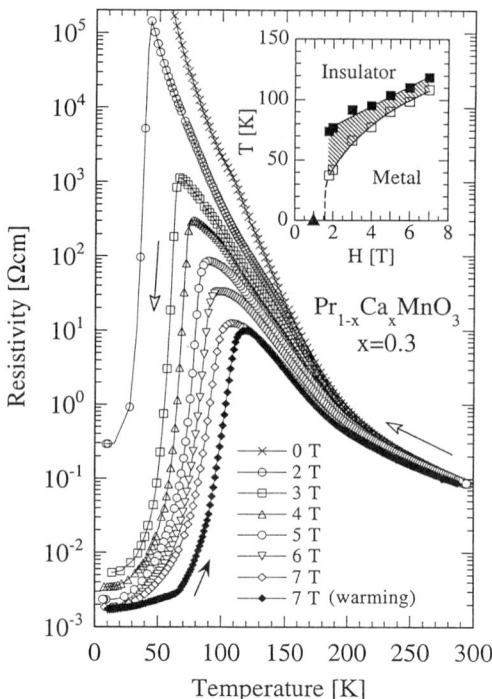

Figure 9. The temperature profile of resistivity for a $Pr_{1-x}Ca_xMnO_3$ ($x = 0.3$) crystal under several magnetic fields [53]. Inset shows the phase diagram determined by resistivity measurements in field-cooling (FC; open squares) and field-cooled-warming (FCW; closed ones) runs. The hatched area indicates the hysteretic region.

Figure 10 shows the magnetic field dependence of resistivity and magnetization taken at 4.2 K and 5 K, respectively [52]. In the magnetization curve of Fig. 10, the transition from the canted AF to the fully spin-polarized (ferromagnetic) state is observed at about 3.5 T in the field increasing process. In accord with the behavior in magnetization, the resistivity shows an insulator to metal transition at $\approx$ 4 T. Because of the first order nature, the transition is irreversible at such a low temperature and the ferromagnetic and metallic state remains even after the magnetic field is removed [53].

The observed phenomenon was further investigated by the neutron diffraction measurement [52,54]. Figure 11 shows the magnetic field dependence of the intensities of the ferromagnetic (top), antiferromagnetic (middle) and orbital superlattice (bottom) reflections which were all taken at 5 K. Corresponding to decreases in the intensities of antiferromagnetic and orbital-superlattice reflections at 4 T, an increase is observed for the ferromagnetic reflection. Hence, the observations shown in Figs. 9 ~ 11 indicate that the field induced transition is accompanied by the collapse of the charge/orbital-ordered state. For the $Pr_{1-x}Ca_xMnO_3$ ($x = 0.3$) crystal, a similar insulator to metal transition is also induced by applying an external pressure [55-57] which increases one-electron bandwidth W (or transfer integral t). Thus, an insulator to metal transition accompanied by collapse of charge/orbital-ordered state is due to an increase in W, which is, in the case of double exchange systems, achieved both by either an external magnetic field (see, II-2) or pressure.

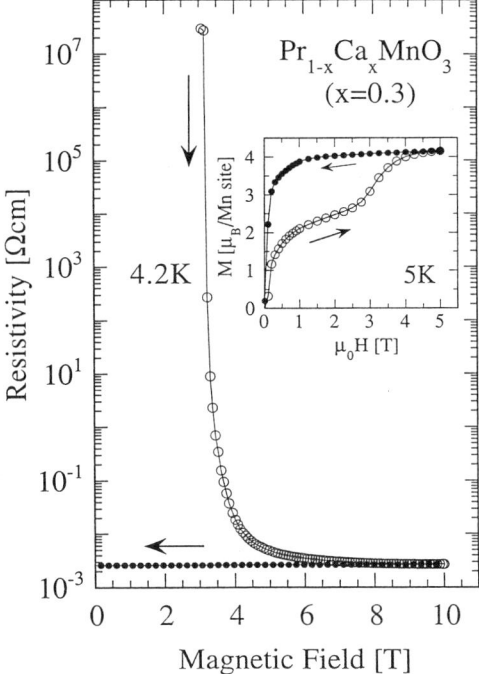

Figure 10. The resistivity at 4.2 K and magnetization at 5 K (inset) as a function of magnetic field for a $Pr_{1-x}Ca_xMnO_3$ ($x = 0.3$) crystal [52]. Both measurements were performed after the sample was cooled under zero field to a prescribed temperature.

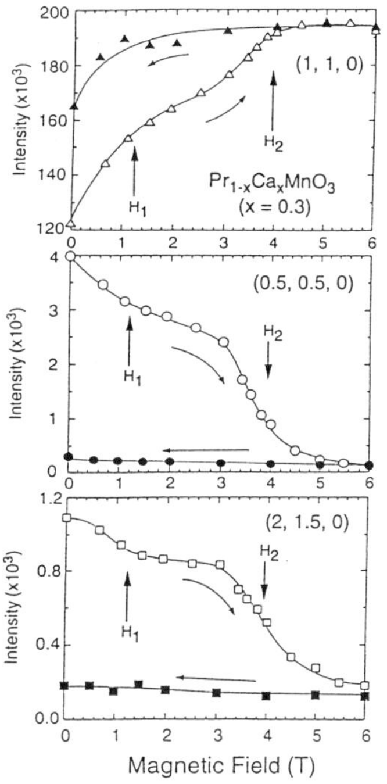

Figure 11. The magnetic field dependence of the intensities of the ferromagnetic (top), antiferromagnetic (middle) and orbital superlattice (bottom) reflections obtained by nuetron diffraction measurements at 5 K [52]. The superlattice reflection arises from the concomitant charge/orbital ordering.

III-3. Variation of the *T-H* Phase Diagram with *x*

In $Pr_{1-x}Ca_xMnO_3$, the dependence of the field induced transition (or collapse of the charge/orbital-ordered state under a magnetic field) on the carrier concentration has systematically been investigated for $0.3 \leq x \leq 0.5$ [51]. Figure 12 shows the variation of temperature profiles of resistivity under $m_0H = 0$, 6 and 12 T with *x*. In the case of $x = 0.5$, the temperature of the charge/orbital ordering transition, which is manifested by a steep increase in resistivity, is lowered with increase in magnetic field. However, the high resistivity due to the charge/orbital ordering recovers below ≈ 160 K even under 12 T. In the case of $x = 0.4$, the resistivity under 6 T shows a sudden decrease at ≈ 60 K and that under 12 T shows no manifestation of charge/orbital ordering but becomes metallic over the whole temperature region. In the case of $x = 0.35$, the resistivity under 6 T rather shows an insulator to metal transition at ≈ 80 K and that under 12 T is quite similar to the comparative data of $x = 0.4$. Namely, Fig. 12 indicates that the robustness of the charge/orbital ordering changes depending on the carrier concentration or doping level.

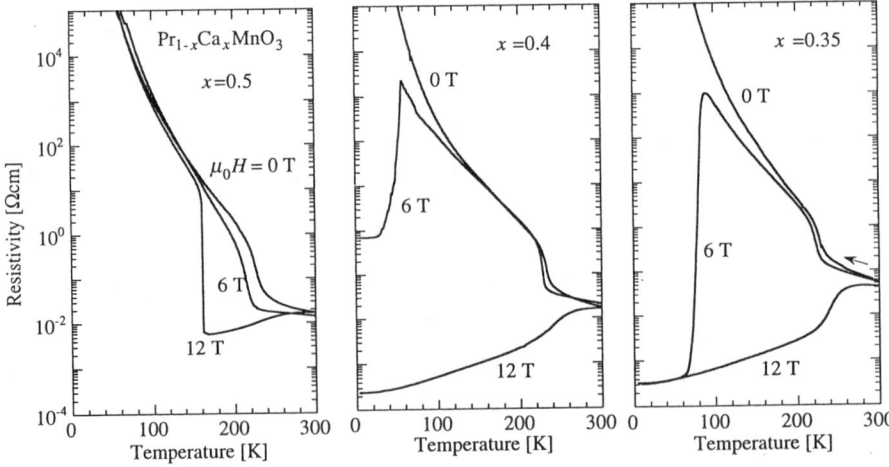

Figure 12. The temperature dependence of resistivity under $\mu_0 H = 0$, 6 and 12 T for $Pr_{1-x}Ca_xMnO_3$ ($x = 0.5$, 0.4 and 0.35) crystals [51]. The resistivity was measured in a field cooling (FC) run.

Such a feature is well demonstrated by the charge/orbital ordering phase diagram plotted on the magnetic field-temperature (H - T) plane as shown in Fig. 13 [52]. In this figure, the phase boundaries have been determined by the measurements of the magnetic field dependence of resistivity (ρ - H) and magnetization (M - H) at fixed temperatures. The charge/orbital ordering for $x = 0.5$ is so strong that the critical field to destruct the charge/orbital-ordered state becomes as large as about 27 T at 4.2 K (Fig. 6), and a similar feature is also seen for $x = 0.45$ [49]. For $x \leq 0.4$, on the other hand, the shrinkage of the charge/orbital-ordered phase-region becomes remarkable. The averaged value (H_{av}) of the critical fields in the H-increasing and -decreasing runs at a constant temperature rather shows a decrease with decrease in temperature below ≈ 175 K ($dH_{av}/dT < 0$). In the case of $x = 0.3$, a collapse of a charge/orbital-ordered state (*i.e.*, an appearance of a FM metallic state) is realized by applying an external magnetic field of only several Tesla when the temperature is set below ≈ 50 K. Another noteworthy aspect in Fig. 13 is the expansion of the field-hysteresis with decreasing temperature, which is characteristic of a first order phase transition as described in Sec. II-2 [25]. Such a variation of the charge/orbital-ordered phase with the carrier concentration has similarly been observed for a further W-reduced system, $Nd_{1-x}Ca_xMnO_3$ [49,58]. The common feature for the modification of the phase diagram with x seems to be correlated with the action of the extra electron-type carriers in the CE-type structure [27], which was already discussed in III-1. Thermodynamically, excess entropy may be brought into the charge/orbital-ordered state by the extra localized carriers and their related orbital degrees of freedom, which are pronounced as x deviates from 0.5. The excess entropy may reduce the stability of the charge-ordered state with decreasing temperature and cause the reduction in the critical magnetic field ($dH_{av}/dT < 0$) as observed in the case of $x \leq$

0.4 in Fig. 13.

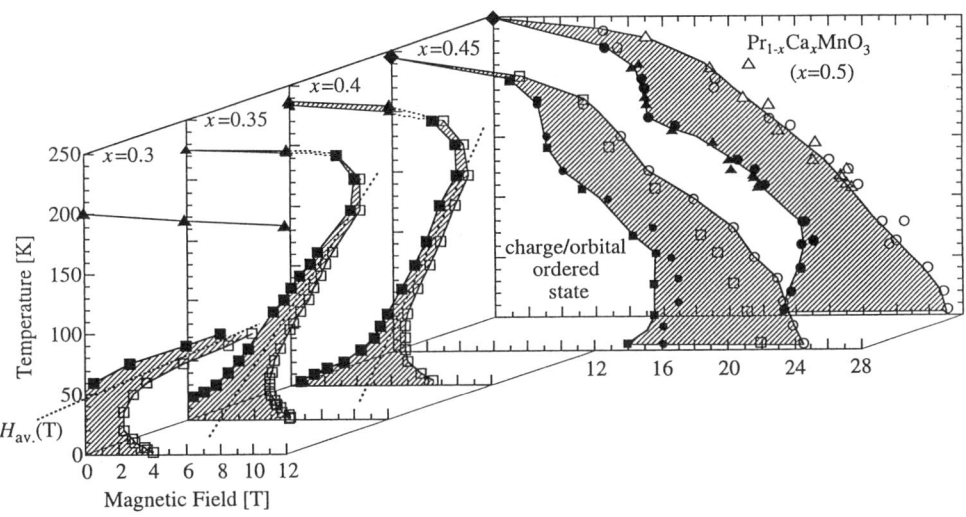

Figure 13. The charge/orbital-ordered state of $Pr_{1-x}Ca_xMnO_3$ ($x = 0.5, 0.45, 0.4, 0.35$ and 0.3), which is plotted on the magnetic field-temperature plane [51]. The phase boundaries for $x = 0.5$ and 0.45 have been determined by means of pulsed high magnetic fields up to 40 T [49]. The hatched area indicates the hysteresis region. In the cases of $x = 0.3, 0.35$ and 0.4, the lines of $H_{av}(T)$ at which the free energies of both the charge/orbital-ordered and the FM metallic states are supposed to be equal are schematically indicated by dashed lines.

III-4. Magnetic Field Induced Insulator-Metal Transition as Investigated by Optical Spectroscopy

To investigate the change in the electronic structure accompanied with the magnetic field induced insulator-metal transition, measurements of optical spectra are useful. We again show in Figure 14 the phase diagram of $Pr_{1-x}Ca_xMnO_3$ ($x = 0.4$) in the temperature and magnetic-field plane. The hatched area indicates a field-hysteresis. As shown in Fig. 14, the $Pr_{1-x}Ca_xMnO_3$ ($x = 0.4$) undergoes the charge ordering transition at $T_{CO} \sim 235$ K under zero magnetic field and an antiferromagnetic transition at $T_N \sim 170$ K. The patterns of spin, charge and orbital ordering below T_N are schematically shown in Fig. 8.

Figure 15 shows the optical conductivity spectrum ($\sigma(\omega)$) of $Pr_{1-x}Ca_xMnO_3$ ($x = 0.4$) at 293 K and 10 K [59]. At 293 K ($> T_{CO}$), almost the same spectrums are observed between the E//c and E//b configurations. With decrease in temperature from 293 K to 10 K, however, the spectrral weight below 0.2 eV is suppressed for both the E//c and E//b polarizations indicating an opening of the charge gap due to the charge ordering. (The spicky structures below 0.06 eV are due to optical phonon modes.) Further, each $\sigma(\omega)$ has a different onset energy (Δ) which is obtained by extrapolating linearly from the rising part of the b and c axes polarized $\sigma(\omega)$ to the abscissa, as shown by the dashed lines in Figure 15. Such an anisotropic feature is related with the modification of the CE-type spin-ordering due to the excess electrons by $|1/2 - x|$, which is described in Sec. III-1 (Fig. 8). The excess electrons are likely to occupy the $3dz^2\text{-}r^2$ orbital of the Mn^{4+} site, and the Δ_c originates from the intersite transition of such excess $3dz^2\text{-}r^2$ electrons. Figure 15 indicates that the effective intersite Coulomb correlation is larger for the in plane than that for the c axis. The gap energy ($\Delta_c \sim$ 0.18 eV) of the density of states is comparable to that of the charge-ordered state of other $3d$ transition metal oxides, e.g., Fe$_3$O$_4$ (~ 0.14 eV) [60] and La$_{2-x}$Sr$_x$NiO$_4$ (~ 0.24 eV) [61].

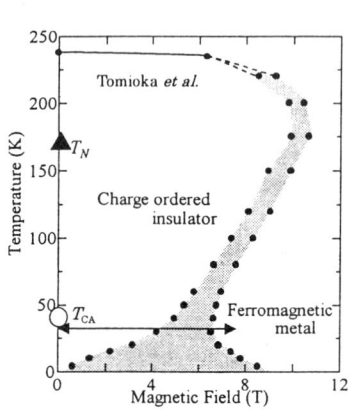
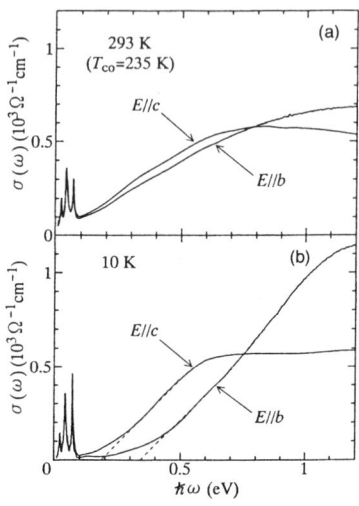

Figure 14. The temperature and magnetic phase diagram of Pr$_{1-x}$Ca$_x$MnO$_3$ ($x = 0.4$). The phase boundaries between the charge-ordered insulator and the ferromagnetic metal were determined by the measurements of the isothermal magnetoresistance and the temperature dependence of resistivity under magnetic fields. The hatched area indicates a field-hysteresis region.

Figure 15. The anisotropy of the optical conductivity in a single crystal of Pr$_{1-x}$Ca$_x$MnO$_3$ ($x = 0.4$) at (a) 293 K and (b) 10 K [59].

In Figure 14, the magnetic field induced insulator-metal transition at 30 K occurs at $\mu_0 H \sim 6.4$ T and 4.2 T in field increasing and decreasing runs, respectively. We show in Figure 16 the variation of the b- and c-polarized reflectivity ($R(\omega)$) with an external magnetic field, which is taken at 30 K. In both configurations, the infrared reflectivity gradually increases as the magnetic field increases up to 6 T (A spikey structure around 0.06 eV is due to the highest-lying oxygen phonon mode.), and abruptly changes to the metallic one between 6 and 7 T. As shown in this figure, the variation of reflectivity with an external magnetic field is over a wide energy region up to 3 eV. Figure 17 shows the b- and c-polarized $\sigma(\omega)$ spectra which is deduced by Kramers-Kronig (KK) analysis in Fig. 16. In consistent with the field dependence of $R(\omega)$ spectrum, the onset of $\sigma(\omega)$ gradually shifts to lower energy with increase in an external magnetic field and the $\sigma(\omega)$ undergoes a large change into a metallic band at 7 T. A similar change in $\sigma(\omega)$ spectrum is also observed in the field decreasing process.

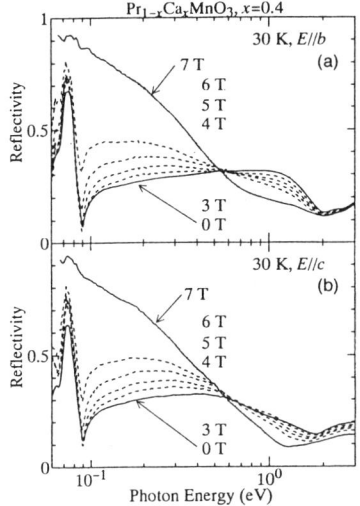
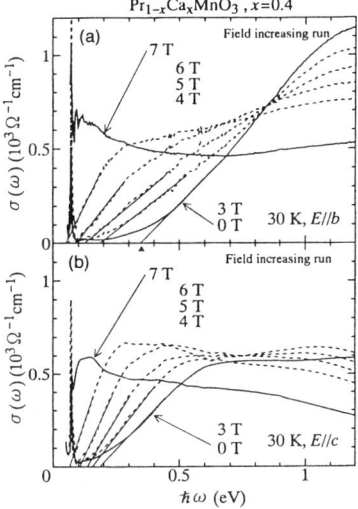

Figure 16. The magnetic-field dependence of (a) b-polarized and (b) c-polarized reflectivity spectra of $Pr_{1-x}Ca_xMnO_3$ ($x = 0.4$) at 30 K. A spikey structure around 0.06 eV is due to the highest-lying oxygen phonon mode [59].

Figure 17. The magnetic-field dependence of (a) b-polarized and (b) c-polarized optical conductivity spectra of $Pr_{1-x}Ca_xMnO_3$ ($x = 0.4$) at 30 K deduced by Kramers-Kronig analysis [59].

III-5. Variation of the T-H Phase Diagram with Bandwidth

In $Pr_{1-x}Ca_xMnO_3$, the charge/orbital-ordered phase still remains at $x = 0.35$ as shown in Fig. 13. The substitution of Ca with the larger cation Sr formulated as $Pr_{0.65}(Ca_{1-y}Sr_y)_{0.35}MnO_3$ is expected to increase the effective one-electron bandwidth W while keeping the doping level constant. Such an increase in W due to an chemical pressure also causes a modification of the charge/orbital-ordered phase in the H - T plane, as in the commensurate case of $x = 0.5$ (Fig. 6). However, a different feature from the latter case is observed.

Figure 18 (a) shows the temperature profile of resistivity with varying y in $Pr_{0.65}(Ca_{1-y}Sr_y)_{0.35}MnO_3$ [62]. For $y \leq 0.2$, the resistivity shows an insulating behavior while for $y > 0.3$ it appears metallic below some critical temperature. For $y = 0.3$, the compound once shows the charge/orbital ordering transition at $T_{CO} \approx 200$ K and with further decrease of temperature it shows an insulator to metal transition at ≈ 65 K. Figure 18 (b) shows the temperature profiles of resistivity for $y = 0.2$. Similarly to the phenomena as shown in Fig. 9, a magnetic field induces the transition from a charge/orbital-ordered insulator to a ferromagnetic metal under $\mu_0 H > 2$ T.

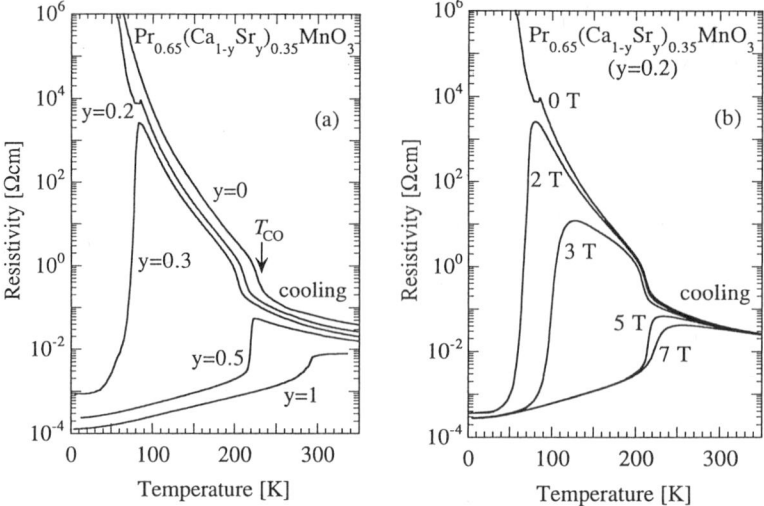

Figure 18. (a) The temperature dependence of resistivity (in the cooling run) for $Pr_{0.65}(Ca_{1-y}Sr_y)_{0.35}MnO_3$ crystals with varying y. The arrow indicates the charge/orbital-ordering transition for $Pr_{0.65}Ca_{0.35}MnO_3$. (b) The temperature dependence of resistivity of a $Pr_{0.65}(Ca_{1-y}Sr_y)_{0.35}MnO_3$ ($y = 0.2$) crystal under several magnetic fields (in the cooling run). The anomaly around 90 K under $H = 0$ is due to the spin-canting transition.

As discussed in Sec. II-2, the transfer integral according to the double exchange model is expressed as that $t_{DE} = t_0 \cos(\theta/2)$ (see, Sec. II-2). Qualitatively, the magnetic field induced insulator to metal transition as shown in Fig. 18 (b) is caused by an increase of t_{DE} in terms of $\cos(\theta/2)$. On the other hand, the W-controlled phenomena shown in Fig. 18 (a) are caused by an increase in t_0. A similar nonmetal to metal transition induced by the chemical pressure has also been reported for different doping levels, e.g. $x = 0.3$ [63-65]. Since the charge/orbital ordering is further weakened in the case of $x = 0.3$, the metallic state due to the destruction of the charge/orbital-ordered state appears for $y \geq 0.16$ [63-65].

Figure 19 shows the magnetic field dependence of resistivity taken at 200 K with varying y. The transition from the high to low (low to high) resistive state takes place in the field increasing (decreasing) run for all the samples, and the critical magnetic field systematically decreases with increase in y. Figure 20 shows the obtained phase diagram of the charge/orbital-ordered state for $Pr_{0.65}(Ca_{1-y}Sr_y)_{0.35}MnO_3$ with varying y. (The phase diagram for $y = 0$ is identical with that of $x = 0.35$ in Fig. 13.) In Fig. 20, the charge/orbital-ordered phase systematically slides toward the lower-H region with increase in y. In the case of $y = 0$, a collapse of a charge/orbital-ordered state under ≈ 6 T takes place below ≈ 65 K, which in the case of $y = 0.3$ occurs even under zero field. As a result of increase in t_0, the charge/orbital ordering for $y = 0.3$ is weakend so that it collapses under zero field. For $y = $

0.3, the disappearance of the charge/orbital ordering accompanied by the I-M transition at ≈ 65 K has been confirmed through the temperature dependence of the intensity of orbital superlattice reflection [62].

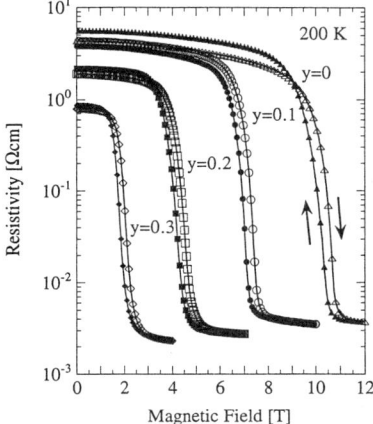

Figure 19. The magnetic field vs. resistivity curves at 200 K for $Pr_{0.65}(Ca_{1-y}Sr_y)_{0.35}MnO_3$ crystals with varying y [62]. The data in the field-increasing and -decreasing runs are denoted as open and closed symbols, respectively.

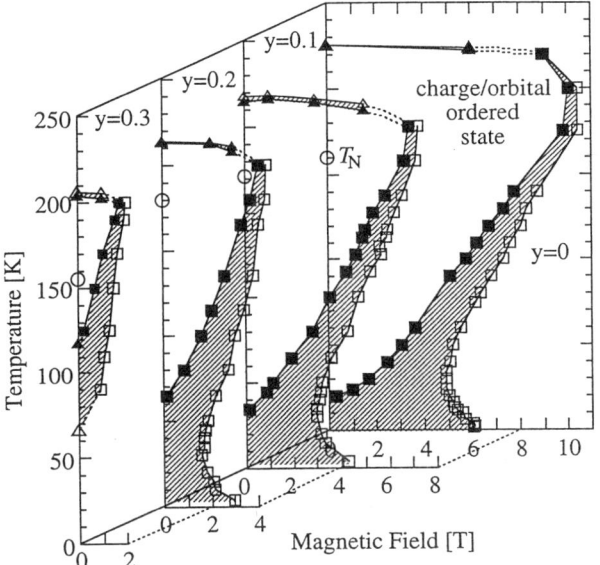

Figure 20. The charge/orbital-ordered state of $Pr_{0.65}(Ca_{1-y}Sr_y)_{0.35}MnO_3$ with varying y. The hysteresis is indicated by the hatched area. Open circles denote the antiferromagnetic Neel temperatures, which were determined by the temperature profiles of magnetization under 100 Oe.

IV. SUMMARY

In the manganese oxides with perovskite-type structure, various instabilities competing with the double exchange interaction play important roles, which depends on the effective one-electron bandwidth and the nominal hole concentration. As one such example, the magnetic field induced metal-insulator phenomena releavant to collapse of the charge/orbital ordering have been described. From the optical spectroscopy, the variation of reflectivity with an external magnetic field in this field induced transition is over a wide energy region up to few eV. The phase control related with the robustness of the charge/orbital ordering has been demonstrated by presenting the phase diagram on the magnetic field-temperature plane.

Acknowledgement(s)

The authors would like to thank Y. Moritomo, T. Kimura, R. Kumai, M. Kasai, T. Okuda, Y. Okimoto, H. Yoshizawa, H. Kawano, R. Kajimoto, N. Miura and M. Tokunaga for their enlightening discussions. Some works described in this chapter was supported by NEDO (New Energy and Industrial Technology Development Organization) of Japan and performed in the Joint Research Center for Atom Technology (JRCAT) under the joint research agreement between the National Institute for Advanced Interdisciplinary Research (NAIR) and the Angstrom Technology Partnership (ATP).

REFERENCES

[1] E.J.W. Verwey, P.W. Haayman and F.C. Romeijin, J. Chem. Phys. **15**, 181 (1947).
[2] C.H. Chen, S-W. Cheong and S.A. Cooper, Phys. Rev. Lett. **71**, 2461 (1993).
[3] P.D. Battle, T.C. Gibb and P. Lightfoot, J. Solid State Chem. **84**, 271 (1990).
[4] J.Q. Li, Y. Matsui, S.K. Park and Y. Tokura, Phys. Rev. Lett. **79**, 297 (1997).
[5] J.M. Tranquada, B.J. Sternlieb, J.D. Axe, Y. Nakamura and S. Uchida, Nature **375**, 561 (1995).
[6] H. Sawada, N. Hamada, K. Terakura and T. Asada, Phys. Rev. B**53**, 12742 (1996).
[7] J.L. G.-Munos, J.R.-Carvajel and P. Lacorre, Europhys. Lett. **20**, 241 (1992).
[8] J.R.-Carvajal, S. Rosenkranz, M. Medarde, P. Laccore, M.T. F.-Diaz, F. Fauth and V. Trounov, Phys. Rev. B**57**, 456 (1998).
[9] K.I. Kugel and D.I. Khomskii, Sov. Phys.-JETP **37**, 725 (1974).
[10] J. Kanamori, J. Appl. Phys. Suppl. **31**, 14S (1960).
[11] G. Matsumoto, J. Phys. Soc. Jpn. **29**, 606 (1970).
[12] I. Solovyev, N. Hamada and K. Terakura, Phys. Rev. Lett. **76**, 4825 (1996).
[13] G.H. Jonker and J.H. van Santen, Physica **16**, 337 (1950).
[14] G.H. Jonker, Physica **22**, 707 (1956).
[15] J.B. Goodenough and J.M. Longo: *Landolt-Börnstein, New Series, Group III, Magnetic and Other Properties of Oxides and Related Compounds*, edited by K.-H. Hellwege and A.M. Hellwege (Springer Verlag, Berlin, 1970), Vol. 4a, p. 126.
[16] C. Zener, Phys. Rev. **82**, 403 (1951).
[17] P.W. Anderson and H. Hasegawa, Phys. Rev. **100**, 675 (1955).

[18] P.-G. de Gennes, Phys. Rev. **118**, 141 (1960).

[19] J.B. Trrance, P. Laccore, A.I. Nazzal, E.J. Ansaldo and Ch. Niedermayer, Phys. Rev. B**45**, 8209 (1992).

[20] A.J. Millis, P.B. Littlewood and B.I. Shraiman, Phys. Rev. Lett. **74**, 5144 (1995).

[21] A.J. Millis, P.B. Littlewood and B.I. Shraiman, Phys. Rev. B**54**, 5405 (1996).

[22] H. Röder, J. Zhang and A.R. Bishop, Phys. Rev. Lett. **76**, 1356 (1996).

[23] P. Schiffer, A.P. Ramirez, W. Bao and S.-W. Cheong, Phys. Rev. Lett. **75**, 3336 (1995).

[24] E.O. Wollan and W.C. Koehler, Phys. Rev. **100**, 545 (1955).

[25] H. Kuwahara, Y. Tomioka, A. Asamitsu, Y. Moritomo and Y. Tokura, Science **270**, 961 (1995).

[26] Z. Jirak, S. Krupicka, V. Nekvasild, E. Pollert, G.Villeneuve and F. Zounova: J. Mag. Mag. Mat. **15-18**, 519 (1980).

[27] Z. Jirak, S. Krupicka, Z.Simsa, M. Dlouha and Z. Vratislav: J. Mag. Mag. Mat. **53**, 153 (1985).

[28] R.D. Shannon, Acta Cryst. **A32**, 751 (1976). The ionic radius of $^{XII}Pr^{3+}$ is obtained by the proportional allotment of the radii of $^{XII}Ce^{3+}$ and $^{XII}Nd^{3+}$.

[29] V. Kiryukhin, D. Casa, J.P. Hill, B. Keimer, A. Vigliante, Y. Tomioka and Y. Tokura, Nature **386**, 813 (1997).

[30] K. Miyano, T. Tanaka, Y. Tomioka and Y. Tokura, Phys. Rev. Lett. **78**, 4257 (1997).

[31] T. Mori, K. Ogawa, K. Yoshida, K. Miyano, Y. Tomioka and Y. Tokura, J. Phys. Soc. Jpn. **66**, 3570 (1997)

[32] K. Ogawa, W. Wei, K. Miyano, Y. Tomioka, and Y. Tokura, Phys. Rev. B**57**, R15033 (1998).

[33] A. Asamitsu, Y. Tomioka, H. Kuwahara and Y. Tokura, Nature **388**, 50 (1997).

[34] P.G. Radaelli, D.E. Cox, M. Marezio, S.-W. Cheong, P.E. Shiffer and A.P. Ramirez, Phys. Rev. Lett. **75**, 4488 (1995).

[35] P.G. Radaelli, D.E. Cox, M. Marezio and S.-W. Cheong, Phys. Rev. B**55**, 3015 (1997).

[36] Y. Moritomo, Y. Tomioka, A. Asamitsu, Y. Tokura and Y. Matsui, Phys. Rev. B**51**, 3297 (1995).

[37] B.J. Sternlieb, J.P. Hill, U.C. Wildgruber, G.M. Luke, B. Nachumi, Y. Moritomo and Y. Tokura, Phys. Rev. Lett. **76**, 2169 (1996).

[38] H. Kawano, R. Kajimoto, H. Yoshizawa, Y. Tomioka, H. Kuwahara and Y. Tokura, Phys. Rev. Lett. **78**, 4253 (1997).

[39] J.B. Goodenough, Phys. Rev. **100**, 564 (1955).

[40] Y. Tokura, A. Urushibara, Y. Moritomo, T. Arima, A. Asamitsu, G. Kido and N. Furukawa, J. Phys. Soc. Jpn. **63**, 3931 (1994).

[41] A. Urushihara, Y. Moritomo, T. Arima, A. Asamitsu, G. Kido and Y. Tokura, Phys.

Rev. B **51**, 14103 (1995).

[42] G. Xiao, E.J. McNiff, Jr, G.Q. Gong, A. Gupta, C.L. Canedy and J.Z. Sun, Phys. Rev. B **54**, 6073 (1996).

[43] Y. Tokura, H. Kuwahara, Y. Moritomo, Y. Tomioka and A. Asamitsu, Phys. Rev. Lett. **76**, 3184 (1996);

[44] H. Kuwahara, Y. Moritomo, Y. Tomioka, A. Asamitsu, M. Kasai and Y. Tokura, Phys. Rev. B **56**, 9386 (1997);

[45] H. Kuwahara, Y. Moritomo, Y. Tomioka, A. Asamitsu, M. Kasai and Y. Tokura, J. Appl. Phys. **81**, 4954 (1996).

[46] H. Kuwahara, Y. Tomioka, Y. Moritomo, A. Asamitsu, M. Kasai, R. Kumai and Y. Tokura, Science **272**, 80 (1996).

[47] M. Kataoka, Czech. J. Phys. **46**, Suppl. S4, (1996).

[48] M. Kataoka and M. Tachiki, Physica B**237-238**, 24 (1997).

[49] M. Tokunaga, N. Miura, Y. Tomioka and Y. Tokura, Phys. Rev. B **57**, 5259 (1998).

[50] H. Kuwahara, T. Okuda, Y. Tomioka, T. Kimura, A. Asamitsu and Y. Tokura, Mat. Res. Soc. Sym. Proc. **494**, 83 (1998).

[51] Y. Tomioka, A. Asamitsu, H. Kuwahara, Y. Moritomo and Y. Tokura, Phys. Rev. B **53**, R1689 (1996).

[52] H. Yoshizawa, H. Kawano, Y. Tomioka and Y. Tokura, Phys. Rev. B **52**, R13145 (1995).

[53] Y. Tomioka, A. Asamitsu, Y. Moritomo and Y. Tokura, J. Phys. Soc. Jpn. **64**, 3626 (1995).

[54] H. Yoshizawa, H. Kawano, Y. Tomioka and Y. Tokura, J. Phys. Soc. Jpn. **65**, 1043 (1996).

[55] H.Y. Hwang, T.T.M. Palstra, S.-W. Cheong and B. Batlogg, Phys. Rev. B **52**, 15046 (1995).

[56] Y. Moritomo, H. Kuwahara, Y. Tomioka and Y. Tokura, Phys. Rev. B **55**, 7549 (1997).

[57] H. Yoshizawa, R. Kajimoto, H. Kawano, Y. Tomioka and Y. Tokura, Rev. B **55**, 2729 (1997).

[58] K. Liu, X.W. Wu, K.H. Ahn, T. Sulchek, C.L. Chien and J.Q. Xiao, Phys. Rev. B **54**, 3007 (1996).

[59] Y. Okimoto, Y. Tomioka, Y. Onose, Y. Otsuka and Y. Tokura, Phys. Rev. B **57**, R9377 (1998).

[60] S.-K. Park, T. Ishikawa and Y.Tokura (unpublished).

[61] T. Katsufuji, T. Tanabe, T. Ishikawa, Y. Fukuda, T. Arima and Y. Tokura, Phys. Rev.B **54**, R14230 (1996).

[62] Y. Tomioka, A. Asamitsu, H. Kuwahara and Y. Tokura, J. Phys. Soc. Jpn. **66**, 302 (1997).

[63] H.Y. Hwang, S.-W. Cheong, P.G. Radaelli, M. Marezio and B. Batlogg, Phys. Rev. Lett. **75**, 914 (1995).
[64] B. Raveau, A. Maignan and Ch. Simon, J. Solid State Chem. **117**, 424 (1995).
[65] A. Maignan, Ch. Simon, V. Caignaert and B. Raveau, J. Mag. Mag. Mat. **152**, 5 (1996).

OPTICAL PROPERTIES OF COLOSSAL MAGNETORESISTANCE MANGANITES

T. W. Noh, J. H. Jung, and K. H. Kim

Department of Physics, Seoul National University, Seoul 151-742, Korea

I. INTRODUCTION

Recently, there have been lots of attentions paid on perovskite manganites due to their interesting physical properties. A parent compound, $LaMnO_3$, has an antiferromagnetic insulating ground state with four $3d$ electrons in a high spin configuration, i.e. $t_{2g}^3 e_g^1$. By substituting La with a divalent cation, Mn^{4+} ions can be introduced. Depending on the hole doping concentration (x) and temperature (T), the compounds show several intriguing phenomena, including a metal-insulator (M-I) transition,[1] a charge and/or orbital ordering,[2-4] a polaron ordering,[5] a magnetic ordering,[6] a magnetic field dependent structural phase transition,[7] and a collapse of charge ordering under a magnetic field.[8]

Near x=0.3, $(R,A)MnO_3$ [R = La or Nd, and A = Ca, Sr, or Ba] shows a M-I transition near a ferromagnetic ordering temperature T_C. Under a magnetic field, this compound shows a very large resistance drop, which is called the "colossal magnetoresistance" (CMR).[9-13] The coexistence of the ferromagnetism and the metallic conduction has been explained by the double exchange (DE) model, suggested by Zener:[14]

$$\widetilde{H} = -t \sum_{<ij>\sigma}(c_{i\sigma}^+ c_{j\sigma} + H.c.) - J\sum_i \sigma_i \cdot S_i, \qquad (1)$$

Physics of Manganites. Edited by Kaplan and Mahanti
Kluwer Academic/Plenum Publishers, 1999

where σ_i and S_i denote spins of itinerant e_g electrons and localized t_{2g} electrons, respectively. J is the Hund's coupling constant between the e_g and the t_{2g} electrons, and t is the bare hopping matrix element between nearest neighbor sites. [This Hamiltonian is often called the "Kondo lattice model".] In the limit of $J \cdot S \to \infty$, the effective Hamiltonian can be written as[15,16]

$$H_{eff} = -t \sum_{<ij>} \cos\frac{\theta_{ij}}{2} c_i^+ c_j, \qquad (2)$$

where θ_{ij} is the angle between i- and j-th spins. Note that, in this effective Hamiltonian, the electron hopping between neighbor sites becomes a maximum when they have parallel spins.

However, it was demonstrated that the DE interaction alone is not sufficient to explain the magnitude of the CMR.[17] Therefore, it was argued that other degrees of freedom should be included. Till now, several candidates have been proposed: the dynamic Jahn-Teller (JT) interaction,[18-20] an orbital degree of freedom,[21] a superexchange interaction,[22] and a disorder.[23,24] Optical measurements are useful methods to investigate such low energy excitations.

In this paper, we will focus our discussion on optical properties which are closely related to the CMR. [Optical properties of layered manganites,[25] and issues related to charge ordering[26] will be omitted.] In Section II, we will review theoretical and experimental works. In Section III, we will present our proposed schematic diagrams for optical transitions occurring in the CMR manganites. In Section IV, we will present details of our experimental works on $(La,Ca)MnO_3$ and $(La,Pr)_{0.7}Ca_{0.3}MnO_3$, and discuss them with our proposed diagrams. In Section IV. 4, we will also include our results on frequency shifts of internal phonon modes near T_C in $La_{0.7}Ca_{0.3}MnO_3$. Finally, we will make a summary in Section V.

II. OVERVIEW

1. Theoretical Works

Furukawa investigated optical conductivity spectra $\sigma(\omega)$ using the Kondo lattice model with classical spins in an infinite dimension.[27] As shown in the inset of Fig. 1, when J is larger than the electron bandwidth W, the e_g band should be splitted into up-spin ($e_g^\uparrow$) and down-spin ($e_g^\downarrow$) bands at $-J$ and $+J$, respectively. Then, $\sigma(\omega)$ should be composed of a Drude peak located at $\omega = 0$ and an interband transition located at $\omega = 2J$, as shown in Fig. 1. [The latter peak is attributed to spin-conserved optical transitions of $e_g^\uparrow(t_{2g}^\uparrow) \to e_g^\uparrow(t_{2g}^\downarrow)$ and $e_g^\downarrow(t_{2g}^\downarrow) \to e_g^\downarrow(t_{2g}^\uparrow)$: namely, transitions between the spin split e_g bands with different t_{2g}

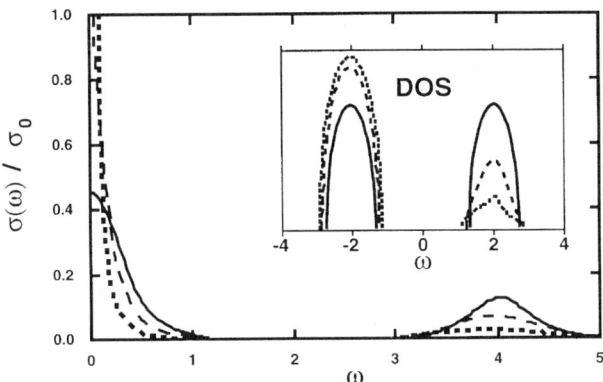

Figure 1. Optical conductivity spectra for $J=2$ in $La_{0.8}Sr_{0.2}MnO_3$. Temperatures are $T=1.05T_C$ (solid curves), $T=0.5T_C$ (dashed curves), and $T=0.25T_C$ (dotted curves). In the inset, the densities of states for up-spin e_g bands are also shown. (Reproduced from Ref. 27.)

spin backgrounds.] As T decreases below T_C, t_{2g} spins are aligned, so magnetization M should increase. Then, the interband transition peak near $\omega = 2J$ decreases, and the Drude free carrier peak increases. Therefore, in this spin split band model which includes the DE interaction only, the spectral weight change should be proportional to M^2.

Taking into account of the orbital degeneracy in the e_g orbitals, Shiba and coworkers investigated effects of t_{2g} spin disorder in $\sigma(\omega)$.[28,29] The conductivity due to an intraband transition within the e_g bands was found to be very similar to Fig. 1. In this model, an interband transition also becomes allowed due to orbital dependence of the transfer integral, which comes from overlap of the Mn e_g and the O $2p$ wave functions. As t_{2g} spin disorder decreases with increasing M, the interband transition in a frequency region between ~$2t$ and $4t$ was predicted to increase. Ishihara and coworkers investigated orbital fluctuations in the low-T ferromagnetic metallic state and their effects on optical properties.[30,31] They found that the orbital degrees of freedom remained disordered even in the ferromagnetic state, and that such an orbital liquid state results in an incoherent absorption band. They found that a Drude peak with a small spectral weight appeared in $\sigma(\omega)$ when a JT interaction was included. Recently, Horsch, Jaklic, and Mack also investigated effects of the orbital degeneracy using a Hamiltonian which is a generalized t-J model with anisotropic interactions and 3-site hopping. In this 'optical t-J model', $\sigma(\omega)$ is characterized by a broad incoherent spectrum with increasing intensity as T is lowered, and a Drude peak with a small weight.[32]

Millis, Mueller, and Shraiman studied a model of electrons interacting with dispersionless classical phonons.[17-19] In addition to Eq. (1), they included Hamiltonian

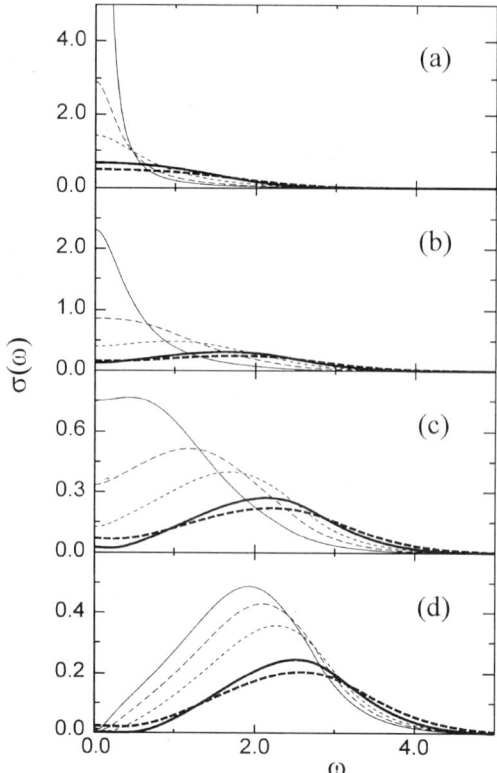

Figure 2. Optical conductivity spectra with different electron-phonon coupling, λ. For the case of $x=0$, with (a) $\lambda=0.71$, (b) $\lambda=1$, (c) $\lambda=1.08$, and (d) $\lambda=1.15$ are shown. Temperatures are $T=0.02$ (light solid line), $T=T_C/2$ (light dashed line), $T=3T_C/4$ (light dotted line), $T=T_C$ (heavy solid line), and $T=2T_C$ (heavy dashed line). Note that in (a) the lowest T is 0.025 not 0.02. (Reproduced from Ref. 19.)

terms representing an electron-phonon coupling and phonon energies. They assumed a JT form for the electron-phonon coupling, and introduced a dimensionless parameter, $\lambda=g^2/kt$, where g and k represent the electron-phonon coupling and the elastic restoring force constant, respectively. Depending on λ, they found a crossover from Fermi liquid to polaron behavior in a limit of $J \cdot S \rightarrow \infty$. Figure 2 shows T-dependent $\sigma(\omega)$ spectra for the case of $x=0$. For a weak coupling $\lambda=0.71$, $\sigma(\omega)$ has a strong Drude peak, which becomes broaden and acquires a T-independent part due to spin scattering as T increases. In a moderate coupling $\lambda=1$, $\sigma(\omega)$ has a Drude form at a low T and, as T increases, a broad peak develops. The latter peak was attributed to transitions between the two JT split levels. And, in a strong coupling $\lambda=1.15$, $\sigma(\omega)$ do not show a Drude peak any longer even at $T=0$. For the general case of $x \neq 0$, midgap transitions should occur between a JT split Mn^{3+} level and a nearest neighbor Mn^{4+} level.

Röder, Zang, and Bishop also investigated the interplay between electron-phonon and

DE interactions using a similar Hamiltonian.[20,33] Under the Lang-Firsov transformation for a quantum phonon problem, they derived an effective Hamiltonian over a phonon vacuum:

$$\tilde{H} = -t\xi(\eta) \sum_{<ij>\sigma} \cos\frac{\theta_{ij}}{2}(c_{i\sigma}^+ c_{j\sigma} + H.c.) + \sum_i D_i. \quad (3)$$

The polaron band narrowing is given by $\xi(\eta)=\exp(-\varepsilon_p\eta^2/\hbar\omega_0)$ with an effective electron-phonon coupling parameter ε_p and a coupled phonon frequency ω_0. Here, η measures a degree of the polaron effect, which is determined self-consistently and dependent on ε_p, ω_0, and x. D_i refers to diagonalized term over the phonon vacuum. Within this Hamiltonian, Röder et al. derived a relation on T_C:

$$T_C(x) = \frac{9}{50}[-e_b(x)]\xi(\eta), \quad (4)$$

where e_b is a band energy, which is proportional to t and depends on x. [As η approaches zero, $e_b(x)$ is roughly proportional to $x(1-x)$.] So, $\xi(\eta)$ becomes important in determining both metallic and insulating properties of CMR materials, as well as the variation of T_C with doping. Even though they did not investigate $\sigma(\omega)$ explicitly, they predicted a continuous crossover from a large polaronic state to a quasi-self-trapped small polaron state near T_C.

2. Experimental Works

Numerous optical investigations on manganites showed that there do exist drastic T-dependent changes in spectral weight,[34-43] which were believed to be closely related to the M-I transition. However, identifications of optical transitions vary among researchers, and origins of the spectral weight changes are not clearly understood yet. Three important cases will be explained in detail.

Moritomo et al. investigated T-dependent absorption coefficient of a $La_{0.6}Sr_{0.4}MnO_3$ thin film in a frequency range of 0.5 ~ 5 eV.[34] As shown in Fig. 3, the spectral weight above 2.2 eV was transferred to a low energy region with decreasing T. The T-independent part, which is denoted as the hatched area in Fig. 3, was considered as a charge-transfer transition from the O $2p$ states to the Mn $3d$ levels. Then, the spectral weights of the T-dependent part above and below 2.2 eV, denoted by S_A and S_B, respectively, were evaluated. [The spectral weight of the T-independent part is denoted by S_0.] They found that S_A/S_0 was proportional to $1-(M/M_S)^2$, so they claimed that S_A was originated from the interband transition between the Hund's rule split bands. This interpretation is consistent with the prediction of the Furukawa's spin split band model.

Okimoto et al. investigated $\sigma(\omega)$ of a $La_{0.825}Sr_{0.175}MnO_3$ single crystal ($T_C \sim 283$ K).[37] As shown in Fig. 4, there was a broad peak centered around 1.5 eV above T_C. With

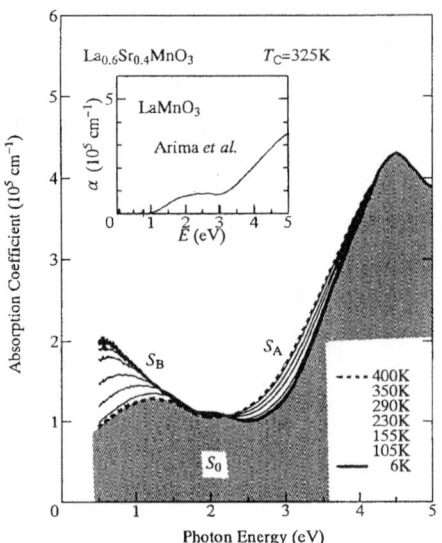

Figure 3. Temperature dependent absorption coefficient for $La_{0.6}Sr_{0.4}MnO_3$ thin film. The temperature independent spectral weight and the temperature dependent higher-, and lower-lying spectral weight are denoted as S_0, S_A, and S_B, respectively. In the inset, room temperature absorption coefficient for $LaMnO_3$ is also shown. (Reproduced from Ref. 34.)

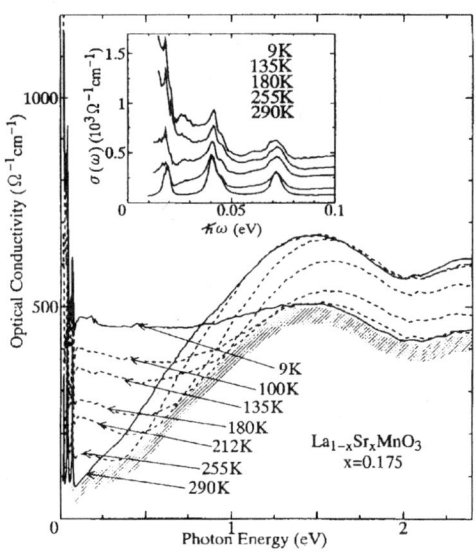

Figure 4. Optical conductivity spectra of $La_{0.825}Sr_{0.175}MnO_3$ ($T_C \sim 238$ K). The hatched curve represents the temperature independent part of the spectra deduced from the envelope of the respective curves. The inset shows a magnification of the far-infrared part. (Reproduced from Ref. 37.)

decreasing T, the spectral weight above 1.0 eV was transferred to lower energy region. They subtracted the T-independent part, which was also considered as the charge-transfer contribution, from their experimental $\sigma(\omega)$. They explained the remaining parts in terms of the spin split band picture: the low and the high frequency parts were regarded as the intraband and the interband transitions between the spin split bands. However, their observed spectral weight for the intraband transition was not proportional to $(M/M_S)^2$. Moreover, as shown in the inset of Fig. 4, there appears a sharp Drude-like peak below 0.05 eV, so it is difficult to consider the spectral weight below 1.0 eV as the Drude peak in the spin split band model. In a later systematic work on (La,Sr)MnO$_3$ single crystals,[38] the same authors observed similar discrepancies, and suggested that another degree of freedom (e.g. orbital ordering and/or electron-lattice interactions) should be taken into account.

Kaplan et al. measured reflectance and transmittance spectra of a Nd$_{0.7}$Sr$_{0.3}$MnO$_3$ thin film.[39] As shown in Fig. 5, $\sigma(\omega)$ showed a broad peak structure near 1.2 eV above T_C. With cooling below T_C, the conductivity at the low energy region increased dramatically and the peak shifted to a lower energy. [Interestingly, they reported no Drude-like peak below T_C, even though the sample was metallic. However, a later work by the same group on an annealed Nd$_{0.7}$Sr$_{0.3}$MnO$_3$ thin film showed a Drude-like peak.[40]] Based on the similarity between this spectra and that in Fig. 2 (d), they assigned the 1.2 eV peak as an optical transition from a JT split Mn^{3+} level to a nearest neighbor Mn^{4+} level.

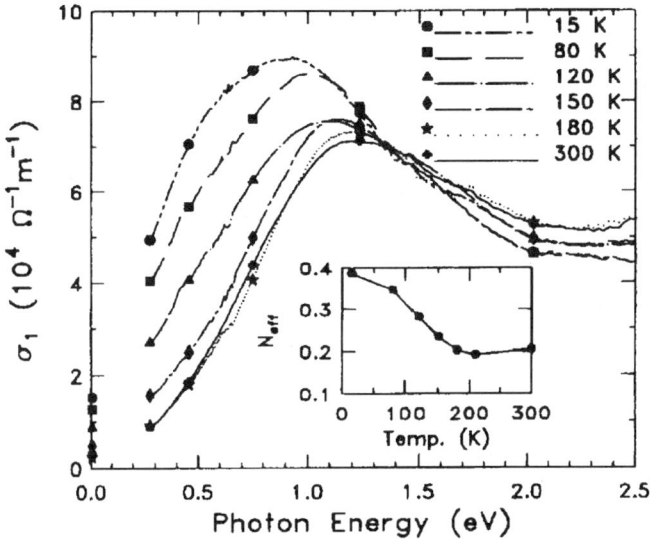

Figure 5. Real part of optical conductivity for Nd$_{0.7}$Sr$_{0.3}$MnO$_3$ thin film. In the inset, effective number of carriers, N_{eff}, up to 1.5 eV at various temperatures is shown. (Reproduced from Ref. 39.)

III. OUR PROPOSED SCHEMATIC DIAGRAMS

As shown in the former section, interpretations on $\sigma(\omega)$ are quite different among experimental groups. However, a correct interpretation is essential to understand physics of the CMR. We think that both the DE and the electron-phonon interactions play important roles in determining $\sigma(\omega)$. In this section, we will present our proposed schematic diagrams.

Fig. 6 (a) shows a schematic diagram of $\sigma(\omega)$ for $T>T_C$. There are four main peaks below 5 eV: (i) a small polaron absorption peak below 1.3 eV, (ii) a peak centered around 1.5 eV, which will be called the "1.5 eV peak", (iii) a broad peak centered around 3.5 eV, which is originated from interband transitions between the Hund's rule split bands, and (iv) a peak due to charge transfer transitions between the O 2p and the Mn d levels.

Fig. 6 (b) shows a schematic diagram of $\sigma(\omega)$ for $T \ll T_C$. As T becomes lower, the peak due to the interband transitions between the Hund's rule split bands decreases. Then, the spectral weight below 1.2 eV should increase, but it develops into two parts, namely a coherent Drude peak and an incoherent mid-infrared (IR) peak. On the other hand, the 1.5 eV peak becomes weaker slightly with decreasing T, and the peak due to the charge transfer transitions is nearly T-independent.

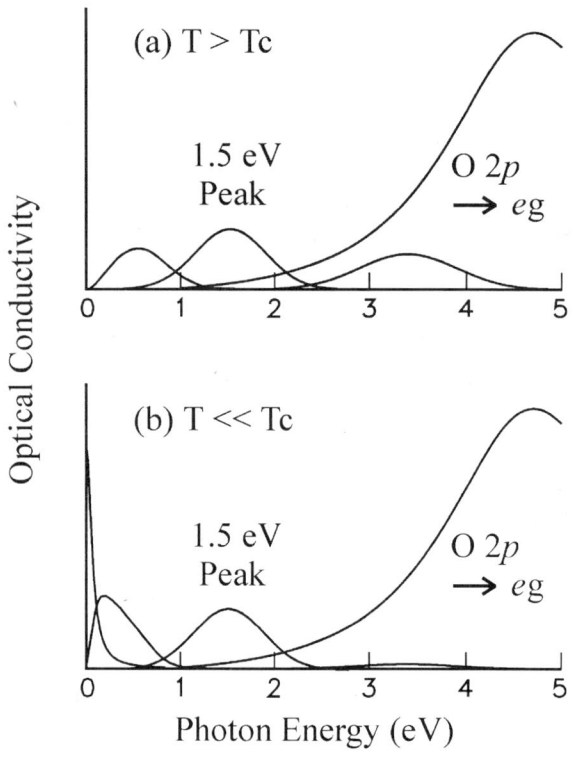

Figure 6. Proposed diagrams of optical conductivity spectra for (a) $T>T_C$ and (b) $T \ll T_C$.

Note that our proposed diagrams include all the optical transitions which are related to the DE and the electron-phonon interactions. In the Furukawa's model, only terms related to the DE interaction, i.e. the Drude peak and the interband transitions between the spin split bands, were considered. On the other hand, in the Millis' model, only the spectral evolution in the low frequency region (i.e. below 1.5 eV in this diagram) due to the electron-phonon interaction was addressed.

There have been some debates on existence and identity of the 1.5 eV peak. For most manganites with hole concentrations of 0.3, the 1.5 eV peak could not be seen clearly. However, from doping dependent conductivity studies on (La,Ca)MnO$_3$, we showed that such a peak should exist.[44] Recently, Machida, Moritomo, and Nakamura observed similar peaks in transmission spectra of $(Nd_{0.25}Sm_{0.75})_{0.6}Sr_{0.4}MnO_3$ and $Sm_{0.6}Sr_{0.4}MnO_3$ films.[35,36] And, to explain T-dependent optical spectra of $Nd_{0.7}Sr_{0.3}MnO_3$, $La_{0.7}Ca_{0.3}MnO_3$, and $La_{0.7}Sr_{0.3}MnO_3$ thin films, Quijada et al. assumed the existence of the 1.5 eV peak.[40] However, there are still debates on the origin of the 1.5 eV peak. Machida et al. attributed it to an optical transition related to the JT clusters.[35,36] Quijada et al. assigned it to an interatomic transition between the JT split Mn^{3+} levels.[40] On the other hand, we assigned it to an intraatomic transition. Details of our experimental results on the 1.5 eV peak will be presented in Section. IV. 2.

IV. DATA AND DISCUSSION

1. Our Experimental Techniques

We measured reflectivity spectra $R(\omega)$ in a wide photon energy region of 5 meV ~ 30 eV.[45] $R(\omega)$ in far- and mid-IR regions were obtained using a conventional Fourier transform spectrophotometer. Above 0.6 eV, grating spectrometers were used. Especially, in the frequency region of 6 ~ 30 eV, we used a synchrotron radiation at Normal Incidence Monochromator Beamline in Pohang Light Source. In order to get T-dependence of $R(\omega)$ from 5 meV to 2.5 eV, we used a liquid He-cooled cryostat.

In our earlier works,[41,42] room temperature data were smoothly connected for lower temperature $R(\omega)$ since the T-dependence was not significant above 2.5 eV. From measured $R(\omega)$, we obtained $\sigma(\omega)$ using the Kramers-Kronig transformation. For these analyses, $R(\omega)$ for a low frequency region were extrapolated with constants for insulating states or using the Hagen-Rubens relation for metallic states.[46] For a high frequency region, reflectivity at 30 eV was extended up to 40 eV and above which ω^{-4} dependence was assumed. This approach limited our investigation of spectral weight changes below 2.0 eV.

To extend our investigation in a wider region, we recently started to use a spectroscopic ellipsometer.[47] In a frequency region of 1.5 ~ 5.0 eV, we were able to obtain T-dependent optical constants directly without relying on the extrapolation techniques. In a frequency region lower than 1.5 eV, we could use the Kramers-Kronig transformation with anchoring techniques.[48] Using this new approach, we were able to obtain T-dependent $\sigma(\omega)$ more accurately and extend our investigation region up to 5.0 eV.

2. Doping-dependent Optical-conductivity Studies of (La,Ca)MnO$_3$

To get a clear understanding on the low energy optical transitions in Fig. 6, we systematically investigated doping dependent $\sigma(\omega)$ of (La,Ca)MnO$_3$ at room temperature.[44] The left pictures in Fig. 7 show $\sigma(\omega)$ of (La,Ca)MnO$_3$ in paramagnetic insulating states. For LaMnO$_3$, $\sigma(\omega)$ show peak structures centered around 2 eV and 5 eV, which are marked with Peaks A and B, respectively. Arima et al. assigned Peaks A and B as charge transfer transitions of O $2p \rightarrow e_g$ and O $2p \rightarrow t_{2g}$, respectively.[49] Using detailed studies based on strength arguments, we found that Peaks A and B should come from $e_g^1 \rightarrow e_g^2$ and O $2p \rightarrow e_g^2$ transitions, respectively.[50] Since the e_g^1 band is located near the top of the O $2p$ band and since there is a strong hybridization between the O $2p$ and the Mn e_g wavefunctions, the hole state in this compound can have some O $2p$ characters which were observed in some other experiments.[51-54] However, we think that it should still have the symmetry of the e_g^1 band and that a significant portion of the hybridized state should be originated from the e_g^1 band.

For a quantitative analysis of the other (La,Ca)MnO$_3$ spectra, we regarded $\sigma(\omega)$ as a sum of a midgap component $\sigma_{ms}(\omega)$ and Lorentz oscillator components $\sigma_L(\omega)$, i.e. $\sigma(\omega)=\sigma_{ms}(\omega)+\sigma_L(\omega)$. The dotted lines in the left pictures of Fig. 7 indicate contributions of the Lorentz oscillators. For the first Lorentz oscillator, its strength and width increase and its peak frequency decreases with increasing x. Such behaviors can be well explained when we assigned its peak as O $2p \rightarrow e_g$ transition. [Note that the transition from the lower to the upper Hund's rule split bands is omitted in this analysis, since its strength seems to be much smaller than that of the O $2p \rightarrow e_g$ transition. However, such a transition can be seen in Moritomo's work,[34] Quijada's work,[40] and our recent studies on La$_{7/8}$Sr$_{1/8}$MnO$_3$.[43]]

The right pictures of Fig. 7 show the midgap components of (La,Ca)MnO$_3$. As mentioned earlier, the peak around 2 eV for LaMnO$_3$ comes from $e_g^1 \rightarrow e_g^2$ transition. With the Ca doping, the peak position shifts to lower energies, and a double peak structure can be clearly seen for 0.5≤x≤0.9. We clearly observed similar double peak structures for La$_{0.7}$Ca$_{0.3}$MnO$_3$, (La,Pr)$_{0.7}$Ca$_{0.3}$MnO$_3$ samples and a Nd$_{0.7}$Sr$_{0.3}$MnO$_3$ single crystal below T_C.[55] Therefore, it is quite natural to analyze $\sigma_{ms}(\omega)$ in terms of two peaks. [A recent Monte Carlo calculation using the two-orbital Kondo model with JT phonons showed such a double peak structure clearly.[56]]

For the x =0.3 sample, the spectrum below 2.0 eV looks like one peak. However, there will be some discrepancies between the optical data and other experimental data, when the broad peak around 1.0 eV is interpreted as one optical transition coming from the a JT split Mn^{3+} level to a nearest neighbor Mn^{4+} level, i.e. a small polaron peak. For examples, a

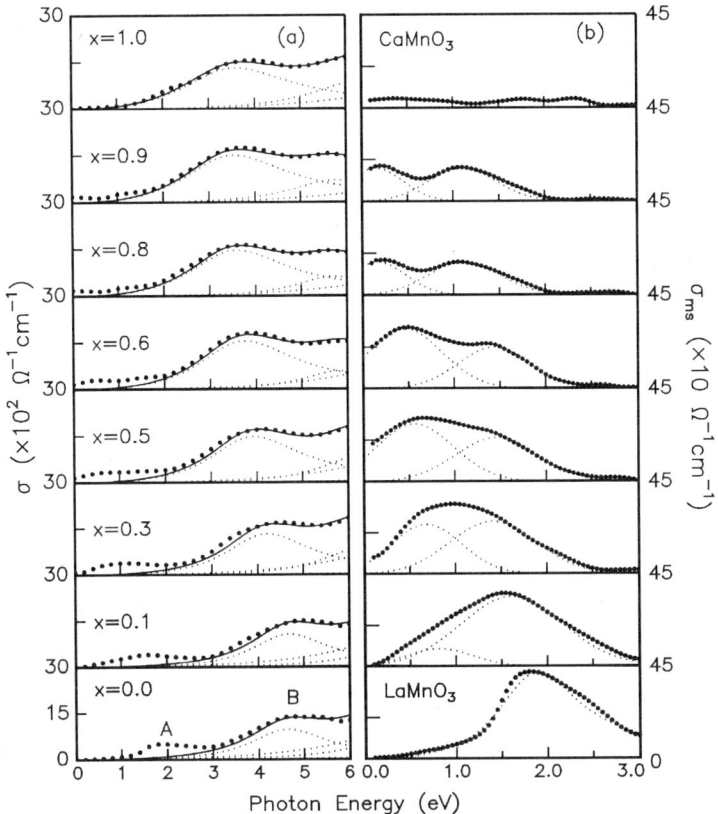

Figure 7. (a) Doping dependent optical conductivity spectra of (La,Ca)MnO$_3$ below 6 eV. The solid circles are experimental data. The dotted and solid lines represent the Lorentz oscillators and sum of their contributions, respectively. Note that the first and the second peaks in LaMnO$_3$ are denoted as Peak A and Peak B, respectively. (b) Optical conductivity spectra of midgap components in (La,Ca)MnO$_3$. The dotted lines represent a Lorentz oscillator for $x=0.0$ and two Gaussian functions for $x \neq 0.0$.

recent transport measurement shows that a small polaron activation energy is about 0.15 eV,[57] which is smaller than 0.25 eV. [It is known that the energy of a small polaron motion measured by optical techniques is about 4 times larger than the activation energy measured by transport measurements.[58]] Moreover, recent structural experiments, such as EXAF and PDF measurements, showed that the local distortion of the Mn-O octahedron becomes weaker as T decreases below T_C.[59,60] However, Fig. 5 shows that the peak near 1.0 eV actually increases at low T.

Within the double peak structure interpretation, the first peak around 0.6 eV can be attributed to a small polaron peak. In this case, the peak position agrees with the transport measurement data. Moreover, this peak will decrease at low T due to a crossover from the small to the large polaron regimes, which will explained in Section IV. 3.

In doped manganites, Millis *et al.* predicted existence of midgap states which come from optical transitions among the Mn *d* levels.[19] They were assumed to be composed of two parts: *i.e.* the small polaron peak and the interatomic transition from a JT split Mn^{3+} lower level to a Mn^{3+} upper level. Using this assignment, Quijada *et al.*[40] interpreted their experimental spectra. However, the interatomic transition will cost on-site Coulomb repulsion energy U and JT splitting energy E_{JT}. Since $U \geq 3.0$ eV and $E_{JT} \geq 1.0$ eV, which were determined from other experiments,[61,62] the peak corresponding to interatomic $Mn^{3+} \rightarrow Mn^{3+}$ transition should be located at a energy higher than 4.0 eV. Moreover, such an assignment cannot explain the Ca doping concentration dependence of the 1.5 eV peak strength.[44]

In our earlier paper, we assigned the 1.5 eV peak come from the intraatomic transition between JT split Mn^{3+} levels.[44] Quijada *et al.*, pointed out that the 1.5 eV peak could not originate from the intraatomic transition, since both the initial and the final states had *d*-symmetry with respect to the same origin. However, we think that there are a couple of points to be considered. First, there is a strong hybridization of the e_g band with the O 2*p* band. Second, recent structure experiments suggests that the Mn-O octahedron is strongly distorted locally and such a distortion becomes weaker as T decreases.[59,60] Using a cluster model of $(MnO_6)^{9-}$, we found that the intraatomic $d \rightarrow d$ transition becomes possible when the symmetry of the Mn-O octahedron becomes lower than orthorhombic. Moreover, this assignment can explain the decrease of the 1.5 eV peak at low T, which were observed by numerous optical measurements.[34-42] However, more systematic studies are required to get a better understanding on the origin of the 1.5 eV peak.

3. Polaron Absorption in $La_{0.7}Ca_{0.3}MnO_3$: A Crossover from Small to Large Polarons

Figure 8 shows the T-dependent $\sigma(\omega)$ of $La_{0.7}Ca_{0.3}MnO_3$ ($T_C \sim 250$ K).[41] As T decreases, the spectral weight is transferred from high to low energies. The crossover energy is around 0.5 eV. Similar spectral weight changes were observed in many other CMR materials, as shown in Figs. 3–5. As shown in Fig. 8 (b), the spectra below 0.5 eV can be characterized by two types of responses, i.e. a sharp Drude peak in a far-IR region and a broad absorption band in a mid-IR region. Then a corresponding conductivity spectrum can be written as a sum of the two contributions: $\sigma(\omega) = \sigma_{drude}(\omega) + \sigma_{mir}(\omega)$. A more detailed picture for the Drude peak is shown in the inset. As T decreases, $\sigma_{mir}(\omega)$ increases initially but becomes saturated around 120 K. However, $\sigma_{drude}(\omega)$ increases continuously without a saturation.

Such T-dependent behaviors of $\sigma_{mir}(\omega)$ are quite different from the predictions of the model of Furukawa. [See Fig. 1.] Therefore, we should strongly rely on the electron-phonon coupling effects suggested by Millis *et al.*[17-19] and Röder *et al.*[20] Above T_C, i.e. in the insulating regime, it is widely accepted that a small polaron plays an important role. And, $\sigma_{mir}(\omega)$ below 0.8 eV at 260 K can be fitted reasonably well with the small polaron model.[63,64] However, below T_C, it is not known clearly whether a large polaron coherent

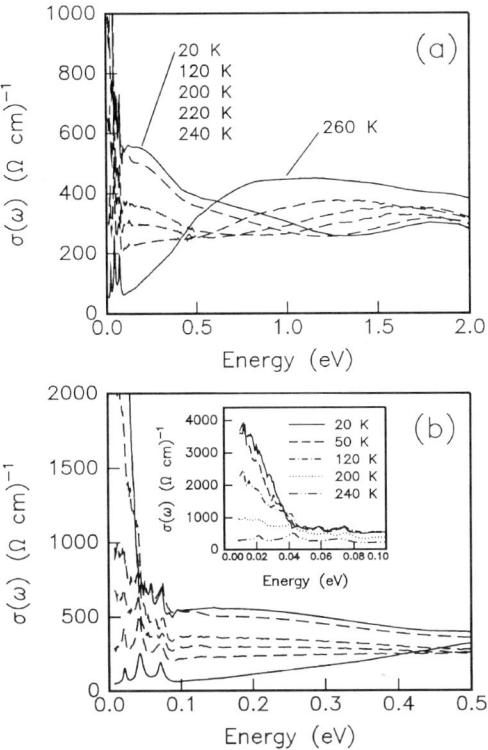

Figure 8. (a) Optical conductivity spectra of $La_{0.7}Ca_{0.3}MnO_3$ below 2.0 eV. (b) Optical conductivity spectra below 0.5 eV. Temperature ranges are the same as those in (a). Inset shows detailed optical conductivity spectra in far-IR region.

motion[20] or a small polaron tunneling[65] will be the origin of the metallic behavior.

Emin investigated frequency responses of large and small polaron absorptions.[66] A coherent band of large polaron should show up at frequencies below characteristic phonon modes and becomes more significant as T decreases. And, its photoionization should bring out an incoherent mid-IR band, which is very asymmetric and shows a long tail above its peak position. On the other hand, a coherent band of small polaron, i.e. polaron tunneling band, should occur at an energy region much lower than the characteristic phonon modes.

Behaviors of both $\sigma_{drude}(\omega)$ and $\sigma_{mir}(\omega)$ at $T \leq 120$ K, shown in Fig. 8, are quite consistent with the features of large polaron absorption, predicted by Emin. As shown in the inset, the Drude peak becomes evident just below the bending phonon mode frequency, i.e. ~ 330 cm^{-1}, and its width does not change too much. Moreover, $\sigma_{mir}(\omega)$ near 20 K is quite asymmetric and shows a long tail above its peak position. From this observation, it can be argued that *a crossover from a small polaron state to a large polaron state* accompanies the M-I transition in $La_{0.7}Ca_{0.3}MnO_3$. [The small to large polaron crossover was also observed by Yoon *et al.* using a diffusive electronic Raman scattering response.[67] And, there are some

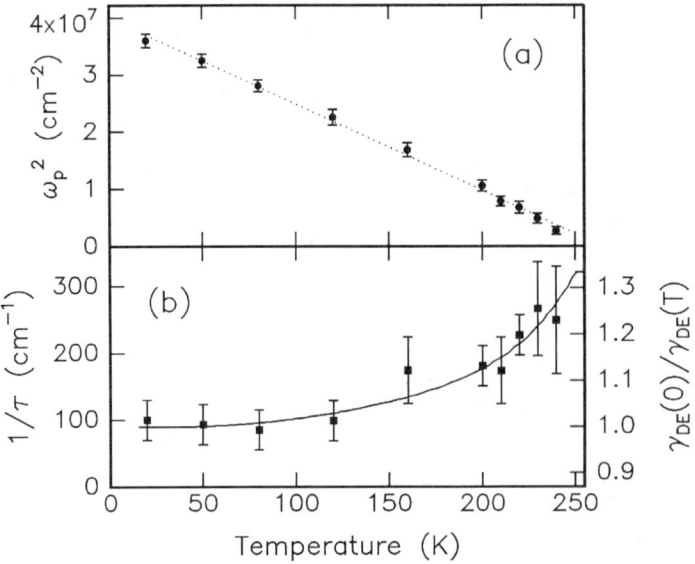

Figure 9. (a) Values of ω_p^2 below T_C. The dotted line is a linear guide for eye. (b) Scattering rates, $1/\tau$, for free carriers (solid squares). The behavior of $1/\gamma_{DE}(T)$ is overlapped as the solid line.

other experimental evidences for this crossover.[68,69]]

In the large polaron state, the coherent band can be described by the simple Drude model. Therefore, we could determine electrodynamic quantities, including plasma frequency ω_p^2 and scattering rate $1/\tau$. The detailed procedure is given in Ref. 41. As shown in Fig. 9 (a), ω_p^2 linearly increases with decreasing T. This behavior agrees with photoemission data which showed that density of states at the Fermi energy increased progressively below T_C.[61] [Recently, Alexandrov and Bratkovsky predicted carrier density collapse in doped charge-transfer insulators with strong electron-phonon coupling as a consequence of bipolaron formation in the paramagnetic phase.[70] However, more experimental studies are required to test this model.] As shown in Fig. 9 (b), $1/\tau$ remains nearly a constant below 120 K and starts to increase above 160 K. It should be noted that this behavior is not consistent with a prediction for the large lattice polaron: i.e. $1/\tau$ should be proportional to T.[66] So, we proposed that the $1/\tau$ behavior might come from the ferromagnetic alignment of spins below T_C. Actually, we found that $1/\tau$ could be scaled with $1/\gamma_{DE}(T) = 1/\langle\cos(\theta/2)\rangle$, shown as a solid line in Fig. 9 (b). This indicates that the polaron state should be also coupled with the spin degree of freedom.

As shown in Fig. 8 (b), the spectral weight of the Drude peak seems to be quite small at overall temperatures, compared with the doped carrier density, i.e. $n \approx 0.3$ hole per Mn. If we assume this value of n at 20 K, the corresponding ω_p^2 value predicts that the effective mass m^* should be about $13 m_e$, where m_e is the band mass. Then, the small values of the

Drude weight could be explained by the large effective mass of the coherent polaron motion especially at low T. In the case of the strongly coupled large polaron, $m^*/m_e = (1+0.02\alpha^4)$, where α represents the Fröhlich coupling constant.[71] Then, α can be estimated to be about 5, which suggests that *the large polaron in* $La_{0.7}Ca_{0.3}MnO_3$ *might be in a strong coupling regime*, which has not been realized in other physical system before. When the characteristic phonon frequency ω_0 was chosen to be about 300 cm^{-1}, the size of the large polaron was estimated to be $(3/\alpha)[\pi \hbar/(4m_e\omega_0)]^{1/2} \approx 7$ Å, which is in agreement with the theoretical work by Röder *et al.*[20] However, note that these parameters were obtained under assumption of the constant n value, and that our optical measurements can provide information only on (n/m^*). More studies are required to determine the values of n and/or m^* more accurately and the coupling strength of the large polaron.

4. A Scaling Behavior of Spectral Weight Change in $(La,Pr)_{0.7}Ca_{0.3}MnO_3$

To get a further insight on the polaron absorption bands, we performed systematic optical measurements on orthorhombically distorted $La_{0.7-y}Pr_yCa_{0.3}MnO_3$ (LPCMO) samples [y=0.0, 0.13, 0.4, 0.5, and 0.7].[42] Hwang *et al.* showed that dc transport properties of LPCMO were affected by a carrier hopping parameter, which could be varied systematically by the Pr doping.[72] [Recently, there are some works which suggest that phase separation should occur for the samples with large *y* values.[73]]

Figure 10 shows $\sigma(\omega)$ of the LPCMO samples. At room temperature, all the samples show very similar optical conductivity spectra: each of $\sigma(\omega)$ has a broad absorption features centered around 1 eV and vanishes as ω approach to zero. These gap-like features are consistent with the fact these samples are in insulating states at room temperature. For $Pr_{0.7}Ca_{0.3}MnO_3$, $\sigma(\omega)$ are nearly T-independent, since it remains in an insulating state down to 15 K. For the samples with y=0.13, 0.4, and 0.5, $\sigma(\omega)$ show quite strong T-dependence. As T decreases below T_C, the spectral features approximately above 0.5 eV decrease and those below 0.5 eV increase. Interestingly, for the y=0.4 sample, such spectral weight changes at all temperatures occur at one balancing point, fixed near 0.5 eV. As y increases, the mid-IR absorption features decrease systematically.

The effective number N_{eff} of carriers below a cutoff frequency, ω_c, was evaluated using the following relation:

$$N_{eff} = \frac{2m}{\pi e^2 n} \int_0^{\omega_c} \sigma(\omega)d\omega. \tag{5}$$

For actual evaluation, the value of $\hbar\omega_c$ was chosen as 0.5 eV. And in evaluating n, T-dependent volume change was considered properly from reported structural data.[74] Figure 11 shows the $N_{eff}(\omega_c)$ vs. T curves for LPCMO. Note that $N_{eff}(\omega_c)$ for each sample increases abruptly near T_C. As the Pr doping increases, $N_{eff}(\omega_c)$ decreases systematically at overall

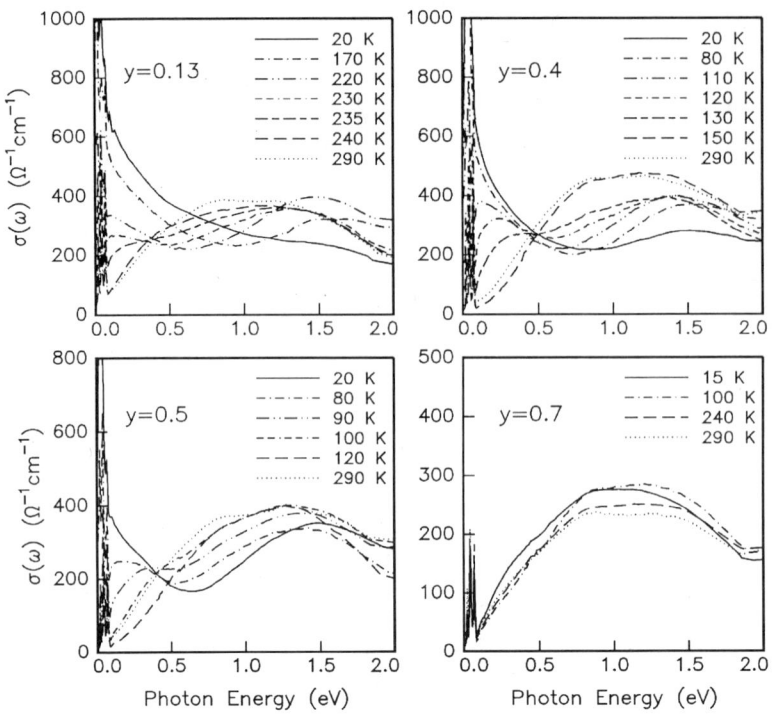

Figure 10. Optical conductivity spectra of $La_{0.7-y}Pr_yCa_{0.3}MnO_3$ below 2.0 eV.

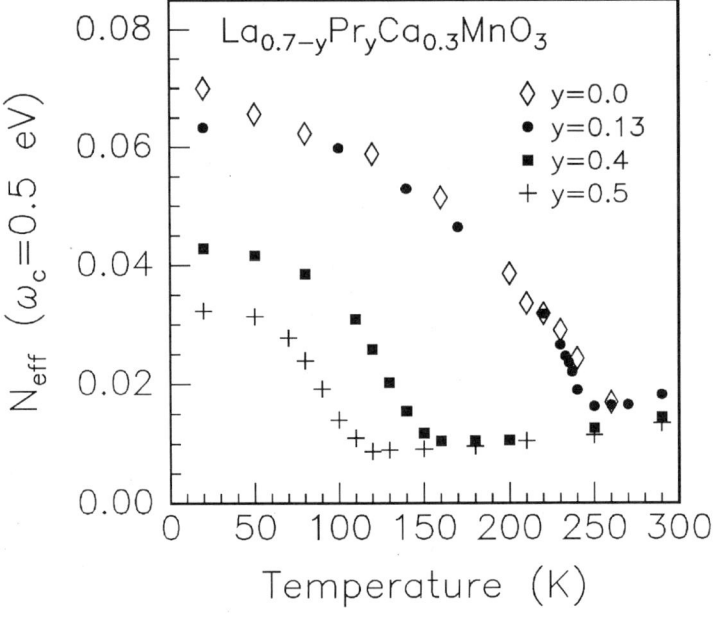

Figure 11. Temperature dependence of N_{eff} (ω_c=0.5 eV) for $La_{0.7-y}Pr_yCa_{0.3}MnO_3$.

temperature. It is interesting to note that the shape of $N_{eff}(\omega_c)$ for each sample is quite similar. And, the increasing rate of $N_{eff}(\omega_c)$ is nearly the same as that of T_C. Therefore, we plot the $N_{eff}(\omega_c)/T_C$ vs. T/T_C curves, shown in Fig. 12.

Interestingly enough, even if detailed spectral weight transfer behaviors in Fig. 11 are quite different, all of experimental values of $N_{eff}(\omega_c)/T_C$ fall into nearly one curve, shown in Fig. 12. The results demonstrate that the $N_{eff}(\omega_c)/T_C$ vs. T/T_C curves in these LPCMO compounds can be described by one scaling function. Moreover, as shown in the inset of Fig. 12, the scaling function is nearly linearly proportional to $\gamma_{DE}(T)$, which is a mean field value for the T-dependent DE bandwidth predicted by Kubo and Ohata.[75]

To explain this scaling behavior, we adopted the theoretical results by Röder et al.[20] who investigated a Hamiltonian including the DE and the JT electron-phonon coupling terms. If we assume that the mid-IR absorption below 0.5 eV, should be proportional to the hopping term in Eq. (3),

$$N_{eff}(\omega,T) \approx t\xi(\eta) <\cos(\theta/2)> \approx t\xi(\eta)\gamma_{DE}(T). \qquad (6)$$

Using Eqs. (4) and (6), $N_{eff}(\omega_c)/T_C$ becomes proportional to $\gamma_{DE}(T)$, when the hole

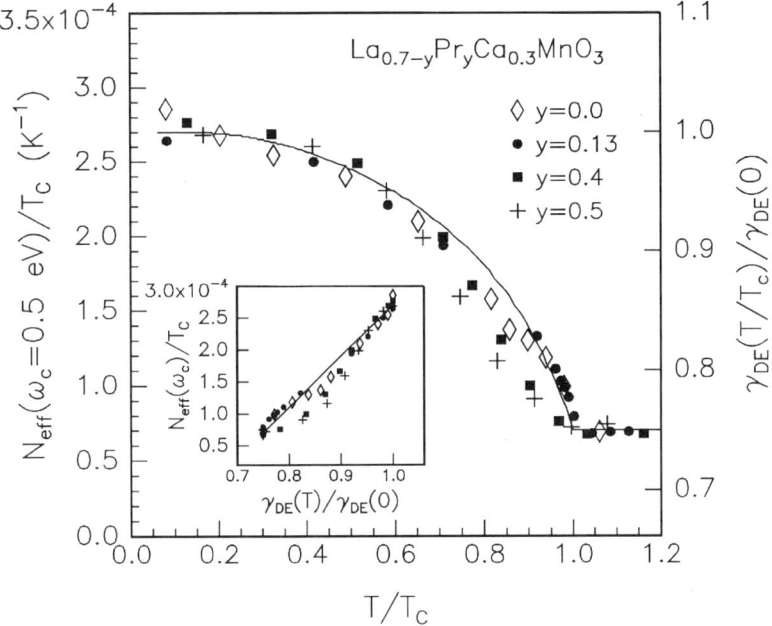

Figure 12. $N_{eff}(\omega_c=0.5\ \text{eV})/T_C$ vs. T/T_C curves of $La_{0.7-y}Pr_yCa_{0.3}MnO_3$. The solid line refers to the behavior of $\gamma_{DE}(T) \equiv <\cos(\theta/2)>$. The inset shows a linear scaling behaviors between $N_{eff}(\omega_c=0.5\ \text{eV})/T_C$ and $\gamma_{DE}(T)$.

concentration can be considered to be fixed, i.e. 0.3 per Mn atom, for all samples of LPCMO.

Note that our experimental data strongly suggest that the kinetic energy term should be scaled linearly with T_C. The variation of t was estimated to be less than 2 % in LPCMO compounds under the tight-binding approximation.[74] The small change in t is difficult to explain the large variation in $N_{eff}(\omega_c)$ and T_C. In the theory by Röder et al, both T_C and the interatomic transfer matrix from Mn^{3+} to Mn^{4+} will be significantly affected simultaneously by the change in $\xi(\eta)$, which comes from the self-amplifying coupling in both DE and JT effects. As far as we know, it is the only model which can explain such a scaling behavior.

5. Phonon Frequency Shift in $La_{0.7}Ca_{0.3}MnO_3$

Phonon modes are sensitive to local lattice distortions,[76,77] so infrared spectroscopy can be used to probe some properties of the local lattice distortions. In cubic perovskite materials, ABO_3, three optic phonons are infrared active.[78] The external mode corresponds to a relative motion of the A site ion with respect to the BO_6 octahedron. And, two internal modes, bending and stretching modes, are sensitive to the B-O-B bond angle and the B-O bond length, respectively.

Figure 13 shows far-IR $\sigma(\omega)$ of $La_{0.7}Ca_{0.3}MnO_3$ at various temperature.[79] Three strong

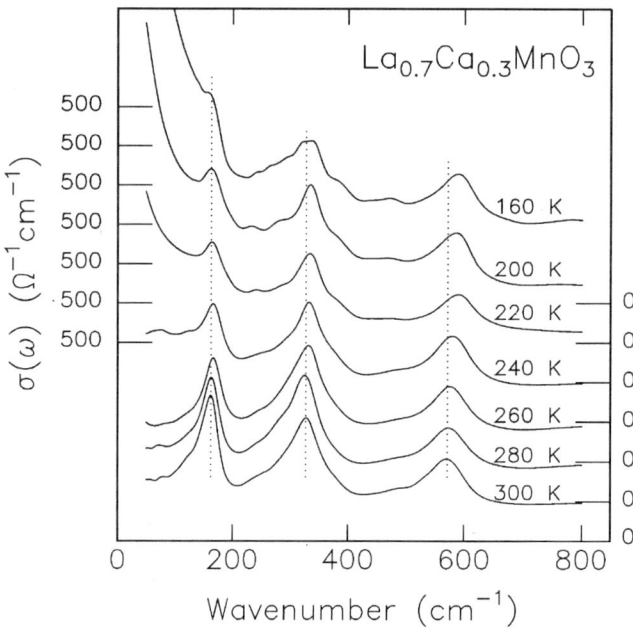

Figure 13. Optical conductivity spectra of a $La_{0.7}Ca_{0.3}MnO_3$. The room temperature positions of TO phonon frequencies are shown as the dotted lines.

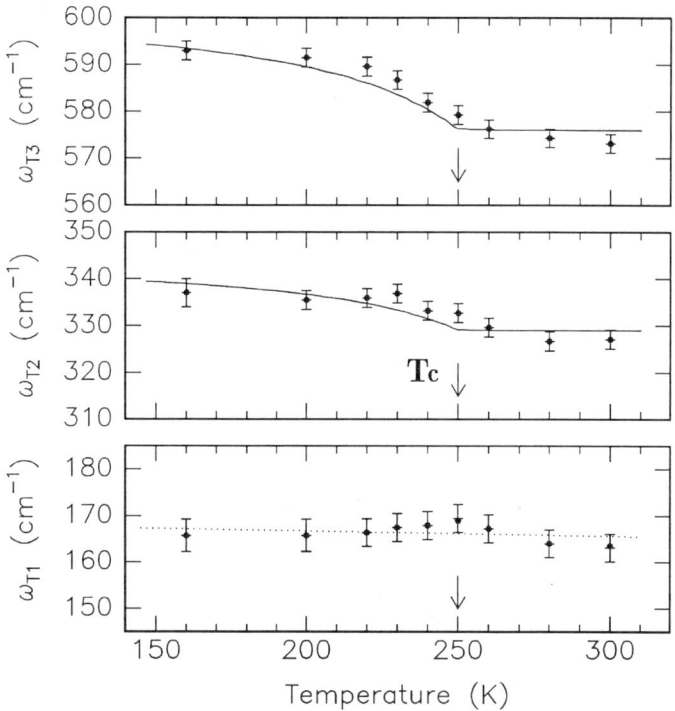

Figure 14. TO phonon frequencies at various temperatures. The solid lines for ω_{T2} and ω_{T3} are predictions of a model by Lee and Min. Refer to Ref. 65.

transverse optic (TO) phonon peaks can be clearly seen. The peaks located around 170, 330, and 580 cm^{-1} correspond to the external, the bending, and the stretching modes respectively. Below T_C, contribution of free carriers can be seen, and the bending and the stretching modes significantly shift to higher frequencies. Note that the phonon frequency of the external mode does not shift significantly.

Values of the external, the bending, and the stretching mode frequencies, which are denoted by ω_{T1}, ω_{T2}, and ω_{T3}, respectively, were obtained by the fitting with the Lorentz oscillators. The TO phonon frequencies obtained from such a fitting are shown in Fig. 14. While ω_{T1} is nearly T-independent, the internal phonon mode frequencies change notably around T_C: ω_{T2} and ω_{T3} shift to higher frequencies by about 10 cm^{-1} and 20 cm^{-1}, respectively. These frequency shifts indicate that the Mn-O-Mn bond angle and the Mn-O distance significantly change near T_C.

The anomalously large frequency shifts of the internal phonon modes indicate that there is a strong electron-phonon coupling in $La_{0.7}Ca_{0.3}MnO_3$. Recently, Lee and Min investigated the polaron transport and lattice dynamics using a model with the DE Hamiltonian and the electron-phonon interaction.[65] They predicted that a phonon frequency could shift due to a change in electron screening near T_C. The theoretical predictions based on this model are shown as solid lines in Fig. 14, and in good agreement with our experimental values. Therefore, this IR work strongly supports that the interplay between the DE and the electron-phonon interactions in $La_{0.7}Ca_{0.3}MnO_3$ becomes very important, which is also consistent with our conclusion in Section IV. 4.

Similar phonon frequency shifts were reported in Raman spectroscopy[80] and sound velocity[81] measurements. And, there were also several reports that internal phonon frequencies were very sensitive to lattice parameters, which could be controlled by doping rare earth elements or by changing substrate-induced stress in film samples.[82,83,84]

V. SUMMARY

In this article, we reviewed theoretical and experimental works on optical properties of colossal magnetoresistance materials. Our detailed studies on (La,Ca)MnO$_3$ at room temperature revealed that there exist two absorption features which can be attributed to a small polaron absorption and an optical transition between the Jahn-Teller split Mn^{3+} levels. When T becomes much lower than T_C, a crossover from a small to a large polaron seems to occur. And, detailed electrodynamics analyses suggested that the large polaron might be in a strong coupling regime, which has not been realized in a real system before. And, there should be couplings between charge, lattice, and spin degree of freedom. Studies on $(La,Pr)_{0.7}Ca_{0.3}MnO_3$ reveal that the total spectral weight of the Drude peak and the mid-infrared band is proportional to T_C. This scaling behavior can be explained by the theoretical model by Röder, Zang, and Bishop, which is based on the double-exchange and the electron-phonon interactions. Furthermore, the internal phonon modes show drastic phonon frequency shifts near T_C, which also demonstrates the importance of the electron-phonon interactions in the colossal magnetoresistance manganites.

ACKNOWLEDGEMENTS

We would like to acknowledge Prof. E. J. Choi, Jaejun Yu, B. I. Min, Y. H. Jeong, J.-G. Park, S. C. Lee, J. S. Kang, S.-J. Oh, M. S. Han, Y. W. Kwon, Y. S. Kwon, K.-T. Park, R. Ramesh, H. D. Drew, and Dr. Y. Chung for their useful discussions. We also appreciate H. J. Lee, Y. S. Lee, S. T. Lee, H. S. Choi, J. Y. Gu, J. S. Ahn, and N. J. Hur for their invaluable help in doing these researches. One of us (TWN) appreciates the hospitality of LG-CIT during his visit when this manuscript was written.

REFERENCES

1. G. H. Jonker and J. H. van Santen, Physica (Utrecht) 16:337 (1950); J. H. van Santen

and G. H. Jonker, *ibid* 16:599 (1950).
2. C. H. Chen and S-W. Cheong, Phys. Rev. Lett. 76:4042 (1996).
3. S. Mori, C. H. Chen, and S-W. Cheong, Nature 392:473 (1998).
4. H. Kuwahara, Y. Tomioka, A. Asamitsu, Y. Moritomo, and Y. Tokura, Science 270:961 (1995).
5. Y. Yamada, O. Hino, S. Nohdo, R. Kanao, T. Inami, and S. Katano, Phys. Rev. Lett. 77:904 (1996).
6. J. B. Goodenough, Phys. Rev. 100:564 (1955).
7. A. Asamitsu, Y. Moritomo, Y. Tomioka, T. Arima, and Y. Tokura, Naurte 373:407 (1995).
8. Y. Tomioka, A. Asamitsu, Y. Moritomo, H. Kuwahara, and Y. Tokura, Phys. Rev. Lett. 74:5108 (1995).
9. S. Jin, T. H. Tiefel, M. McCormack, R. A. Fastnacht, R. Ramesh, and L. H. Chen, Science 264:413 (1994).
10. R. von Helmolt, J. Wecker, B. Holzapfel, L. Schultz, and K. Samwer, Phys. Rev. Lett. 71: 2331 (1993).
11. J. Y. Gu, K. H. Kim, T. W. Noh, and K.-S. Suh, J. Appl. Phys. 78:6151 (1995).
12. J. Y. Gu, C. Kwon, M. C. Robson, Z. Trajanovic, K. Ghosh, R. P. Sharma, R. Shreekala, M. Rajeswari, T. Venkatesan, R. Ramesh, and T. W. Noh, Appl. Phys. Lett. 70:1763 (1997).
13. J. Y. Gu, S. B. Ogale, M. Rajeswari, T. Venkatesan, R. Ramesh, V. Radmilovic, U. Dahmen, G. Thomas, and T. W. Noh, Appl. Phys. Lett. 72:1113 (1998).
14. C. Zener, Phys. Rev. 82:403 (1951).
15. P. W. Anderson and H. Hasegawa, Phys. Rev. 100:675 (1955).
16. P. G. de Gennes, Phys. Rev. 118:141 (1960).
17. A. J. Millis, P. B. Littlewood, and B. I. Shraiman, Phys. Rev. Lett. 74:5144 (1995).
18. A. J. Millis, R. Mueller, and B. I. Shraiman, Phys. Rev. B 54:5389 (1996).
19. A. J. Millis, R. Mueller, and B. I. Shraiman, Phys. Rev. B 54:5405 (1996).
20. H. Röder, J. Zang, and A. R. Bishop, Phys. Rev. Lett. 76:1356 (1996).
21. R. Maezono, S. Ishihara, and N. Nagaosa, Phys. Rev. B 57:R13993 (1998).
22. S. Ishihara, J. Inoue, and S. Maekawa, Phys. Rev. B 55:8280 (1997).
23. C. M. Varma, Phys. Rev. B 54:7328 (1996).
24. L. Sheng, D. Y. Xing, D. N. Sheng, and C. S. Ting, Phys. Rev. B 56:R7053 (1997).
25. T. Ishikawa, T. Kimura, T. Katsufuji, and Y. Tokura, Phys. Rev. B 57:R8079 (1998).
26. Y. Okimoto, Y. Tomioka, Y. Onose, Y. Otsuka, and Y. Tokura, Phys. Rev. B 57:R9377 (1998).
27. N. Furukawa, J. Phys. Soc. Jpn. 64:3164 (1995).
28. P. E. de Brito and H. Shiba, Phys. Rev. B 57:1539 (1998).
29. H. Shiba, R. Shiina, and A. Takahashi, J. Phys. Soc. Jpn. 66:941 (1997).
30. S. Ishihara, M. Yamanaka, and N. Nagaosa, Phys. Rev. B 56:686 (1997).

31. S. Ishihara and N. Nagaosa, J. Phys. Soc. Jpn. 66:3678 (1997).
32. P. Horsch, J. Jaklic, and F. Mack (Report No. cond-mat/9807255).
33. J. Zang, A. R. Bishop, and H. Röder, Phys. Rev. B 53:R8840 (1996).
34. Y. Moritomo, A. Machida, K. Matsuda, M. Ichida, and A. Nakamura, Phys. Rev. B 56:5088 (1997).
35. A. Machida, Y. Moritomo, and A. Nakamura, Phys. Rev. B (in press).
36. A. Machida, Y. Moritomo, and A. Nakamura (submitted).
37. Y. Okimoto, T. Katsufuji, T. Ishikawa, A. Urushibara, T. Arima, and Y. Tokura, Phys. Rev. Lett. 75:109 (1995).
38. Y. Okimoto, T. Katsufuji, T. Ishikawa, T. Arima, and Y. Tokura, Phys. Rev. B 55:4206 (1997).
39. S. G. Kaplan, M. Quijada, H. D. Drew, D. B. Tanner, G. C. Xiong, R. Ramesh, C. Kwon, and T. Venkatesan, Phys. Rev. Lett. 77:2081 (1996).
40. M. Quijada, J. Černe, J. R. Simpson, H. D. Drew, K. H. Ahn, A. J. Millis, R. Shreekala, R. Ramesh, M. Rajeswari, and T. Venkatesan (Report No. cond-mat/9803201).
41. K. H. Kim, J. H. Jung, and T. W. Noh, Phys. Rev. Lett. 81:1517 (1998).
42. K. H. Kim, J. H. Jung, D. J. Eom, T. W. Noh, Jaejun Yu, and E. J. Choi (Report No. cond-mat/9804284).
43. J. H. Jung, K. H. Kim, H. J. Lee, J. S. Ahn, N. J. Hur, T. W. Noh, M. S. Kim, and J.-G. Park (submitted).
44. J. H. Jung, K. H. Kim, T. W. Noh, E. J. Choi, and Jaejun Yu, Phys. Rev. B 57:R11043 (1998).
45. J. H. Jung, K. H. Kim, D. J. Eom, T. W. Noh, E. J. Choi, Jaejun Yu, Y. S. Kwon, and Y. Chung, Phys. Rev. B 55:15489 (1997).
46. K. H. Kim, J. Y. Gu, H. S. Choi, D. J. Eom, J. H. Jung, and T. W. Noh, Phys. Rev. B 55:4023 (1997).
47. H. S. Choi, J. Bak, J. S. Ahn, N. J. Hur, T. W. Noh, and J. H. Cho, J. Korean Phys. Soc. 33:185 (1998).
48. Edward D. Palik, *Handbook of Optical Constants of Solids,* Academic Press, London (1985).
49. T. Arima and Y. Tokura, J. Phys. Soc. Jpn. 64:2488 (1995).
50. J. H. Jung, K. H. Kim, H. J. Lee, T. W. Noh, E. J. Choi, and Y. Chung, J. Korean Phys. Soc. 31:L549 (1997).
51. H. L. Ju, H.-C. Sohn, and K. M. Krishnan, Phys. Rev. Lett. 79:3230 (1997).
52. T. Saitoh, A. E. Bocquet, T. Mizokawa, H. Namatame, A. Fujimori, M. Abbate, Y. Takeda, and M. Takano, Phys. Rev. B 51:13942 (1995).
53. T. Saitoh, A. Sekiyama, K. Kobayashi, T. Mizokawa, A. Fujimori, D. D. Sarma, Y. Takeda, and M. Takano, Phys. Rev. B 56:8836 (1997).
54. A. Chainani, M. Mathew, and D. D. Sarma, Phys. Rev. B 47:15397 (1993).
55. H. J. Lee and T. W. Noh (unpublished).

56. S. Yunoki, A. Moreo, and E. Daggoto (Report No. cond-mat/98007149).
57. Guo-meng Zhou (private communication).
58. P. A. Cox, *Transition Metal Oxides*, Clarendon Press, Oxford (1992).
59. C. H. Booth, F. Bridges, G. H. Kwei, J. M. Lawrence, A. L. Cornelius, and J. J. Neumeier, Phys. Rev. Lett. 80:853 (1998).
60. S. J. L. Billinge, R. G. DiFrancesco, G. H. Kwei, J. J. Neumeier, and J. D. Thompson, Phys. Rev. Lett. 77:715 (1996).
61. J.-H. Park, C. T. Chen, S-W. Cheong, W. Bao, G. Meigs, V. Chakarian, and Y. U. Idzerda, Phys. Rev. Lett. 76:4215 (1996).
62. C.-H. Park, D. S. Dessau, T. Saitoh, Z.-X. Shen, Y. Moritomo, and Y. Tokura (submitted).
63. R. Mühlstroh and H. G. Reik, Phys. Rev. 162:703 (1967).
64. A. S. Alexandrov and S. N. Mott, *Polarons and bipolarons*, World Scientific, Singapore (1995).
65. J. D. Lee and B. I. Min, Phys. Rev. B 55:12454 (1997).
66. D. Emin, Phys. Rev. B 48:13691 (1993).
67. S. Yoon, H. L. Liu, G. Schollerer, S. L. Cooper, P. D. Han, D. A. Payne, S-W. Cheong, and Z. Fisk, Phys. Rev. B 58:2795 (1998).
68. D. Louca, T. Egami, E. L. Brosha, H. Röder, and A. R. Bishop, Phys. Rev. B 56:R8475 (1997).
69. J.-S. Zhou, J. B. Goodenough, A. Asamitsu, and Y. Tokura, Phys. Rev. Lett. 79:3234 (1997).
70. A. S. Alexandrov and A. M. Bratkovsky (Report No. cond-mat/ 9806030).
71. G. D. Mahan, *Many-Particle Physics*, Plenum, New York (1990).
72. H. Y. Hwang, S-W. Cheong, P. G. Radaelli, M. Marezio, and B. Batlogg, Phys. Rev. Lett. 75:914 (1995).
73. J. B. Goodenough (private communication).
74. P. G. Radaelli, G. Iannone, M. Marezio, H. Y. Hwang, S-W. Cheong, J. D. Jorgensen, and D. N. Argyriou, Phys. Rev. B 56:8265 (1997).
75. K. Kubo and N. Ohata, J. Phys. Soc. Jpn. 33:21 (1972).
76. T. Riste, *Electron-Phonon Interaction and Phase Transitions*, Plenum, New York (1977).
77. J. B. Goodenough, A. Wold, R. J. Arnott, and N. Menyuk, Phys. Rev. 124:373 (1961).
78. M. D. Fontana, G. Metrat, J. L. Servoin, and F. Gervais, J. Phys. C 16:483 (1984).
79. K. H. Kim, J. Y. Gu, H. S. Choi, G. W. Park, and T. W. Noh, Phys. Rev. Lett. 77:1877 (1996).
80. J. C. Irwin, J. Chzanowski, and J. P. Franck (submitted).
81. A. P. Ramirez, P. Schiffer, S-W. Cheong, C. H. Chen, W. Bao, T. T. M. Palstra, P. L. Gammel, D. J. Bishop, and B. Zegarski, Phys. Rev. Lett. 76:3188 (1996).
82. L. Kebin, L. Xijun, Z. Kaigui, Z. Jingsheng, and Z. Yuheng, J. Appl. Phys. 81:6943 (1997).

83. A. V. Boris, N. N. Kovaleva, A. V. Bazhenov, A. V. Samoilov, N.-C. Yeh, and R. P. Vasquez, J. Appl. Phys. 81:5756 (1997).
84. A. J. Millis, T. Darling, and A. Migliori, J. Appl. Phys. 83:1588 (1998).

POLARONS IN MANGANITES; NOW YOU SEE THEM NOW YOU DON'T

Simon. J. L. Billinge

Department of Physics and Astronomy and
Center for Fundamental Materials Research,
Michigan State University,
East Lansing, MI 48824-1116.

INTRODUCTION

In this paper we will give a brief review of the experimental evidence for the presence of lattice polarons in colossal magnetoresistant manganite materials, concentrating on our own work on the system $La_{1-x}Ca_xMnO_3$. We will describe how local structural distortions appear and disappear as a function of doping and temperature and relate this to the nature, localized or delocalized, of the doped carriers.

The possible importance of the lattice degrees of freedom to the physics of the manganites was recently pointed out by Millis *et al.*[1] based firstly on the observation that the pure double exchange model did not give quantitative agreement with experimental results, and secondly that in this mixed valent system the Jahn-Teller distortion is present[2] and could have a profound effect on the properties. This has stimulated a theoretical effort[1,3-5] and a large number of experiments designed to investigate the existence of polarons. More recently it has come to light that the charge distribution may be inhomogeneous on a microscopic length-scale in these materials. Phases of ordered charge-stripes have been seen at certain concentrations.[6] Also, there are theoretical predictions that doped holes will phase separate into charge rich and charge poor regions.[7] The importance of orbital ordering is also now in the spotlight.[8-11]

Because of the strong coupling of the charges to the lattice through the Jahn-Teller distortion, the above mentioned effects are apparent in the atomic structure and probes of local structure are sensitive to, and can give information about, them. The importance of *local* structural information is that, in general, the charge density modulations (stripes or otherwise) are not long-range ordered and may also be fluctuating in time. In that case the structural distortions will not be apparent in the average structure, but will be present in the local structure.

The paper will be organized as follows. First we give a brief overview of the general experimental evidence for polarons in these materials. We then concentrate on local structural evidence. We concentrate on our own work using the atomic pair distribution function (PDF) analysis technique to study the $La_{1-x}Ca_xMnO_3$ system; however, we also mention related local structural work from extended x-ray absorption fine structure (XAFS) and PDF studies. We first describe the PDF technique. We then describe the temperature dependence of the local structure in the colossal magnetoresistant (CMR) region of the phase diagram, the doping dependence of local structure and, finally, we describe the interesting marginal behavior in the $x = 0.5$ doped material.

The reader is referred to a number of excellent recent reviews covering the experimental[12] and theoretical[4] situations in the manganites, as well as other articles in this volume.

EXPERIMENTAL EVIDENCE FOR POLARONS IN MANGANITES

We briefly review some of the experimental evidence for the existence of polarons in the perovskite manganites. Firstly, activated transport is observed at high temperature in the doped manganites[13–15] and the data are well fit with by adiabatic small polaron hopping.[15] Also, in the ferromagnetic metallic phase at lower-T the resistivity is found to be proportional to $\exp(-m(h,t)/m_0)$, where m is the magnetization[16,17] which suggests that the transport, even in the ferromagnetic phase, is controlled by magnetic polarons.[4] The thermopower at high temperature is also typically polaronic[15,18–21] and an unusual Hall effect of the wrong sign[22,23] is well (and self-consistently) explained by the presence of small polarons.[22] The local magnetic moment measured using electron paramagnetic resonance[24] (EPR) also indicates the presence of magnetic clusters with an effective moment of ~ 30 close to T_c. An isotope effect in the EPR indicates that the polarons have a lattice component.[25] The connection of the lattice to charge transport was also indicated by ion channeling measurements[26] which report a strong correlation between the resistivity and the temperature dependence of atomic displacements.

Optical conductivity measurements can show excitations which originate from polarons. Early results showed that no Drude peak is apparent at high temperature, but appears in the ferromagnetic metallic phase.[27,28] These authors also saw a broad feature above 1 eV which Kaplan et al.[28] interpreted as coming from a charge transition from an Mn^{3+} to an adjacent Mn^{4+} site, the excitation being related to the polaron binding energy. A crossover from small polarons at high-T to large polarons at low-T was also inferred from an Infrared reflectivity study of phonon frequency shifts at T_c.[29] More recently, the broad midinfrared features have been analyzed more quantitatively and assigned to large polaron bands in the metallic state.[30]

Direct evidence for lattice polarons from local structural studies will be elaborated more in the next section. They include PDF studies,[31–34] and x-ray absorption fine structure (XAFS) studies.[35–41] Other techniques giving information about local order have also been applied. Muon spin resonance studies indicate an inhomogeneous moment distribution and the muon relaxation varying in a manner similar to spin-glasses.[42] High resolution transmission electron microscopy is now showing direct evidence for charge-stripe ordering in these materials,[6,43–45] and the observation of charge ordering superlattice peaks in neutron diffraction[46,47] shows beyond doubt that the lattice responds to the charge order since the neutron superlattice peak is coming from a structural distortion accompanying charge order. Diffraction studies of the average structure[48,49] are also consistent with the presence of local structural distortions and disordered Jahn-Teller distortions. Small angle neutron scattering[50] has also been used to observe microscopic inhomogeneities in the spin density.

Finally, the observation of a large isotope effect on T_c[51,52] is direct evidence that the lattice is involved in the ferromagnetic transition.

LOCAL STRUCTURAL EVIDENCE FOR POLARONS IN MANGANITES

Atomic Pair Distribution Function Method

We have used the Atomic Pair Distribution Function Method for analyzing neutron powder diffraction data. This method reveals information about the local, and intermediate range (on a nanometer length-scale), atomic structure directly. In this section we describe how the PDF is obtained experimentally and how it is interpreted, giving a few illustrative examples.

The atomic pair distribution function, $G(r)$, can be obtained from powder diffraction data through a sine Fourier transformation:[53]

$$G(r) = 4\pi r[\rho(r) - \rho_0] = \frac{2}{\pi}\int_0^\infty Q[S(Q) - 1]\sin Qr\, dQ, \qquad (1)$$

where $\rho(r)$ is the microscopic pair density, ρ_0 is the average number density, $S(Q)$ is total structure function which is the normalized scattering intensity, and Q is the magnitude of the scattering vector, $Q = |\mathbf{k} - \mathbf{k_0}|$. For elastic scattering, $Q = 4\pi \sin\theta/\lambda$, where 2θ is the scattering angle and λ is the wavelength of the scattering radiation.

The PDF is a measure of the probability of finding an atom at a distance r from another atom. It has been used extensively for characterizing the structure of disordered materials such as glasses and liquids. However, the same approach can be applied to study crystalline materials.[54-56] In the past this was rarely the method of choice for studying well ordered materials. However, a need to characterize disorder in new materials, coupled with improved technologies for carrying out the experiments, is now making this approach more interesting. The advent of powerful synchrotron x-ray and pulsed neutron sources are allowing high quality data to be routinely collected over wide ranges of Q, and modern high speed computing is allowing efficient data analysis and modeling to be carried out. The experimental determination of PDFs has been described in detail elsewhere[53,56,58,59] and we will not discuss it here.

An illustrative example of the capabilities of this technique is provided by solid C_{60}. The famous "bucky ball" C_{60} molecule is a rigid, well defined, soccer ball shaped molecule which is shown in Fig. 1. These molecules form into a cubic solid, also shown in the Figure. At room temperature the balls are spinning. What we expect to see in the PDF is the following. The balls themselves are rigid and well defined objects; the correlations between the atoms on the balls are well defined and should give sharp peaks in the PDF. However, the balls are spinning independently of each other and there are no orientational correlations between the balls at this temperature. This means that a particular atom on one ball does not have a specific relationship to an atom on another ball and there are no inter-ball atom-atom correlations. In the PDF there should be no sharp peaks from the inter-ball correlation function. Since the diameter of the ball is 7.1 Å the longest intra-ball correlation possible is 7.1 Å and there should be no peaks in the PDF beyond this distance. This is exactly what is seen in the data shown in Fig. 1. The upper panel on the right shows the diffraction data

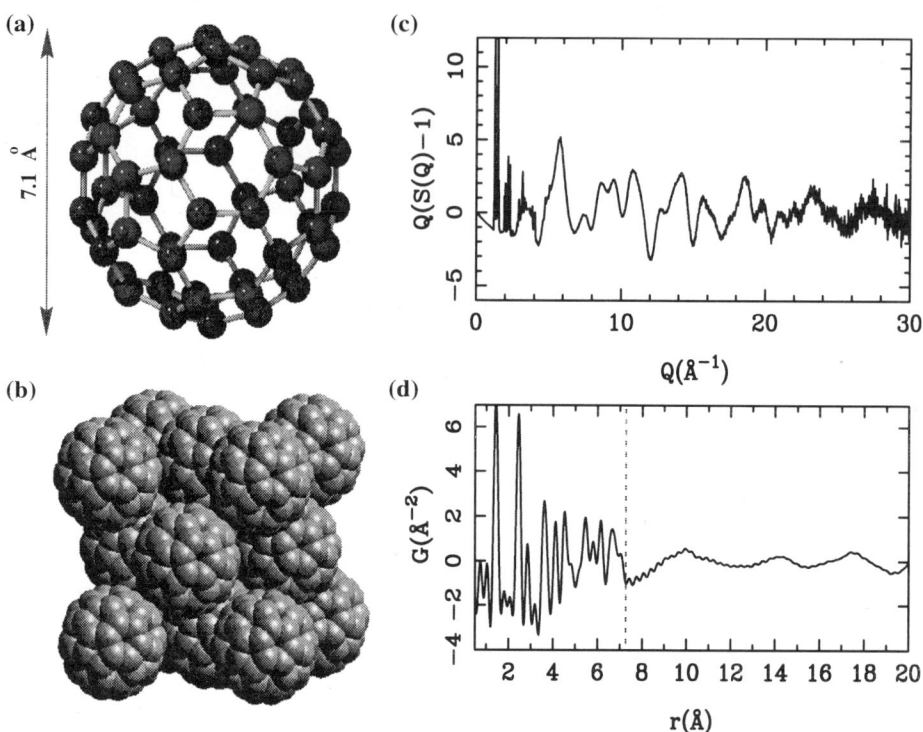

Figure 1. PDF of solid C_{60}. Data are neutron diffraction data collected at room temperature using the GLAD diffractometer at the Intense Pulsed Neutron Source at Argonne National Laboratory. On the left the C_{60} molecule is shown and below it the arrangement of the balls in the solid. The upper panel on the right shows the diffraction data plotted as $i(Q) = Q[S(Q) - 1]$. The resulting PDF is shown in the bottom panel on the right.

in the form of $Q[S(Q) - 1]$. Bragg peaks are apparent in the low-Q region. Beyond $Q \sim 5$ Å^{-1} there is only diffuse scattering which is coming from the balls themselves. The Fourier transform of these data is shown below. It is clear that $G(r)$ from these data behaves exactly as we expect. Within the same function, we are recovering the different short and intermediate range order in this material.

The small amount of structure evident in the inter-ball region of $G(r)$ can be explained as follows. The broad bumps centered at 10 Å, 14 Å and 18 Å come about because of the arrangement of the molecules themselves in space. This is the PDF you would expect to get from such an arrangement of balls of diameter 7.1 Å with a uniform atomic density smeared over their surface. The bumps appear at the nearest-neighbor, second-neighbor and third-neighbor distances of the cubic arrangement. The high frequency oscillations, which are also evident in Fig. 1(d), are a mixture of termination ripples coming from the sharp peaks of the intra-molecular PDF, and random noise in the data. The termination ripples are well controlled systematic errors in the data which come about because we Fourier transform a finite range data-set. We take account of these termination ripples in our modeling.[60]

The bucky ball example has a particular relevance to the manganite research. In the case of the manganites, information about the state of localization of the doped carriers is contained in the presence or absence of local Jahn–Teller distortions of the MnO$_6$ octahedra. These distorted octahedra are, in general, orientationally disordered and can also fluctuate dynamically. This is in complete analogy with the bucky ball case. We want to determine the size and shape of the individual MnO$_6$ octahedra, regardless of how they are arranged in space and whether or not they are orientationally ordered. It is clear from the C$_{60}$ example that the PDF will reveal exactly this information if we consider the intra-octahedral PDF. These are the peaks in the manganite PDF below $r = 3.0$ Å.

We obtain quantitative structural information from the PDF by comparing experimentally determined PDFs to PDFs calculated from model structures. An example is shown in Fig. 2. Model parameters such as atom positions are then varied in such a way as to improve the agreement between the calculated and measured PDFs. This is full-profile refinement technique which is analogous to Rietveld refinement of powder diffraction data[61] but which yields the local structure directly. A detailed description of the modeling procedure is given in Ref. 60 and less completely in Ref. 55.

Temperature Dependence of the Local Structure

In this section we consider the temperature dependence of the local structure of La$_{1-x}$Ca$_x$MnO$_3$ as the sample passes through the metal-insulator transition. This work is described in more detail in Ref. 31.

The main result is shown in Fig. 3 The PDF peak-height of the peak at $r = 2.75$ Å is plotted as a function of temperature for 3 different samples. Two of them $x = 0.21$ and $x = 0.25$, lie in the CMR region of the phase diagram and change from a paramagnetic insulator at high temperature to a ferromagnetic metal at low temperature. The temperature of the insulator-metal transition (T_{IM}) is indicated by arrows in the figure and was determined as the maximum in the resistivity. The resistivity measurements are shown as insets in the Figure. What is clear from the Figure is that the PDF peak sharpens in response to the IM transition. The sharpening is smooth and continuous and takes place over a wide range of temperature below T_{IM} with a clear onset at T_{IM}. The sample with no IM transition does not show any anomalous sharpening beyond what is expected from normal thermal effects.

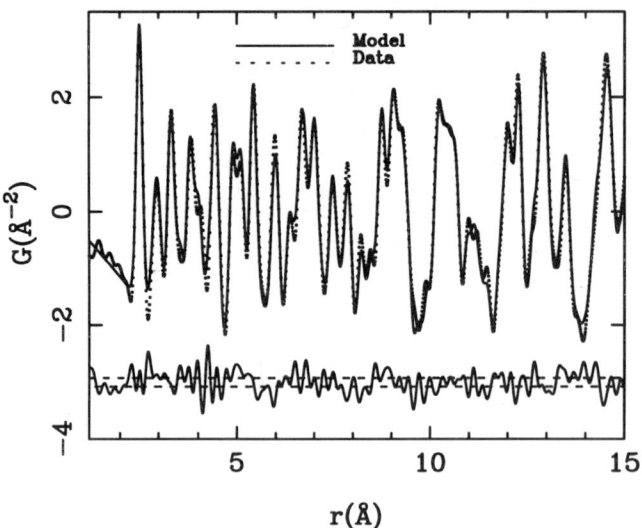

Figure 2. Example of a model-PDF (solid line) fit to data (dotted line). Below is the difference curve. The dashed lines indicate the random errors at the level of ±1 standard deviation. The data are from $YBa_2Cu_3O_{7-\delta}$ collected on HIPD at MLNSC, Los Alamos at 10 K. Reproduced from Ref. 60 with permission.

The PDF peak sharpening is interpreted in the following way. The width of the PDF peak gives the width of the *distribution* of bond lengths of the underlying atom-atom pair; in this case the O-O bonds on the MnO_6 octahedron. A sharper peak means a narrower distribution of bond lengths. In this case it means that in the ferromagnetic metallic (FM) phase the Jahn-Teller distortion is disappearing and the local structure is becoming more ordered. The implication is that the high-temperature paramagnetic insulating phase is polaronic but the polarons are disappearing as the sample enters the FM phase.

The change in the local structure as the sample went through the IM transition was modeled as a breathing mode type distortion of amplitude $\delta = 0.12$ Å occuring around one in four Mn sites and was ascribed to the formation of small lattice polarons as the sample went into the PI phase. Figure 4 shows two data-PDFs from below and above the IM transition, (i), and two model-PDFs without, and with, the breathing mode distortion, (ii). The *change* in the structure is evident in the difference curve plotted at the bottom of the Figure as open circles. What is clear is that the changes in the model-PDF (solid line in the difference curve) do a very satisfactory job of reproducing the structural changes seen in the sample. What is notable about the agreement between the two difference curves is the number of features, over such a large range of r, which are well explained by the distortion. This comes about from a model which has only a single parameter: the amplitude of the displacement of the breathing mode distortion. This is one of the strengths of the PDF technique: that data over a wide range of r are available to be fit in a self consistent manner and our interpretation does not have to rely on fitting a single feature.

We would like to point out here that because of the way that the modeling was carried out the atomic displacements which give rise to the breathing mode distortion

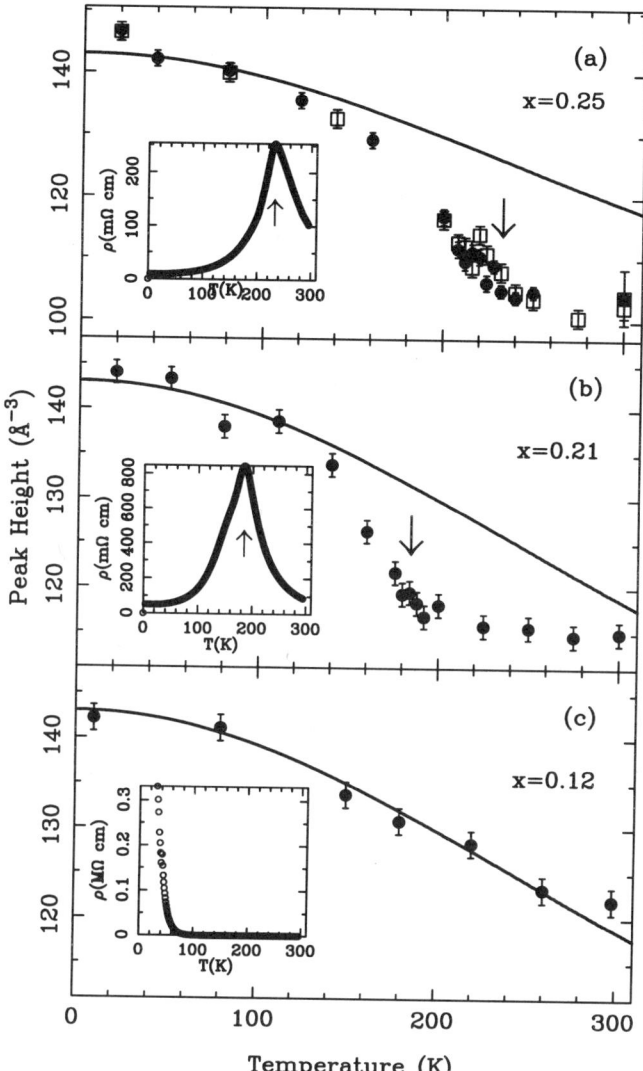

Figure 3. PDF peak height vs temperature of the PDF peak at $r = 0.275$ Å. This is from O-O correlations on the MnO_6 octahedra. The peak is seen to sharpen in response to the Insulator-metal transition. (a) $x = 0.25$ sample, (b) $x = 0.21$ sample, (c) $x = 0.12$ sample. No anomalous sharpening is evident in the $x = 0.12$ sample which has no metal-insulator transition. Resistivity data are shown in the insets. Arrows mark the temperatures of the metal-insulator transitions. Reproduced from Ref. 31 with permission.

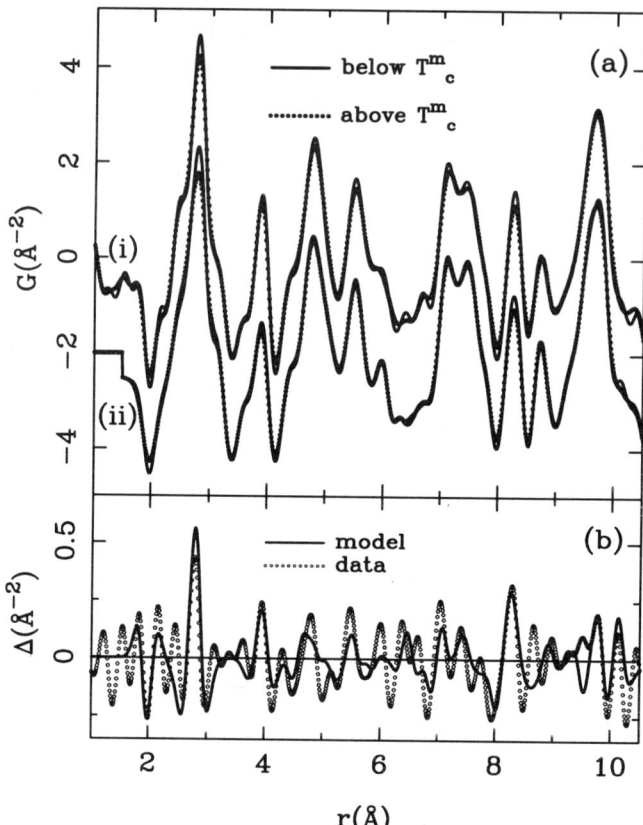

Figure 4. A comparison of PDFs with, and without, the breathing mode distortion which appears at the MI transition. (a) Data-PDFs are shown in (i) from above (circles) and below (solid line) T_{MI}. Model-PDFs are shown in (ii) with (circles) and without (solid line) the breathing mode distortion. (b) Difference curves from the PDFs in (a) showing the *change* in the structure: open circles - data, solid line - models. The changes in the PDF produced by introducing the breathing mode distortion into the model reproduce the changes in the data on going through the IM transition very well. Reproduced from Ref. 31 with permission.

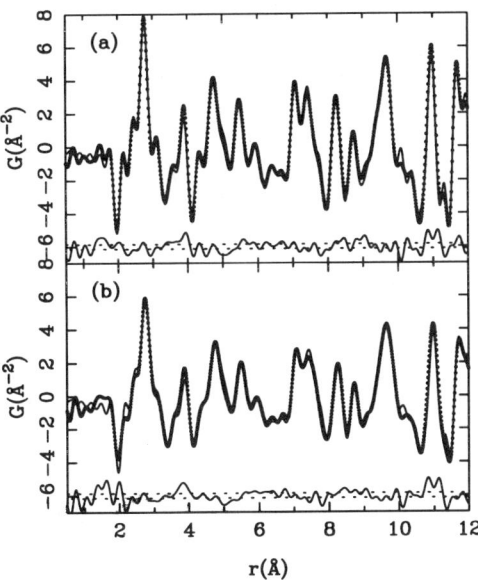

Figure 5. Best fit of the crystallographic model to the $x = 0.25$ data. (a) 10 K data, and (b) 220 K data. symbols are the data, solid line is the model-PDF. The dashed line around the difference curve is the expected level of uncertainty due to random counting statistics at the level of 2 standard deviations..

automatically result in a Jahn-Teller type distortion appearing on the remaining Mn sites. This means that the structural changes occuring at the IM transition are also consistent with a JT distortion *appearing* on 3/4 of the Mn sites as the sample goes from the low-temperature metallic phase to the high-temperature insulating phase.

The structural distortions modeled above disappear smoothly below T_{IM} over a wide temperature range. This raises an important question: is the structure locally distorted at the lowest temperature or do local structural distortions persist at the lowest temperature. The best way to answer this is to take the crystal structure of the material and see how well it manages to fit the PDF at low temperature. Good agreement between the crystal structural model and the PDF implies an absence of local disorder. The fit is shown in Fig. 5(a) for the $x = 0.25$, 10 K, data set. There is good agreement suggesting that the level of disorder is small.

There is further evidence of this if we concentrate on the intra-octahedral part of the PDF at low-r. This region of the PDF is expanded in Fig. 6(a). The crystallographic model is orthorhombic and there are three independent Mn-O bond lengths. However, in $La_{1-x}Ca_xMnO_3$ for $x = 0.25$ the orthorhombic distortion is very small and the 3 bond-lengths are essentially the same. In our modelling they refined to 1.96 Å, 1.97 Å and 1.97 Å, respectively. Thus, in the crystallographic model we have used to fit the data, the Jahn-Teller distortion has essentially completely gone. It has been pointed out previously[1] that if the orbitals are dynamic, or disordered, a local Jahn-Teller distortion can persist even when the average one disappears. This is clearly seen in PDF[33,62] and XAFS[36,37] data. The expected position of the long Mn-O bond is indicated by an arrow on the Figure. The position of the arrow was set by the position of the long Mn-O bond

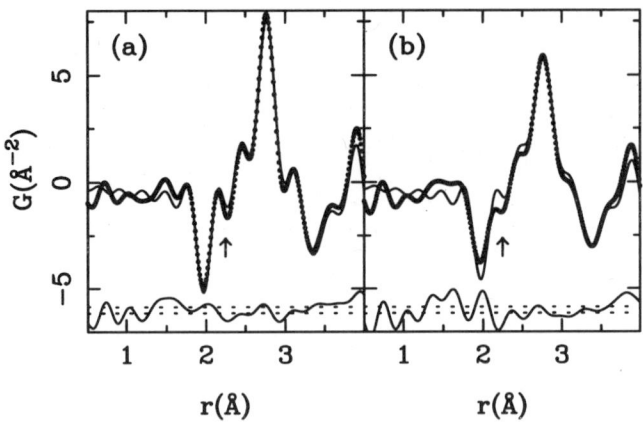

Figure 6. Same as Fig. 5 but the low-r region has been expanded. Data are from the $x = 0.25$ sample, (a) 10 K and (b) 220 K.

which is clearly observable in our data-PDF from an undoped sample of $LaMnO_{3.006}$ and is in good agreement with the long Mn-O bond seen crystallographically for the stoichiometric sample[63]. If a *local* JT distortion exists we might expect to see intensity close to the position of this arrow. Indeed there is a feature at this position. However, the modelling, shown as the solid line, shows that this feature is well explained as a termination ripple coming from the sharp peaks on either side of it. It might be argued that there is some additional intensity in the feature around 2.2 Å; however, this is small and puts an upper limit of a few percent of the number of sites which contain a JT distortion. Otherwise, the first Mn-O peak is well fit by the model. Furthermore, the presence of significant termination ripples around this peak indicates that it is a very sharp, resolution limited, peak lending further credence to the idea that there is just a single Mn-O bond length in this sample.

This modeling argues strongly that at the lowest temperature and well into the ferromagnetic region, (i) the structure is not disordered: the local structure agrees with the average structure, and (ii) there is no local Jahn-Teller distortion. This is consistent with the idea that the doped charge carriers are delocalized at low temperature and high enough doping and is in agreement with LDA calculations which suggest that when charges delocalize in the ferromagnetic doped compound, the Jahn-teller distortion disappears completely.[64]

This result is in qualitative agreement with XAFS[37,40,41] and independent PDF[33] results. The XAFS results suggest from an empirical peak-height analysis that residual disorder exists maybe up to 30%, but the same data analyzed for peak width seems to indicate that this may be an overestimate of the residual disorder (Ref. 37, Figure 8). The work by Lanzara *et al.*[40] also suggests that there is a small residual Jahn-Teller distortion although this is not observed in another XAFS study which suggests that the octahedra are locally regular in the FM phase[41]. It should be noted that the resolution of the distribution function obtained from XAFS is considerably lower than that of our PDFs and is only obtained from the XAFS data in an indirect way.[40] We will attempt to fit the model of Lanzara *et al.*[40] for the distorted MnO_6 octahedron to the PDF obtained directly from the diffraction data to look for evidence of the residual Jahn-

Teller effects they describe. The PDF work for the system $La_{1-x}Sr_xMnO_3$[33] reports that the local JT distortion disappears, but at a higher doping level than in the calcium doped case.

The common point emerging from all these studies seems to be that there is a change in character of the carriers, occurring smoothly with temperature, from delocalized at low temperature, to localized small polarons at high temperature. The IM transition coincides with the disappearance of the delocalized carriers. The detailed nature of the ground-state of the charge at low temperature; large polaron or fully delocalized, is not completely resolved at this point.

So what is the nature of the electronic state at intermediate temperatures? Local structural probes have something to say about this also. The situation is illustrated in Figs. 5(b) and 6(b). Intermediate, temperature (220 K) data are shown, again with a converged crystallographic model superimposed. The intra-octahedra region of the PDF is emphasized in Fig. 6(b). How does the MnO peak structure change as temperature is raised? First, it is clear that the peak broadens. This is evident from the fact that the termination ripples, which are so clear at 10 K (Fig. 6(a)), have been suppressed: the peak is no longer resolution limited but has a significant intrinsic width. The model peak is also broader, reflecting this fact, although the broadening in the model peak is due to increased thermal motion since the 3 Mn-O average bond lengths are all equal at 1.97 Å. What is apparent from Fig. 6(b) is that extra intensity is appearing in the data at 2.2 Å, the position of the long Mn-O bond in the undoped material. Again, the arrow in the figure is at the position of the long Mn-O bond in the undoped sample. There appears to be a coexistence of fully relaxed, fully Jahn-Teller distorted, octahedra and undistorted octahedra. If we make a correspondence between undistorted octahedra and delocalized carriers, and between distorted octahedra and localized carriers, the implication is that there is a coexistence of localized and delocalized carriers. We are in the process of trying to quantify this more precisely at present. This is similar to the observations of Lanzana et al.,[40] although again, complete correspondence between the XAFS and PDF data awaits a more quantitative study on the PDF data.

Doping Dependence of the Local Structure

As we have discussed, a response of the local structure is seen as the IM transition is crossed on cooling in the CMR region of the phase diagram. It is also possible to cross an IM transition as a function of doping at low temperature in the $La_{1-x}Ca_xMnO_3$ and $La_{1-x}Sr_xMnO_3$[12] (among others) systems. Is there a qualitative change in the local structure in response to crossing the IM transition in this way? Initial evidence from our data suggested that the local response is very similar crossing vs temperature or crossing vs doping.[65] The changes that are seen in the PDF between an $x = 0.21$ and an $x = 0.12$ sample at low temperature, are extremely similar to those observed between the $x = 0.21$ sample at 225 K and 140 K. This is shown in Fig. 7. It is clear that the two difference curves in Fig. 7(a) are highly similar. This means that the structural differences between each pair of data-sets are very similar. The triangles are differences between PDFs from the same sample ($x = 0.21$) collected above and below T_{IM}; the solid line shows differences between PDFs of two different samples ($x = 0.12$ and $x = 0.21$) collected at 10 K. Again, this spans the IM transition but this time as a function of doping. There is no reason to expect temperature dependent and doping dependent structure changes to be the same per se in this material. To underscore this fact, panel (b) of Figure has the difference between the $x = 0.12$ and $x = 0.21$ samples at 220 K (solid line) superimposed over the same temperature

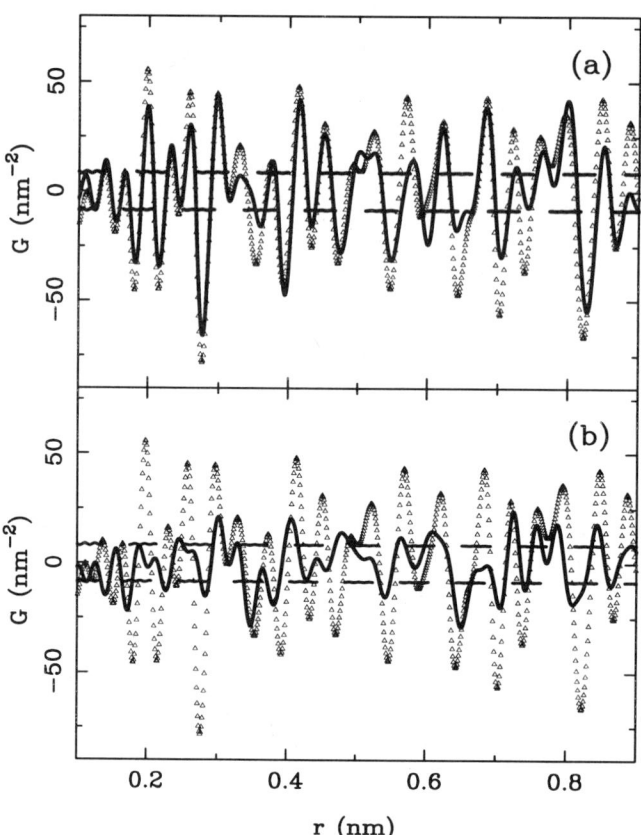

Figure 7. Difference curves between two PDFs showing local structural *differences*. (a) Triangles show the change in the local structure between $x = 0.21$ PDFs above and below T_{IM} ($G(225K) - G(140K)$). The solid line is the difference between $x = 0.21$ and $x = 0.12$ ($G(x = 0.12) - G(x = 0.21)$) at 10 K. (b) Same as (a) except the solid line is the difference between $x = 0.21$ and $x = 0.12$ ($G(x = 0.12) - G(x = 0.21)$) at 220 K. In this case both samples are in the insulating phase. Changes in the PDF are highly correlated when the sample crosses an IM transition regardless of whether it crosses it as a function of temperature or doping (see panel (a)). Reproduced from Ref. 65 with permission.

dependent difference curve shown in panel (a). This time the difference curve does not span the IM transition since both $x = 0.12$ and $x = 0.21$ samples are in the paramagnetic insulating phase at this temperature, and in this case the difference curves are largely uncorrelated. This suggests that it is the transition from insulator to metal that triggers the structure change and a different local structure is associated with the delocalized metallic state than the localized (polaronic) insulating state. It also suggests that small lattice polarons are present in the ferromagnetic insulating phase which exists between $x \sim 0.02$ and $x \sim 0.2$ at low temperature, similar to those in the paramagnetic insulating phase at high temperature in the CMR region.

Again, these results are similar to the XAFS results of Booth *et al.* in which the local structure in the FM phase is seen to be better ordered than in any of the insulating phases. As with our early PDF studies, the number of data points on the x-axis is sparse and it is not clear if the sharpening of the Mn-O bond distribution function (seen directly in the PDF and indirectly in the fourier transform of the XAFS data) occurs abruptly at the IM transition or evolves smoothly with doping.

A more complete PDF study[33] of doping dependence has been carried out on the strontium doped system, $La_{1-x}Sr_xMnO_3$. In this case there is seen to be no abrupt change in the Mn-O bond-length distribution in response to the IM transition. These authors integrated the area under the first PDF peak which was assumed to contain only the short Mn-O bonds (of length ~ 1.96 Å) in the Jahn-Teller distorted octahedra. The number evolved from 4 for $LaMnO_3$ to 6 for $x \sim 0.35$. This implies that the local Jahn-Teller distortion persists until $x = 0.35$ at low temperature in this material which is at higher doping than in the case of the Ca doped material. The data points have large error bars on them reflecting the difficulty in carrying out this integration, but the evolution of the number of short bonds appears to be smooth through the IM transition which occurs at $x = 0.17$ for this system. The behavior is qualitatively similar to that seen in both the XAFS,[37] and our own work, in that the number of short Mn-O bonds is increasing and the atomic probability distribution therefore becoming more sharply peaked as the sample becomes more highly doped and metallic.

Louca *et al*[33] argued that the fact that the number of short bonds evolves smoothly through the IM transition (there is no change in slope in response to the IM transition, though the data are somewhat sparse in the region of the IM transition), that the electronic transition must be percolative. Regions of metallic material are forming at lower dopings but the sample becomes metallic only when these regions can percolate the sample.

We wanted to see if the behavior in the Ca doped system is similar. We report here preliminary results from a recent experiment on the doping dependence of the local structure in the Ca doped system. Instead of integrating the area under the Mn-O PDF peak, we monitored the peak height of the short Mn-O peak at $r = 1.96$ Å in the PDF. The metal-insulator transition was found from resistivity measurements to be at $x = 0.18$ in our samples, a little lower than observed in other studies.[66] The behavior of the PDF peak height as a function of doping is shown in Fig. 8. The height of the peak has been corrected for the change in average scattering length of the sample as its composition changes. The peak height is seen to evolve smoothly through the IM transition with no apparent change in slope at the transition, similar to the observation in the Sr doped system.[33] It would thus appear that in the Ca doped system also, the IM transition is percolative. At low temperature, local Jahn-Teller distorted sites are disappearing consistently with doping until, in this case, $x \sim 0.25$ where they are completely gone. Again, if we associate the observation of Jahn-Teller distorted sites with the presence of localized polaronic charges, These persist at low temperature

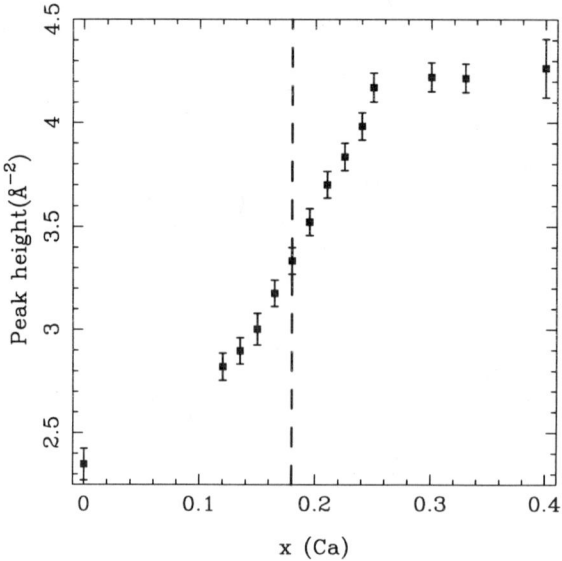

Figure 8. Evolution of the height of the Mn-O PDF peak at $r = 1.96$ Å as a function of doping (x). The peak comes from the Mn-O nearest neighbor correlation All data were taken at $T = 10$ K.

until $x = 0.25$ but are smoothly disappearing over the entire doping range, even in the ferromagnetic insulating phase. The most likely explanation would seem to be a coexistence of fully localized charges (on a single site) with "delocalized" (maybe only over a few atomic sites but maybe more) charges. The metallic regions presumably increase in size and number with doping until they are able to percolate through the sample at which point it becomes a macroscopic conductor. This can be taken as evidence for two-fluid behavior[67] (in the sense that there is a coexistence of localized and delocalized carriers) and for charge phase separation.[7]

$x = 0.5$: a Marginal Case

The special doping fraction of $x = 0.5$ is particularly interesting because it shows behavior marginal between ferromagnetic metallic for $x < 0.5$ and antiferromagnetic, charge ordered and insulating for $x > 0.5$. Not surprisingly the sample shows a rich mixture of magnetic and resistive behaviors. The PDF reveals that there is a competition between charge localization and delocalization as we have seen before, but that charge-ordering sets in to change the delicate energy balance and favor a localized ground-state.

The sample goes through three transitions on cooling. At 210 K it goes from a paramagnet to a ferromagnet. Around 160 K the charges order (CO). This is seen by the appearance of superlattice peaks in the diffraction pattern shown in the inset to Figure 9(a). For example, the superlattice peak can be seen as a shoulder on the (002) nuclear peak at $Q = 1.633$ Å^{-1} ($d = 3.846$Å). As the temperature decreases it moves to higher-Q until it reaches $Q = 1.667$ Å^{-1} ($d = 3.769$Å). This is the $[\frac{1}{2} - \epsilon, 0, 0]$ superlattice peak identified by Radaelli et al.[47] where ϵ is a temperature dependent incommensurability parameter. At low temperature (when the superlattice peak is

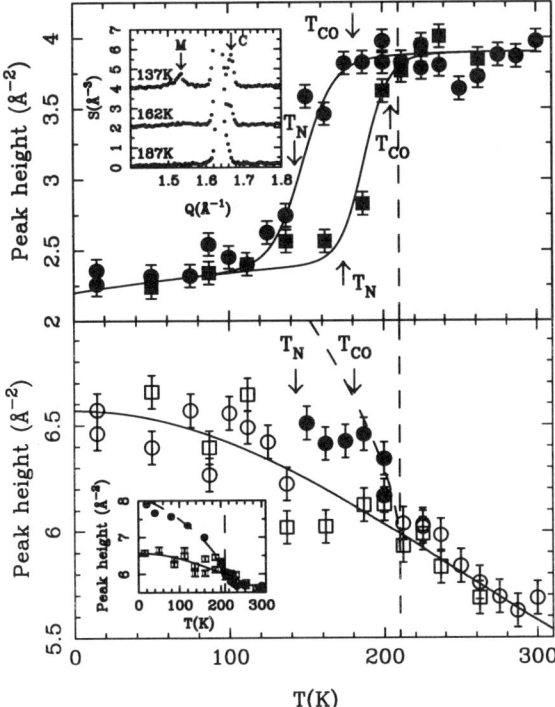

Figure 9. (a) height of PDF peak at $r = 16.2$ Å vs. temperature. The circles show the data measured on cooling and the squares are data collected on warming. The inset shows the appearance of superlattice peaks in the neutron diffraction data. Those marked with a C are charge ordering peaks and those marked with an M are magnetic peaks. (b) height of the PDF peak at $r = 2.6$ Å vs temperature. circles are data collected on cooling and squares are data collected on warming. The inset shows the same data as is shown in the main panel but with peak-height data from the $x = 0.25$ sample superimposed. Please see the text for a description of labels and annotations.

resolved in our data) ϵ is zero. Antiferromagnetic magnetic ordering does not occur until a lower temperature than CO and in our sample occurs between 150 K and 137 K at which point the magnetic superlattice peaks are easily identified in the inset to Figure 9(a). These transitions are also seen on warming but the AF and CO transitions are hysteretic and occur at higher temperatures than on cooling. The approximate temperatures where they occur are indicated by arrows in Figure 9(a) as determined from the appearance of superlattice peaks in the diffraction data. Again, the AF and CO transitions occur at different temperatures from each other.

Thus, at high temperature the sample's behavior is similar to the $x < 0.5$ behavior with polaronic paramagnetic insulating behavior giving over to ferromagnetism and more metallic behavior. At low temperature the sample resembles that of the $x > 0.5$ region with insulating antiferromagnetic and charge ordered behavior. Actually, our sample never becomes metallic in the FM region; however the resistivity curve bends down from the exponentially diverging behavior evident at higher temperature. We

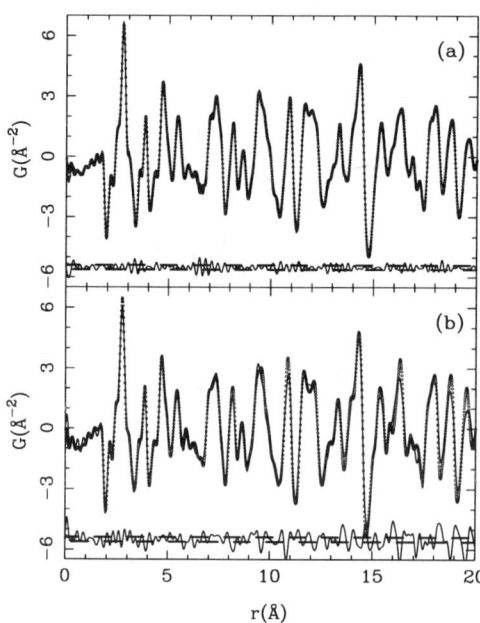

Figure 10. (a) PDFs measured at 112K, solid line while warming and circles while cooling. The difference is plotted underneath, dashed lines indicate two sigmas of statistical uncertainty. The data are reproduced within the statistical uncertainty. (b) As (a) but the data were measured at 162 K on cooling and warming. In this case the cooling data are charge-disordered and the warming data are charge-ordered.

refer to this region as the ferromagnetic incipient metallic (FIM) region.

We first discuss what happens in the intermediate range structure at the CO transition. Figure 10 shows the PDFs from our $x = 0.5$ sample at two temperatures: (a) 112 K and (b) 162 K. The two curves shown in each panel are from data collected on cooling (dots) and warming (solid line). At 112 K the sample is in the CO phase on cooling and warming and, as expected, the PDFs reproduce each other very well. This demonstrates the level of reproducibility expected from our measurements, which is well explained by the random errors of the measurement indicated by dashed lines in the difference curve. At 162 K the sample is charge ordered on warming, but not on cooling. Thus, Fig. 10(b) indicates local structural changes associated with charge melting. In the high-r region (above $r \sim 8$ Å) than in the low-r region. From this we deduce that the local octahedral distortions associated with the localized charges are present at all temperatures. However, when the sample goes through the charge melting transition the orbitals disorder and the average structure goes to a higher symmetry phase ($Pnma$ from $P2_1/m$) because of spatial and/or temporal fluctuations in the orbital and charge ordering. What is evident in the PDF is that at 162 K this disordering transition is already evident above ~ 8 Å and so the disorder is quite short-ranged. It is also evident that in the high-temperature phase the PDF is more sharply peaked. This is the opposite of what one expects from normal temperature effects. In this case it comes about because the high-temperature phase is higher symmetry giving rise to sharper PDF peaks.

To illustrate this effect we have considered a single PDF peak at 16.2 Å. The height of this peak is plotted in Fig. 9(a). The height of the peak is clearly correlated with the appearance of charge order in the sample as observed in the diffraction pattern. The hysteresis of the transition is clearly evident. The PDF is clearly sensitive to the CO transition. However, there is not a large response of the low-r peaks to the CO transition. This is shown in Fig. 9(b) where the peak height of the O-O peak on the MnO_6 octahedron, located at $r = 2.75$ Å, is shown. This strongly suggests that polaronic species survive above the CO temperature but are becoming disordered.

We now look at the behavior of the low-r O-O peak in more detail as this is a sensitive measure of the state, localized or delocalized, of the charge. The peak heights on cooling and warming reproduce each other very well, except in the hysteretic region. What is interesting here is that in this region, the PDF peak height deviates upwards *on cooling* and stays low on warming. This is opposite to normal hysteretic behavior where we expect the high-temperature behavior to persist to lower temperatures because of supercooling effects and the low temperature effects to persist to higher temperature because of superheating. The deviation of the PDF peak height in the hysteretic region cannot be explained as a simple response to the other changes happening in the structure. A deviation of this peak upwards on cooling is exactly the behavior we saw in the lower doped samples as they went through the IM transition. To show that this is the same response, but not as dramatic, we compare the peak height dependence of this sample with that of the $x = 0.25$ sample. This is shown in the inset. The $x = 0.25$ data have have been rescaled so that the peak-height of these peaks lines up at high-temperature. They do not otherwise have the same absolute peak height in this region because the number of Jahn-Teller distorted octahedra is not the same in the two samples. We have fit a function to the peak-height increase of the $x = 0.25$ data and this is shown as a dashed line in the figure. Details of the fit will be described elsewhere. This dashed line is reproduced in the main panel. What appears to be happening is that below 210 K, when the sample enters the FIM phase, the O-O PDF peak height is increasing in a similar way to that observed in the lower-doped materials as charge carriers begin to delocalize. However, at ~ 180 K charge ordering sets in and the delocalization process is interrupted. At this point the PDF peak height breaks off from the delocalization curve. However, some carriers are still delocalized and the peak is still above the solid line which is an extrapolation of the fully polaronic peak height. At this point the CO is incommensurate. This is because some of the carriers are actually delocalized. It is the polaronic carriers which are charge ordering and the density of *polaronic* carriers is less than 0.5/Mn. The charge ordering stabilizes the localized state with respect to the delocalized state and as the temperature is further lowered the delocalized carriers are gradually relocalized and the PDF peak height rejoins the localized curve. This is reflected in the incommensurability of the CO peak which gradually approaches the commensurate position (Fig. 9(a) inset and Ref. 47) as the number of *localized* carriers approaches the nominal 0.5/Mn. The PDF peak-height rejoins the line indicating fully localized carriers at the same point where the sample becomes antiferromagnetic and insulating again.

Thus, we see that there is a competition between the CO and the charge delocalization. The CO stabilizes the localized state at the expense of the delocalized one and suppresses the delocalization process which is able to go to completion at lower doping. On warming in our $x = 0.5$ sample, because of the hysteresis the CO does not melt before the sample becomes paramagnetic and insulating and so the upturn in the PDF peak height is not apparent.

SUMMARY

We have reviewed the evidence in local structural measurements for the existence of polarons in the phase diagrams of $La_{1-x}(Ca,Sr)_xMnO_3$ with greater emphasis being given to our own work on the Ca system. The existence or absence of localized carriers seems to be closely associated with the observation of a local Jahn-Teller distortion which essentially disappears in the metallic phase when charges are delocalized. Localized carriers coexist with delocalized carriers throughout most of the ferromagnetic metallic phase as two fluids, presumably in a microscopically phase separated form. As a function of temperature carriers progressively delocalize on cooling into the FM phase. When the IM transition is crossed as a function of doping the response of the local structure is less apparent and the change in the macroscopic resistivity is presumably coming about because the microscopic regions delocalized carriers percolate the sample.

The $x = 0.5$ sample tries to delocalize in a similar way as the lower doped materials. However, charge ordering of the localized carrier component sets in and stabilizes the localized carriers with respect to the delocalized carriers. This suppresses the delocalization process. The delocalized carriers are gradually relocalized as the temperature is lowered until all the carriers are localized and the sample becomes insulating and antiferromagnetic at low temperature.

ACKNOWLEDGEMENTS

I would like to acknowledge the help of the dynamic group of people who contributed directly to this work: R. G. DiFrancesco, E. Božin, M. F. Hundley, G. H. Kwei, J. J. Neumeier and J. D. Thompson. I would like to acknowledge stimulating discussions with A. R. Bishop, C. Booth, T. Egami, D. Louca, M. F. Thorpe, The work at MSU was supported by NSF grant DMR-9700966 and by the Alfred P. Sloan Foundation. Neutron data were collected at the Intense Pulsed Neutron Source at Argonne National Laboratory which is funded by the USDOE, BES-Materials Science under contract W-31-109-Eng-38, and at the Manuel Lujan Neutron Scattering Center (MLNSC) at Los Alamos National Laboratory which is funded by the US Department of Energy under contract W-7405-ENG-36.

REFERENCES

[1] A. J. Millis, P. B. Littlewood, and B. I. Shraiman, *Phys. Rev. Lett.* 74:5144 (1995).
[2] J. B. Goodenough, *Phys. Rev.* 100:564 (1955).
[3] H. Röder, J. Zang, and A. R. Bishop, *Phys. Rev. Lett.* 76:1356 (1996).
[4] A. R. Bishop and H. Röder, *Current Opinion in Solid State and Materials Science* 2:244 (1997).
[5] A. J. Millis, 53:8434 (1996).
[6] C. H. Chen and S.-W. Cheong, *Phys. Rev. Lett.* 76:4042 (1996).
[7] S. Yunoki, J. Hu, A. L. Malvezzi, A. Moreo, N. Furukawa, and E. Dagotto, *Phys. Rev. Lett.* 80:845 (1998).
[8] R. Maezono, S. Ishihara, and N. Nagaosa, *Phys. Rev. B* 57:R13993 (1998).
[9] S. Ishihara, J. Inoue, and S. Maekawa, *Phys. Rev. B* 55:8280 (1997).
[10] S. Ishihara, M. Yamanaka, and N. Nagaosa, *Phys. Rev. B* 56:686 (1997).
[11] S. Maekawa, *J. Magn. Magn. Mater.* 177-181:850 (1998).
[12] A. P. Ramirez, *J. Phys: Condens. Matter* 9:8171 (1997).

[13] G. M. Jonker and J. H. van Santen, *Physica (Utrecht)* 16:337 (1950).

[14] R. M. Kuster, D. A. Singleton, R. McGreevey, and W. Hayes, *Physica B* 155:362 (1989).

[15] M. Jaime, M. B. Salamon, M. Rubinstein, R. E. Treece, J. S. Horwitz, and D. B. Chrisey, *Phys. Rev. B* 54:11914 (1996).

[16] M. F. Hundley, M. Hawley, R. H. Heffner, Q. X. Jia, J. J. Neumeier, J. Tesmer, J. D. Thompson, and X. D. Wu, *Appl. Phys. Lett.* 67:860 (1995).

[17] B. Martinez, J. Fontcuberta, A. Seffar, J. L. Garcia-Munoz, S. Pinol, and X. Obradors, *Phys. Rev. B* 54:10001 (1996).

[18] V. H. Crespi, L. Lu, Y. X. Jia, K. Khazeni, A. Zettl, and M. L. Cohen, *Phys. Rev. B* 53:14303 (1996).

[19] A. Asamitsu, Y. Moritomo, and Y. Tokura, *Phys. Rev. B* 53:R2952 (1996).

[20] R. Mahendiran, S. K. Tiwary, A. K. Raychaudhuri, R. Mahesh, and C. N. R. Rao, *Phys. Rev. B* 54:R9604 (1996).

[21] M. F. Hundley and J. J. Neumeier, *Phys. Rev. B* 55:11511 (1997).

[22] M. Jaime, H. T. Hardner, M. B. Salamon, M. Rubinstein, P. Dorsey, and D. Emin, *Phys. Rev. Lett.* 78:951 (1997).

[23] P. Matl, N. P. Ong, Y. F. Yan, Y. Q. Li, D. Studebaker, T. Baum, and G. Doubinina, *Phys. Rev. B* 57:10248 (1998).

[24] S. B. Oseroff, M. Torikachvili, J. Singley, S. Ali, S. Cheong, and S. Schultz, *Phys. Rev. B* 53:6521 (1996).

[25] A. Shengelaya, G.-M. Zhao, H. Keller, and K. A. Müller, *Phys. Rev. Lett.* 77:5296 (1996).

[26] R. P. Sharma, G. C. Xiong, C. Kwon, R. Ramesh, R. L. Greene, and T. Venkatesan, *Phys. Rev. B* 54:10014 (1996).

[27] Y. Okimoto, T. Katsufuji, T. Ishikawa, A. Urushibara, T. Arima, and Y. Tokura, *Phys. Rev. Lett.* 75:109 (1996).

[28] S. G. Kaplan, M. Quijada, H. Drew, D. B. Tanner, G. C. Xiong, R. Ramesh, C. Kwon, and T. Venkatesan, *Phys. Rev. Lett.* 77:2081 (1996).

[29] K. H. Kim, J. Y. Gu, H. S. Choi, G. W. Park, and T. W. Noh, *Phys. Rev. Lett.* 77:1877 (1996).

[30] K. H. Kim, J. H. Jung, and T. W. Noh, *Phys. Rev. Lett.* 81:1517 (1998).

[31] S. J. L. Billinge, R. G. DiFrancesco, G. H. Kwei, J. J. Neumeier, and J. D. Thompson, *Phys. Rev. Lett.* 77:715 (1996).

[32] S. J. L. Billinge, R. G. DiFrancesco, M. F. Hundley, J. D. Thompson, and G. H. Kwei, unpublished.

[33] D. Louca, T. Egami, E. L. Brosha, H. Röder, and A. R. Bishop, *Phys. Rev. B* 56:R8475 (1997).

[34] T. Egami and D. Louca, Electron Lattice Coupling in Manganites and Cuprates, to appear in *J. Superconductivity*.

[35] A. T. Boothroyd, A. Mukherjee, and A. P. Murani, *Phys. Rev. Lett.* 77:1600 (1996).

[36] C. H. Booth, F. Bridges, G. J. Snyder, and T. H. Geballe, *Phys. Rev. B* 54:R15606 (1996).

[37] C. H. Booth, F. Bridges, G. H. Kwei, J. M. Lawrence, A. L. Cornelius, and J. J. Neumeier, *Phys. Rev. B* 57:10440 (1998).

[38] T. A. Tyson, J. Mustre de Leon, S. D. Conradson, A. R. Bishop, J. J. Neumeier, H. Röder, and J. Zang, *Phys. Rev. B* 53:13985 (1996).

[39] C. Meneghini, R. Cimino, S. Pascarelli, S. Mobilio, C. Raghu and D. D. Sarma *Phys. Rev. B* 56:3520 (1996).

[40] A. Lanzara, N. L. Saini, M. Brunelli, F. Natali, A. Bianconi, P. G. Radaelli, and S. Cheong, *Phys. Rev. Lett.* 81:878 (1998).

[41] G. Subías, J. Garcia, J. Blasco, M. G. Proietti, *Phys. Rev. B* 57:748 (1998).

[42] R. H. Heffner, L. P. Le, M. F. Hundley, J. J. Neumeier, G. M. Luke, K. Kojima, B. Nachumi, Y. J. Uemura, D. E. MacLaughlin, and S.-W. Cheong, *Phys. Rev. Lett.* 77:1869 (1996).

[43] A. P. Ramirez, P. Schiffer, S.-W. Cheong, C. H. Chen, W. Bao, T. T. M. Palstra, P. L. Gammel, D. J. Bishop, and B. Zegarski, *Phys. Rev. Lett.* 76:3188 (1996).

[44] C. H. Chen, S. Cheong, and H. Y. Hwang, *J. Appl. Phys.* 81:4326 (1997).

[45] S. Mori, C. H. Chen, and S.-W. Cheong, *Nature* 392:473 (1998).

[46] P. G. Radaelli, D. E. Cox, M. Marezio, S.-W. Cheong, P. E. Schiffer, and A. P. Ramirez, *Phys. Rev. Lett.* 75:4488 (1995).

[47] P. G. Radaelli, D. E. Cox, M. Marezio, and S.-W. Cheong, *Phys. Rev. B* 55:3015 (1997).

[48] P. Dai, J. Zhang, H. A. Mook, S.-H. Liou, P. A. Dowben, and E. W. Plummer, *Phys. Rev. B* 54:R3694 (1996).

[49] P. G. Radaelli, G. Iannone, M. Marezio, H. Y. Hwang, S.-W. Cheong, J. D. Jorgensen, and D. N. Argyriou, *Phys. Rev. B* 56:8265 (1997).

[50] J. M. de Teresa, M. R. Ibarra, P. A. Algarabel, C. Ritter, C. Marquina, J. Blasco, J. Garcia, A. del Moral, Z. Arnold, *Nature* 386:256 (1997).

[51] G.-M. Zhou, K. Conder, H. Keller, and K. A. Müller, *Nature* 381:676 (1996).

[52] J. P. Franck, I. Isaac, W. Chen, J. Chrzanowski, and J. C. Irwin, *Phys. Rev. B* 58:5189 (1998).

[53] B. E. Warren, *X-ray Diffraction*, Dover, New York, (1990).

[54] T. Egami, *Mater. Trans.* 31:163 (1990).

[55] S. J. L. Billinge and T. Egami, *Phys. Rev. B* 47:14386 (1993).

[56] T. Egami,

[57] in *Local Structure from Diffraction*, edited by S. J. L. Billinge and M. F. Thorpe, page 1, New York, 1998, Plenum.

[58] H. P. Klug and L. E. Alexander, *X-ray Diffraction Proceedures for Polycrystalline Materials*, Wiley, New York, 2nd edition, (1974).

[59] D. Grimley, A. C. Wright, and R. N. Sinclair, *J. Non-Cryst. Solids* 119:49 (1990).

[60] S. J. L. Billinge, in *Local Structure from Diffraction*, edited by S. J. L. Billinge and M. F. Thorpe, page 137, New York, 1998, Plenum.

[61] H. M. Rietveld, *J. Appl. Crystallogr.* 2:65 (1969).

[62] S. J. L. Billinge, unpublished.

[63] J. B. A. A. Elemans, B. Van Laar, K. R. Van Der Veen, and B. O. Loopstra, *J. Solid State Chem.* 3:238 (1971).

[64] W. E. Pickett and D. J. Singh, *Phys. Rev. B* 53:1146 (1996).

[65] R. G. DiFrancesco, S. J. L. Billinge, G. H. Kwei, J. J. Neumeier, and J. D. Thompson, *Physica B* 241-243:421 (1998).

[66] P. Schiffer, A. P. Ramirez, W. Bao, and S.-W. Cheong, *Phys. Rev. Lett.* 75:3336 (1995).

[67] M. Jaime and M. B. Salamon, unpublished.

OXYGEN ISOTOPE EFFECTS IN MANGANITES: EVIDENCE FOR (BI)POLARONIC CHARGE CARRIERS

Guo-meng Zhao[1,2], H. Keller[1], R. L. Greene[2], and K. A. Müller[1]

[1]Physik-Institut der Universität Zürich,
CH-8057 Zürich, Switzerland

[2]Center for Superconductivity Research
Physics Department, University of Maryland
College Park, MD 20742, USA

1 INTRODUCTION

The isotope effects observed in conventional superconductors show that lattice vibrations play an essential role in bringing about superconductivity. In particular, one finds that the superconducting transition temperature T_c varies as $T_c \propto 1/M^\alpha$ ($\alpha \simeq 0.5$) when the isotopic mass M of the material is varied. If lattice vibrations were not important in this phenomenon there is no reason why T_c should change with the mass of the ions. Thus, the isotope effect provides crucial insight into the microscopic theory of superconductivity in conventional superconductors.

The isotope effects observed in high-temperature superconductors (HTSC) should also place strong constraints on the microscopic pairing mechanism of high-temperature superconductivity. It has been shown that the isotope effects in HTSC are very unusual (see Ref. [1] - [11]). The oxygen isotope shift of the superconducting transition temperature T_c is small in the optimally doped cuprate superconductors, but the shift is very large, and even much larger than the BCS prediction as the doping is reduced towards the deeply underdoped regime. The large isotope effects in the underdoped cuprate superconductors may suggest that lattice vibrations also play an important role in bringing about high-temperature superconductivity. However the isotope effects cannot be simply explained by the conventional theory of superconductivity. Zhao and coworkers have initiated studies of the oxygen isotope effects on both the penetration depth and the carrier concentration in some cuprate superconductors [6, 5, 4, 8]. The results indicate that the effective supercarrier mass depends strongly on the oxygen isotope mass in the underdoped region, while the carrier concentration is independent of the oxygen isotope mass [6, 4, 8]. These isotope-effect results might suggest that polaronic carriers exist and condense into supercarriers in HTSC, which may give experimental support to the theory of (bi)polaronic superconductivity in HTSC [12].

On the other hand, little attempt had been made to investigate the isotope effects in magnetic materials before Zhao and coworkers [13] reported an observation of a giant oxygen isotope shift of the ferromagnetic transition in the colossal magnetoresistive manganite $La_{0.8}Ca_{0.2}MnO_{3+y}$. This is partly because in most materials, magnetic phenomena at room temperature and below are essentially unaffected by lattice vibrations. The atoms can in general be considered as infinitely heavy and static in theoretical descriptions of magnetic phenomena, so there should be no isotope effects in magnetic materials. However, the magnetic properties in the manganese-based perovskites are beyond what one might expect from the conventional theory of magnetism. The magnetic properties of these materials depend very strongly on the oxygen isotope mass, which suggests that lattice vibrations are important to the basic physics of manganites. In this paper we will begin with a brief review of some theoretical and experimental progress in manganite research in Section 2. We will introduce the concepts of polarons and bipolarons in Section 3. In Section 4, we will continue with a detailed review of various oxygen isotope effects observed in several manganite systems, and their implications for the microscopic mechanisms of the ferromagnetic ordering, charge-ordering and colossal magnetoresistance. The concluding remarks will be given in Section 5.

2 EXPERIMENTAL AND THEORETICAL PROGRESS IN MANGANITE RESEARCH

The doped manganites $Ln_{1-x}A_xMnO_3$ (where Ln is a trivalent rare earth ion and A is a divalent ion) have been found to exhibit some remarkable features. The undoped parent compound $LaMnO_3$ (with Mn^{3+}) is an insulating antiferromagnet [14]. When Mn^{4+} ions are introduced by substituting a divalent ion (e.g., Ca) for La^{3+}, the materials undergo a transition from a high-temperature paramagnetic and insulating state to a ferromagnetic and metallic ground state for $0.2 \leq x \leq 0.5$ [15]. For $x \geq 0.5$, the materials exhibit an insulating, charge-ordered and antiferromagetic ground state. The temperature at which the insulator-metal transition occurs can be increased by applying a magnetic field. As a result, the electrical resistance of the material can be decreased by a factor of 1000 or more [16, 17], if the temperature is held in the region of the transition. This phenomenon is now known as colossal magnetoresistance (CMR).

The physics of manganites has primarily been described by the double-exchange model [18, 19]. Crystal fields split the Mn 3d orbitals into three localized t_{2g} orbitals, and two higher energy e_g orbitals which are strongly hybridized with the oxygen p orbitals. Each manganese ion has a core spin of $S = 3/2$, and a fraction $(1 - x)$ have extra electrons in the e_g orbitals with spin parallel to the core spin due to strong Hund's exchange. The electron can hop to an adjacent Mn site with unoccupied e_g orbitals without loss of spin polarization, but with an energy penalty that varies with the angle between the core spins. This double-exchange model accounts qualitatively for ferromagnetic ordering and carrier mobility that depends on the relative orientation of Mn moments which near T_C will therefore be strongly dependent on the applied field. However, Millis, Littlewood and Shraiman [20] have pointed out that double-exchange alone cannot fully explain the data of $La_{1-x}Sr_xMnO_3$. They suggested that lattice-polaronic effects due to strong electron-phonon coupling (arising from a strong Jahn-Teller effect) should be involved. The basic argument [21] is that in the high-temperature paramagnetic state the electron-phonon coupling constant λ is large and the carriers are polarons, while the growing ferromagnetic order increases the bandwidth and thus decreases λ sufficiently for metallic behavior to occur below the Curie temperature T_C. The observed giant oxygen isotope effects [13, 22, 23] along with some other experimental results [24, 25, 26] confirm the polaronic nature of charge carriers in manganites. However, low temperature optical [27], electron-energy-loss (EELS) [28] and photoemission spectroscopies [29] are not consistent with those theoretical models [20, 21, 30]. A broad incoherent spectral feature [27] and a pseudogap in the excitation spectrum [29] were observed while the coherent Drude weight appears to be too small for that of a metal [27]. Recent EELS experiment [28] confirmed the early photoemission and O 1s x-ray absorption spectroscopies [31] of $La_{1-x}Sr_xMnO_3$, which showed that the doped holes are mainly of oxygen p character. However, a substantial mixture between the oxygen p orbitals and manganese e_g orbitals has not been ruled out by these experiments. Thus, the present understanding of the physics in manganites is far from complete, and further theoretical and experimental studies are essential.

3 POLARONS AND BIPOLARONS

The concept of polarons was first introduced by Landau (1933). If an electron is placed into the conduction band of an ionic crystal, the electron is "trapped by digging its own hole" due to a strong Coulombic interaction of the electron with its surrounding positive ions. The electron together with the lattice distortions induced by itself is called a polaron (lattice polaron). Lattice polarons are not 'bare' charge carriers, but are the carriers which are dressed by lattice distortions. Later on, the concept of polarons was treated in much great detail. One of the examples is the Holstein's treatment where an electron is trapped by self induced deformation of two-atomic molecules (Holstein polaron) [32, 33]. In this case, the polaron moves by thermally activated hopping at high temperatures with a diffusion coefficient $\omega a^2 \exp[-(E_p/2-t)/k_BT]$, where a is the lattice constant, ω is the vibration frequency, E_p is the polaron binding energy and t is the bare hopping integral. At low temperatures, the motion of the polarons is coherent, and the polaron behaves like a heavy particle with the effective bandwidth

$$W_{eff} \propto \exp(-\gamma E_p/\hbar\omega), \tag{1}$$

where γ is a dimensionless constant ($0 \leq \gamma \leq 1$) depending on E_p/t. Höck and coworkers [34] generalized the Holstein model to a system with a strong Jahn-Teller effect. The Jahn-Teller polarons can be formed when the Jahn-Teller stabilization energy E_{JT} is large compared to the conduction bandwidth. In this case, the polaron binding energy E_p can be simply replaced by E_{JT}. Since E_p is independent of the isotope mass in the harmonic approximation and $\omega \propto 1/\sqrt{M}$, Eq. 1 indicates that the effective bandwidth of polarons depends on the mass of ions. In the manganite $LaMnO_3$, E_{JT} was estimated to be about 0.5 eV [20] and in La_2CuO_4, $E_{JT} \simeq 1.2$ eV (Ref. [35]). Hence one might expect from Eq. 1 that the isotope effect on W_{eff} should be substantial in both manganites and cuprates.

The concept of a small onsite bipolaron was introduced by Anderson [36], and by Street and Mott [37]. When the attractive interaction between two polarons is larger than the Coulombic repulsion between them, the two polarons can be combined into a real-space pair which is called bipolaron. The onsite bipolaron consists of two polarons sitting on the same unit cell, while the intersite bipolaron includes two polarons sitting on the different unit cells. The onsite bipolaron is very heavy, so a small disorder will make the bipolaron immobile. On the other hand, the effective mass of intersite bipolarons is not so heavy, and thus mobile bipolarons could exist in a real material. The intersite bipolarons were identified in the compound Ti_4O_7 (Ref. [38]). At low temperatures, they form a charge-ordered state and are immobile. The recent normal-state susceptibility measurement in the cuprate superconductors $La_{2-x}Sr_xCuO_4$ [39]

indicates that intersite bipolarons coexist with free carriers in the normal-state. The detailed isotope experiments in this system [8] demonstrate that the mobile intersite bipolarons condense into a superfluid in the deeply underdoped samples.

4 VARIOUS OXYGEN ISOTOPE EFFECTS IN MANGANITES

4.1 Oxygen Isotope Shift of the Curie Temperature. A first observation of the oxygen isotope effect on the Curie temperature was made in $La_{1-x}Sr_xMnO_{3+y}$ system by Zhao and Morris (1995) [40]. In Fig. 1, the normalized magnetizations for the ^{16}O and ^{18}O samples of $La_{0.9}Sr_{0.1}MnO_{3+y}$ are plotted as a function of temperature. The oxygen isotope shifts of T_C were determined from the differences between the midpoint temperatures on the transition curves of the ^{16}O and ^{18}O samples. It is clear that the ^{18}O sample has a lower T_C than the ^{16}O sample by ~ 6.7 K. It should be noted that since the value of y is substantial (> 0.05) when samples are prepared below 1100 °C, the Curie temperatures in these samples are much higher than that for the corresponding single crystal samples where y is close to zero [41]. Actually the extra oxygen in the above chemical formula is related to the existence of cation vacancies.

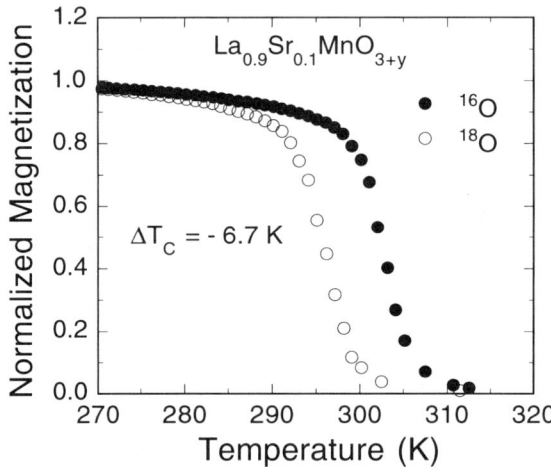

Figure 1: Oxygen isotope effect on the Curie temperature of $La_{0.9}Sr_{0.1}MnO_{3+y}$. The figure is reproduced from Ref. [40].

On the other hand, the oxygen-isotope shift of T_C in $La_{0.8}Ca_{0.2}MnO_{3+y}$ is very large [13], as seen from Fig. 2. The samples with the same isotope mass have the same T_C, while the samples with a heavier oxygen isotope mass (about 95% of ^{18}O) have a much lower T_C. The relative isotope shift of T_C is as large as 10%. Such a large oxygen isotope shift of the ferromagnetic transition is very unusual since lattice vibrations were believed to play no role in the magnetic interactions of most magnetic materials. It is a clear-cut experiment to establish what many have suspected - that atomic motion must be included in any viable description of the manganites [20]. It is also the first experiment in condensed matter physics, which demonstrates that there can be a giant isotope shift of a magnetic transition temperature.

In order to show that the observed isotope effect is intrinsic, we did oxygen isotope back-exchange ($^{16}O \rightarrow ^{18}O$; $^{18}O \rightarrow ^{16}O$). In Fig. 3, we show the normalized magnetization for the ^{16}O and ^{18}O samples of $La_{0.8}Ca_{0.2}MnO_{3+y}$ before and after isotope back-

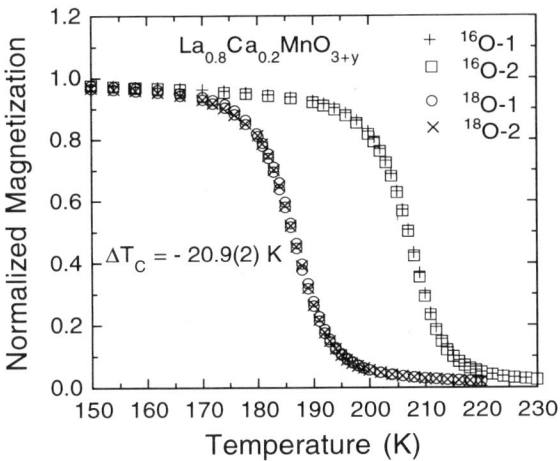

Figure 2: Oxygen isotope effect on the Curie temperature of $La_{0.8}Ca_{0.2}MnO_{3+y}$. The figure is reproduced from Ref. [13].

exchange. The symbol (+) denotes the ^{16}O sample which has been back-exchanged from the original ^{18}O sample (denoted by open square). The symbol (×) represents the ^{18}O sample which has been back-exchanged from the original ^{16}O sample (denoted by open circle). It is evident that the T_C of the ^{16}O (^{18}O) sample goes back completely to that of the original ^{18}O (^{16}O) sample after the isotope back-exchange. This clearly indicates that the shift of T_C is caused only by changing the oxygen isotope mass.

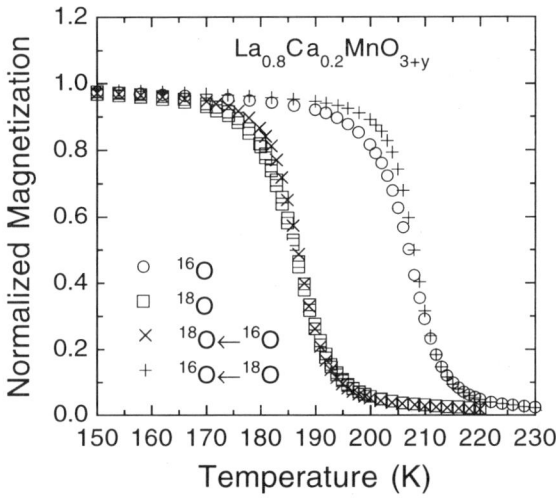

Figure 3: Oxygen isotope back-exchange result for $La_{0.8}Ca_{0.2}MnO_{3+y}$. The figure is reproduced from Ref. [13].

The oxygen isotope exponent is defined as $\alpha_O = - (\Delta T_C/T_C)/(\Delta M_O/M_O)$, where T_C and M_O are for the ^{16}O sample. With this definition, we obtain $\alpha_O = 0.85$ for $La_{0.8}Ca_{0.2}MnO_{3+y}$, and 0.14 for $La_{0.9}Sr_{0.1}MnO_{3+y}$. We have also investigated the oxygen isotope effect on the Curie temperature in other manganite systems. The results are summarized in Table 1. The magnitudes of α_O for $(La_{0.25}Nd_{0.75})_{0.7}Ca_{0.3}MnO_3$ were

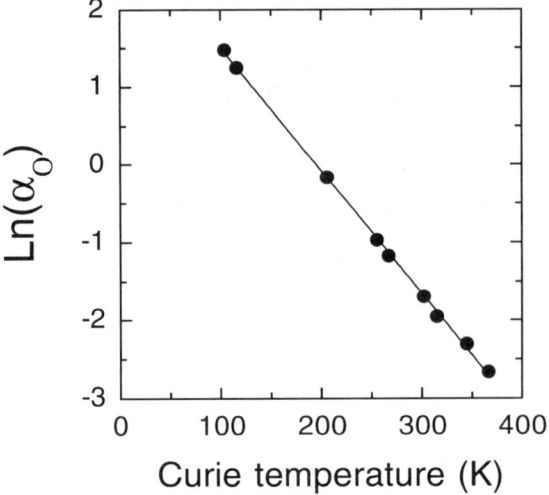

Figure 4: $\ln(\alpha_O)$ vs the Curie temperatures of the ^{16}O samples listed in Table 1.

Table 1: Summary of the oxygen isotope effects in manganites

	$T_C(^{16}O)$ (K)	α_O
$La_{0.67}Sr_{0.33}MnO_{3+y}$ [a]	367	0.070
$La_{0.66}Ba_{0.34}MnO_{3+y}$ [a]	345	0.10
$La_{0.85}Sr_{0.15}MnO_{3+y}$ [b]	315	0.14
$La_{0.90}Sr_{0.10}MnO_{3+y}$ [b]	302	0.18
$La_{0.67}Ca_{0.33}MnO_{3+y}$ [c]	267	0.31
$(LaMn)_{0.945}O_3$ [c]	255	0.38
$La_{0.8}Ca_{0.2}MnO_{3+y}$ [d]	206	0.85
$(La_{0.25}Nd_{0.75})_{0.7}Ca_{0.3}MnO_3$(warm up) [e]	116	3.5
$(La_{0.25}Nd_{0.75})_{0.7}Ca_{0.3}MnO_3$(cool down) [e]	104	4.4

[a]Reference [42]
[b]Reference [40]
[c]Reference [22]
[d]Reference [13]
[e]Reference [43]

estimated from both warm-up and cool-down measurements under a pressure of about 10 kbar (Ref. [43]). In Fig. 4, we plot $\ln(\alpha_O)$ vs the Curie temperatures of the ^{16}O samples listed in Table 1. It can be seen that all the data points fall on a straight line. So the data can be fit by an equation

$$\alpha_O = 21.9\exp(-0.016T_C). \qquad (2)$$

The above simple empirical relation between α_O and T_C is quite unexpected. This relation implies that α_O has no significant dependence on the concentration of the Mn^{4+}

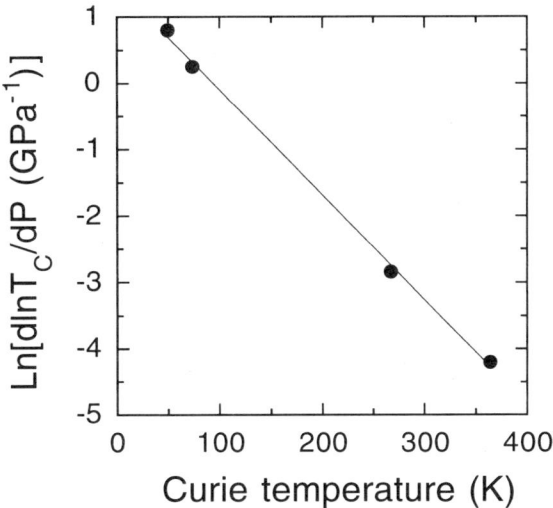

Figure 5: $\ln(d\ln T_C/dP)$ vs Curie temperature (see text for the data). The quantity $d\ln T_C/dP$ is the pressure-effect coefficient in the low pressure region.

when it is in the region of 20-35%, as the case of the samples listed in Table 1. From a simple argument based on the double-exchange model and the small polaron theory discussed above, one would expect that $T_C \propto \exp(-2\alpha_O)$ (Ref. [13]). It is clear that this simple picture cannot quantitatively explain Eq. 2.

In Fig. 5 we show the pressure-effect coefficient $\ln(d\ln T_C/dP)$ vs T_C for a fixed Mn^{4+} of about 30%. The data are listed as follows: $d\ln T_C/dP = 2.23$ GPa^{-1} at $T_C = 49$ K and $d\ln T_C/dP = 1.29$ GPa^{-1} at $T_C = 73$ K for $(La_{0.25}Nd_{0.75})_{0.7}Ca_{0.3}MnO_3$ (Ref. [44]); $d\ln T_C/dP = 0.06$ GPa^{-1} at $T_C = 267$ K for $La_{0.67}Ca_{0.33}MnO_3$ (Ref. [45]); $d\ln T_C/dP = 0.015$ GPa^{-1} at $T_C = 364$ K for $La_{0.7}Sr_{0.3}MnO_3$ (Ref. [41]). The data can be fit by an equation

$$d\ln T_C/dP = 4.4\exp(-0.016T_C). \tag{3}$$

Combining Eq. 2 and Eq. 3, one has $\alpha_O = 5(d\ln T_C/dP)$. Thus, the isotope exponent is simply proportional to the pressure-effect coefficient. Such a simple correlation between the pressure coefficient and isotope exponent implies that the major effect of the pressure is to increase the phonon frequency. An increase of the phonon frequency by the pressure enhances the ferromagnetic exchange interaction while a decrease of the phonon frequency by increasing the oxygen isotope mass reduces the ferromagnetic exchange interaction. Furthermore, this simple relation between α_O and $d\ln T_C/dP$ can be used to check whether the observed oxygen isotope effect is intrinsic. Franck et al., have recently shown that the oxygen isotope exponent in $La_{0.8}Ca_{0.2}MnO_{3+y}$ is reduced to about 0.4 after the two oxygen-isotope exchanged samples were annealed in argon gas [46]. They considered the isotope effect observed in the argon-annealed samples to be intrinsic. At the same time, they suggested that the giant oxygen isotope shift of T_C in $La_{0.8}Ca_{0.2}MnO_{3+y}$ observed by Zhao and coworkers [13] was caused by the existence of extra oxygen in the samples. Their argument is not sound for the following reasons. First of all, the oxygen isotope shift of T_C is not large in $La_{0.9}Sr_{0.1}MnO_{3+y}$ where there is substantial extra oxygen (see above), while the oxygen isotope shift of T_C is colossal (>40 K) in $(La_{0.25}Nd_{0.75})_{0.7}Ca_{0.3}MnO_3$ where there is negligible extra oxygen [43]. Thus the isotope shift has no correlation with the amount of extra oxygen in the samples. Secondly, there has been no experimental proof that the argon-annealing procedure can ensure the same oxygen contents for the two isotope samples. Actually the argon-annealing is a nonequilibrium procedure, so the reduction rate depends on the diffusion coefficient which is isotope-mass dependent (a heavier isotope mass corresponds to a smaller diffusion coefficient, and thus a slower reduction rate). Only if a thermal equilibrium state is reached during diffusion, can the oxygen contents for the two isotope samples be the same. This has been proved in many previous isotope experiments (e.g., Ref. [4, 6]). Since the pressure effect $d\ln T_C/dP$ in $La_{0.8}Ca_{0.2}MnO_{3+y}$ is three times as large as that in $La_{0.67}Ca_{0.33}MnO_{3+y}$ (Ref. [45]), it is not unreasonable that the observed α_O (~0.85) in $La_{0.8}Ca_{0.2}MnO_{3+y}$ is about three times as large as that (~ 0.3) in $La_{0.67}Ca_{0.33}MnO_{3+y}$ (see Table 1).

4.2 Oxygen Isotope Effect from EPR Measurements. Electron Paramagnetic Resonance (EPR) is a powerful technique in condensed matter physics, which allows one to study electron-phonon interactions, static and dynamic magnetic correlations on a microscopic level. The temperature dependence of the EPR linewidth includes some

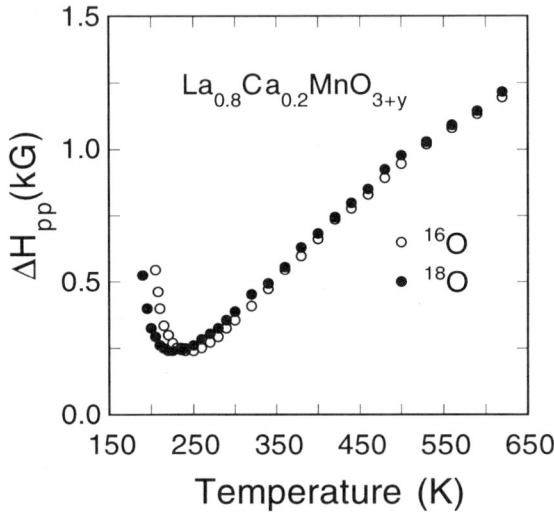

Figure 6: Temperature dependence of the peak-to-peak EPR linewidth ΔH_{pp} for ^{16}O and ^{18}O samples of $La_{0.8}Ca_{0.2}MnO_{3+y}$. The figure is reproduced from Ref. [47].

information about conduction electrons, and the EPR intensity is normally proportional to the static susceptibility. In Ti_4O_7 where intersite bipolarons were identified, the EPR intensity follows a Curie-Weiss law as expected, while the EPR linewidth is thermally activated with an activation energy of about 0.08 eV (Ref. [38]). The temperature-dependent part of the linewidth may be directly related to a Korringa-type relaxation of the center through the single polarons whose density is thermally activated.

We performed EPR measurements [47] on the oxygen isotope exchanged samples of $La_{0.8}Ca_{0.2}MnO_{3+y}$. The EPR linewidth for the two isotope samples is significantly different below a temperature T_{min} where the EPR linewidth has a local minimum (see Fig. 6). Since the isotope shift of T_{min} is similar to the isotope shift of T_C, the sharp upturn of the EPR linewidth below T_{min} might be related to the formation of short-range ferromagnetically-ordered clusters. Well above T_{min}, the EPR linewidths for the two isotope samples have a small difference.

If one assumes that bipolaronic charge carriers also exist in this material, one might expect that the temperature-dependent part of the EPR linewidth should be thermally activated as in the case of Ti_4O_7. It has been shown that [48] the temperature independent part of the EPR linewidth is given by $\Delta H_o = (\Delta H_{dip})^2/H_{ex}$, where ΔH_{dip} is the full dipolar width ($\sim 10^4$ G), H_{ex} is the exchange field which is proportional to T_C ($H_{ex} \sim T_C \sim 10^6$ G). This relation implies that the temperature-independent linewidth of the ^{18}O sample should be 10% larger than the ^{16}O sample (since T_C of the ^{18}O sample is 10% lower than the ^{16}O sample). The absolute value of ΔH_o should be of the order of 100 G (Ref. [48]). In Fig. 7, we plot the temperature dependence of the EPR linewidth above 300 K. Below 300 K, the ferromagnetic clusters start to form as discussed below. The solid lines in the figure are the best fit curves to the data. The equation used for fitting is given by $\Delta H_{pp} = \Delta H_o + AT^{-1.2}\exp(-E_a/T)$. The fit parameters for the ^{16}O sample are $\Delta H_o = 84(15)$ G, $A = 2.11(4) \times 10^7$ GK$^{1.2}$, and $E_a = 0.114(15)$ eV. The fit parameters for the ^{18}O sample are $\Delta H_o = 129(23)$ G, $A = 2.11(7) \times 10^7$ GK$^{1.2}$, and $E_a = 0.116(26)$ eV. The value of ΔH_o for the ^{18}O sample seems to be larger than for the

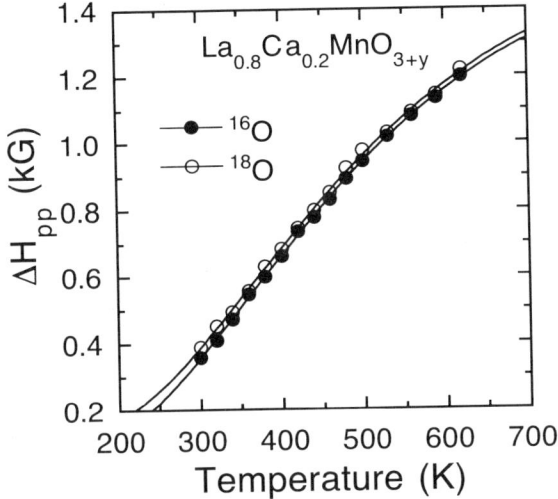

Figure 7: Temperature dependence of the EPR linewidth above 300 K for $La_{0.8}Ca_{0.2}MnO_{3+y}$. The data can be fit by $\Delta H_{pp} = \Delta H_o + AT^{-1.2}\exp(-E_a/T)$. The data are taken from Ref. [47].

^{16}O sample, as expected. The parameters A and E_a are the same for the two isotope samples within the experimental uncertainty. It is important to note that the activation energy E_a deduced here is nearly the same as the value (0.121 eV) deduced from the resistivity data of a thin film of $La_{0.8}Ca_{0.2}MnO_{3+y}$ (Ref. [49]), and that the value of E_a is nearly equal to the energy gap (0.108 eV) measured from tunneling experiments [49]. These results lead us to suggest that bipolaronic charge carriers should also exist in manganites as in the case of Ti_4O_7 (Ref. [38]). Since the bipolarons are heavy and nearly immobile, they have a negligible contribution to the double-exchange. Therefore, the aniferromagnetic superexchange interaction is dominant in the regions where the bipolaronic charge carriers reside, while the ferromagnetic double-exchange interaction dominates in the regions where the mobile polarons sit. The singlet bipolaronic state is not stable with respect to both external magnetic field and ferromagnetic ordering in analogy to the pair-breaking effect in superconductors. Then the density of polaronic carriers increases when an external magnetic field is applied and/or a ferromagnetic ordering takes place. This picture can naturally explain the CMR effect as well as the metal-insulator transition near T_C.

In Fig. 8a we plot the inverse EPR intensity $1/I$ vs temperature. It is clear that the EPR intensity depends strongly on the oxygen isotope mass. Although the unusual

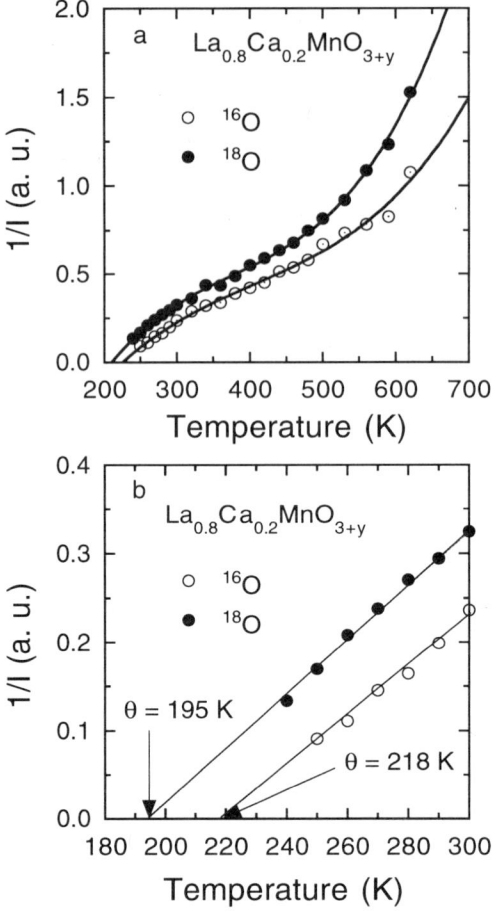

Figure 8: Temperature dependence of the inverse integral intensity of EPR signal for ^{16}O and ^{18}O samples of $La_{0.8}Ca_{0.2}MnO_{3+y}$. The data are taken from Ref. [47].

temperature dependence of the EPR intensity and its dependence on the oxygen isotope mass could be explained by the so called "bottleneck" effect [47], a conventional interpretation (i.e., I is proportional to the static susceptibility) may be more relevant. In Fig. 8b we show $1/I$ below 300 K. The data can be fit by $I \propto 1/(T-\theta)$ with $\theta = 218$ K for the ^{16}O sample and 195 K for the ^{18}O sample. For both isotope samples, the θ value is almost the same as the T_C value, which justifies the conventional explanation. Above 300 K, the θ value appears to increase with increasing temperature. This is quite reasonable since the density of the mobile polaronic carriers (which are responsible for double-exchange) should generally increase with increasing temperature. On the other hand, a formation of ferromagnetic clusters near T_C tends to increase the density of the polaronic carriers (since the ferromagnetic ordering can destabilize the bipolaronic state as discussed above). These competing effects might keep the density of the polaronic carriers nearly unchanged at temperatures from 300 K to T_C. This may explain why the θ value is nearly a constant below 300 K.

4.3 Oxygen Isotope Effect on the Thermal-Expansion Coefficient.

We have also carried out thermal-expansion measurements [22] on the oxygen-isotope substituted manganites $(La_{1-x}Ca_x)_{1-y}Mn_{1-y}O_3$ with a Mn^{4+} concentration of ~33%. The linear thermal-expansion coefficient $\beta(T)$ exhibits an asymmetric peak at the Curie temperature T_C and depends strongly on the oxygen isotope mass as shown in Fig. 9. For $(LaMn)_{0.945}O_3$ (see Fig. 9a), the ^{18}O sample has lower T_C than the ^{16}O sample by ~12 K while the thermal-expansion coefficient jump $\Delta\beta(T_C)$ of the ^{18}O sample is larger than that of the ^{16}O sample by 24(3)%. For $La_{0.67}Ca_{0.33}MnO_3$ (see Fig. 9b), the oxygen

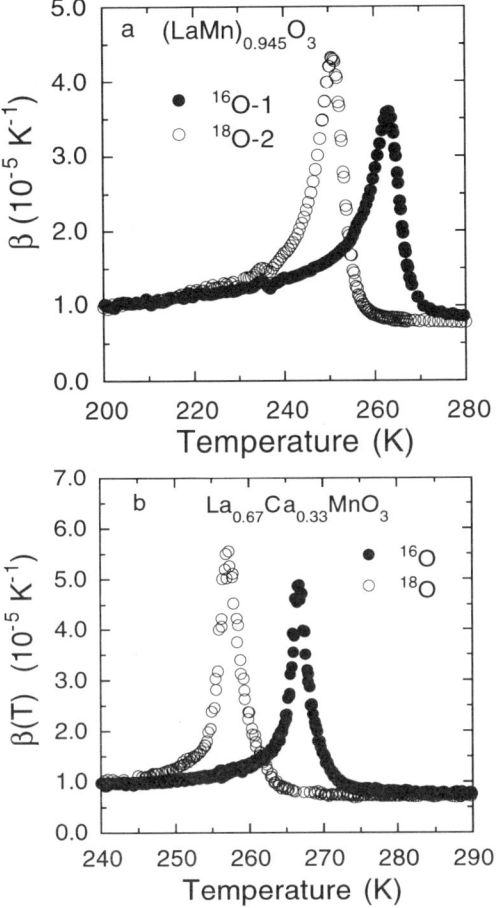

Figure 9: The linear thermal-expansion coefficient $\beta(T)$ for the ^{16}O and ^{18}O samples of $(LaMn)_{0.945}O_3$ (a) and $La_{0.67}Ca_{0.33}MnO_3$ (b). The figure is after Ref. [22].

isotope shift of T_C is about 10 K while the thermal-expansion coefficient jump $\Delta\beta(T_C)$ of the ^{18}O sample is larger than that of the ^{16}O sample by 16(3)%. Although the magnitude of the jump depends on the sharpness of the transition, the relative change in the jump is independent of the transition width provided that two samples have the same transition width. This is the case for the two isotope samples, as seen from the magnetization curves shown in Fig. 10. One can see that the ferromagnetic transition of the samples is very sharp, and that the transition widths of the two isotope samples are the same, i.e., the two curves are the same but for a parallel shift. Therefore, the two isotope samples have the same transition width, which ensures that the observed large isotope effect on $\Delta\beta(T_C)$ is intrinsic.

The asymmetric peak in the thermal-expansion coefficient observed for these samples and no hysteresis in $\beta(T)$ within the accuracy of our measurements (~ 0.1 K) may indicate a second-order ferromagnetic transition. For a second-order phase transition, there is a thermodynamic relation

$$d\ln T_C/dP = 3\Delta\beta(T_C)/\Delta C_P(T_C), \qquad (4)$$

where C_P is the specific heat. For $La_{0.67}Ca_{0.33}MnO_3$, $\Delta\beta(T_C) = 4.2 \times 10^{-5}$ K^{-1} (see Fig. 9b), $\Delta C_P = 50$ J/moleK [50], and unit volume = 60 Å^3. Substituting the above values into Eq. 4, we obtain $d\ln T_C/dP = 0.07$ GPa^{-1}, in excellent agreement with the measured value (~ 0.06 GPa^{-1} [45]). Thus the ferromagnetic transition in $La_{0.67}Ca_{0.33}MnO_3$ seems to be of second order.

We now turn to the isotope dependence of $\Delta\beta(T_C)$. From Eq. 4, we see that there must be a corresponding dependence of $d\ln T_C/dP$ on the oxygen mass M_O provided that ΔC_P is independent of M_O. This should be the case, since $\Delta C_P(T_C)$ arises from

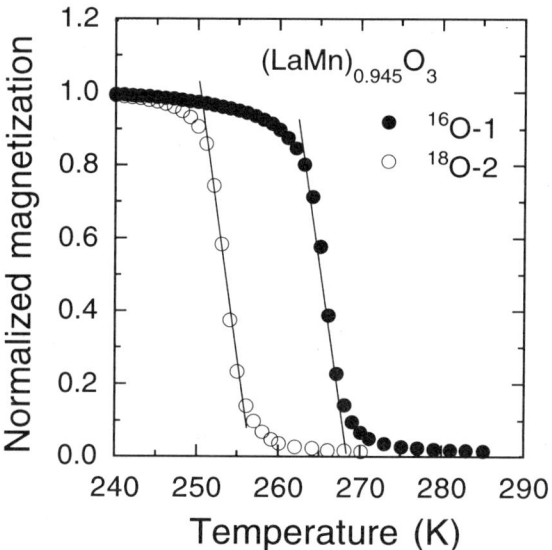

Figure 10: Temperature dependence of the low-field magnetization for the ^{16}O and ^{18}O samples of $(LaMn)_{0.945}O_3$. The figure is after Ref. [22].

entropy changes due to a combination of ferromagnetic ordering of the Mn spins, and delocalization of charge carriers [50, 51], neither of which contributions should be influenced by M_O. So the oxygen isotope effect on $\Delta\beta(T_C)$ is caused by the M_O dependence of $d\ln T_C/dP$. The M_O dependence of $d\ln T_C/dP$ can be understood in terms of the empirical relation: $d\ln T_C/dP \propto \exp(-0.016T_C)$, as discussed above (also see Ref. [22]). This relation implies that the relative change in $d\ln T_C/dP$ is equal to - 0.016ΔT_C. So a decrease of T_C by 12 or 10 K due to an increase of the oxygen isotope mass will lead to a relative increase of $d\ln T_C/dP$ by 19 or 16%, which is in fair agreement with our measured value of 24(3) or 16(3)% for the relative change in $\Delta\beta(T_C)$. Therefore, this provides a quantitative explanation for the observed colossal oxygen-isotope effect on $\Delta\beta(T_C)$.

4.4 Metal-Insulator Transition Induced by Oxygen Isotope Substitution.

As discussed above, the manganese-based perovskites $Ln_{1-x}A_xMnO_3$ exhibit many remarkable features. With a deacrease of temperature, a variety of phase transformations occur depending on the doping and average ionic radius $<r_A>$ of the cations in the $Ln_{1-x}A_x$ site. The ferromagnetic transition temperature for the $x = 1/3$ samples de-

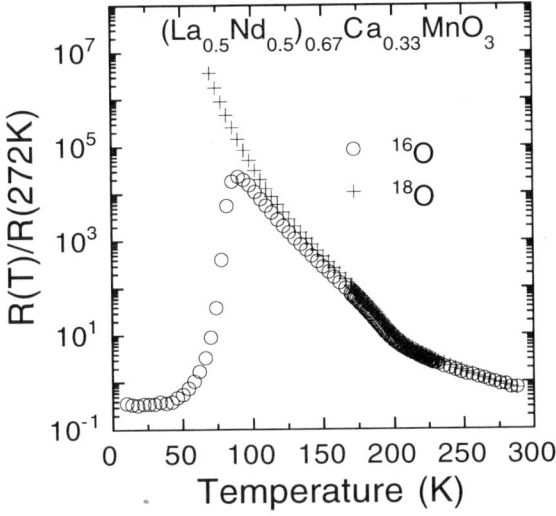

Figure 11: The temperature dependence of the resistivity (normalized to the resistivity at 272 K) for the ^{16}O and ^{18}O samples of $(La_{0.5}Nd_{0.5})_{0.67}Ca_{0.33}MnO_3$.

creases with a decrease of $<r_A>$. When $<r_A>$ is less than a critical value, the ground state is not a metallic and ferromagnetic state, but an insulating and charge-ordered state. The compound $(La_{0.5}Nd_{0.5})_{0.67}Ca_{0.33}MnO_3$ is just near this critical point, and shows a metallic and ferromagnetic state when the sample is prepared in ^{16}O gas environment.

We have carried out the oxygen isotope substitution for this chosen compound $(La_{0.5}Nd_{0.5})_{0.67}Ca_{0.33}MnO_3$. We discovered [23] a novel crossover from a metallic to an insulating ground state by simply replacing the ^{16}O with the ^{18}O isotope. In Fig. 11, we show the temperature dependence of the zero-field resistivity (normalized to the resistivity at 272 K) for the ^{16}O and ^{18}O samples of $(La_{0.5}Nd_{0.5})_{0.67}Ca_{0.33}MnO_3$. The data were taken upon cooling down, while in Ref. [23] the data were taken upon warming up. It is surprising that the ^{16}O sample shows a metallic ground state while the ^{18}O sample exhibits an insulating ground state (the resistance of the ^{18}O sample at 5 K is too large to measure). Somewhat later, the anomalous effect was confirmed by a Russian group [52] in the compound $La_{0.175}Pr_{0.525}Ca_{0.3}MnO_3$.

In the present experiment the sample quality is important because the resistivity of a polycrystalline sample depends on the grain boundaries. For a dense sample, the resistivity should be mainly contributed from the interior of the grains, and may represent an intrinsic electrical property of the material. In order to confirm that this is indeed the case, we show in Fig. 12, the dependences of the magnetization on temperature (Fig. 12a), and on magnetic field (Fig. 12b) for the two isotope samples.

The temperature dependence of the magnetization was measured in a magnetic field of 10 mT after the samples were cooled in the field to 5 K. It is clear that the dependences

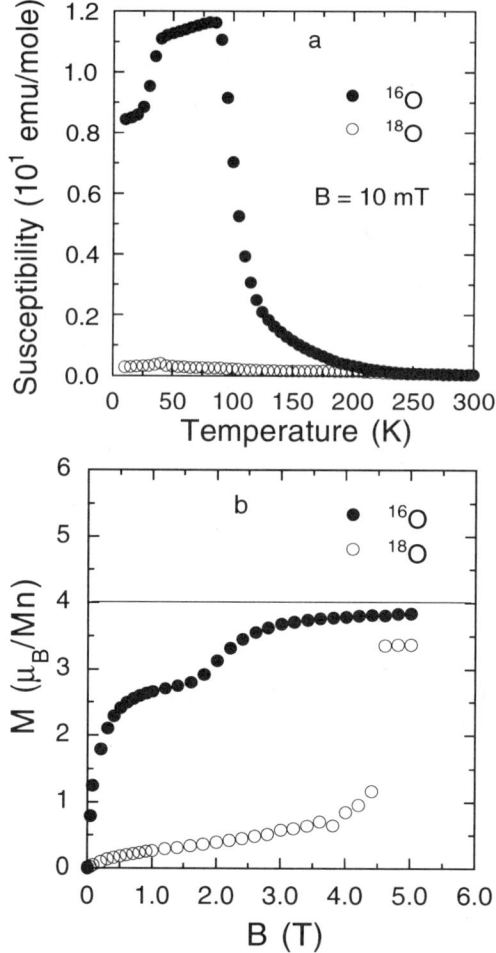

Figure 12: The dependences of the magnetization on temperature (a), and on magnetic field at 5 K (b) for the ^{16}O and ^{18}O samples of $(La_{0.5}Nd_{0.5})_{0.67}Ca_{0.33}MnO_3$. The figure is reproduced from [23].

of the magnetization on both temperature and magnetic field are very different for the two isotope samples. The ^{16}O sample undergoes a ferromagnetic transition at about 140 K, while the ^{18}O sample does not. The ferromagnetic ground state for the ^{16}O sample is consistent with a metallic state at low temperatures. Another anomalous feature shown in Fig. 12b is that the ^{18}O sample has a small moment below a magnetic field of about 4.5 T, then the magnetization undergoes a sudden jump towards a saturated moment. This result also implies that the ground state of the ^{18}O sample is not ferromagnetic in zero field, in agreement with the resistivity data above.

So far there are no theories that can explain the novel crossover from a metallic to an insulating state by simply changing the oxygen isotope from ^{16}O to ^{18}O. Nevertheless, there seems to be a consensus that a strong Jahn-Teller effect in manganites leads to the formation of lattice-polarons, and that most experimental results can be explained by double-exchange and lattice-polaronic effects [30, 21]. On the basis of double-exchange and Jahn-Teller lattice coupling to charge carriers, Millis *et al.* [21] have shown that a metal-insulator transition will occur when the electron-phonon coupling constant λ exceeds a critical value. The coupling constant λ is proportional to $E_{JT}/\bar{t}$, where E_{JT} is the Jahn-Teller stabilization energy, $\bar{t}$ is the hopping integral which can be renormalized by the spin-disorder due to double-exchange. Thus this model can explain an insulator-metal transition as the temperature is reduced through T_C below which the spins become ordered, increasing $\bar{t}$ and thereby reducing λ. However, this model cannot explain the crossover from the metallic to insulating ground state by changing the oxygen mass. This is because the coupling constant λ ($\propto E_{JT}/\bar{t}$) is independent of the oxygen mass. It is worth noting that Millis *et al.* have assumed spatially homogeneous solutions in their calculation [21], which may not be true in the case of extreemly strong electron-phonon interactions.

Both thermal-expansion and small-angle neutron experiments [53] show that the ^{16}O sample of $(La_{0.5}Nd_{0.5})_{0.67}Ca_{0.33}MnO_3$ undergoes a phase segregation into insulating antiferromagnetic domains and metallic ferromagnetic domains at low temperatures. However, the ^{18}O sample has a negligibly small fraction of ferromagnetic domains. Therefore, it is likely that the ^{16}O sample has more mobile carriers (presumely polaronic carriers which are responsible for the ferromagnetic exchange) than the ^{18}O sample. Recently, Zhou *et al.*, have indeed shown that the ^{16}O sample has more mobile carriers than the ^{18}O sample when T_C is low [43]. This is quite reasonable because heavier carriers (e.g., in samples with a heavier isotope mass) are more easily localized by impurity potentials. Thus, this scenario appears to be able to explain this novel isotope effect.

4.5 Oxygen Isotope Shift of the Charge-Ordering Transition. The real-space ordering of charge carriers in crystals is one of the most interesting phenomena in condensed matter physics. Such a charge-ordering state has been observed mostly in transition-metal based oxides, such as Ti_4O_7 (Ref. [54]), $La_{2-x}Sr_xNiO_4$ (Ref. [55]), $(La,Pr,Nd)_{0.5}Ca_{0.5}MnO_3$ (Ref. [56, 58]), $(Pr,Nd)_{0.5}Sr_{0.5}MnO_3$ (Ref. [59, 60]), etc. In particular, the charge-ordering in manganites is rather exotic. As an example, the charge-ordering state in the manganites $Nd_{0.5}Sr_{0.5}MnO_3$ and $Pr_{0.5}Sr_{0.5}MnO_3$ can be destroyed by a small magnetic field (< 10 Tesla) [59, 60]. This implies that the charge-ordering state in these systems is not so stable, which appears to be in contradiction with a large energy gap observed from both photoemission [61] and tunneling [62] experiments. Theoretically, it has been proposed that the long-range Coulomb repulsive interaction among conduction carriers might be responsible for the charge-ordering in these systems [63, 64, 65].

Our isotope experiments do not support these theoretical models. There is a large oxygen isotope effect on the charge-ordering transition temperature in $Nd_{0.5}Sr_{0.5}MnO_3$ and $La_{0.5}Ca_{0.5}MnO_3$ systems [57]. In Fig. 13, we show the temperature dependence of the normalized magnetizations for the ^{16}O and ^{18}O samples under a magnetic field of 1 mT (Fig. 13a), and 2.5 T (Fig. 13b). The normalized magnetization is defined as $(M(T)-M_L)/(M_H-M_L)$, where M_L and M_H are the magnetizations at the lowest and highest temperatures shown in Fig. 13, respectively. With this normalization procedure, the curves for the ^{16}O and ^{18}O samples are parallel shifted. The transition from a high magnetization to a low magnetization state is a signature of the transition from a ferromagnetic to a charge-ordering (CO) state [59, 60]. Here we define the mid-point temperature on the transition curve as the charge-ordering temperature (T_{CO}). With this definition, we find that the charge-ordering temperature for the ^{18}O sample is higher

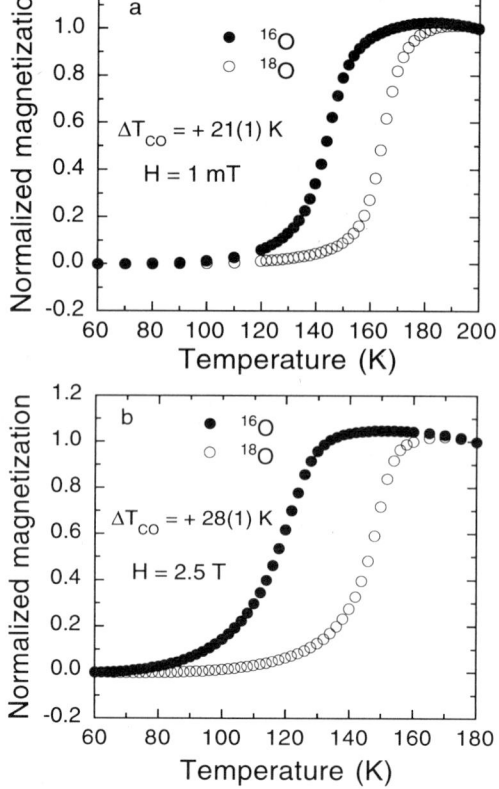

Figure 13: Oxygen isotope effect on the charge-ordering temperature of $Nd_{0.5}Sr_{0.5}MnO_3$: Temperature dependence of the normalized magnetization for the ^{16}O and ^{18}O samples under magnetic fields of 1 mT (a); 2.5 T (b).

than for the ^{16}O sample by 21 K under $H = 1$ mT, by 28 K under $H = 2.5$ T, and 43 K under $H = 5.4$ T. This is the first observation of a colossal negative oxygen isotope shift of the charge-ordering temperature.

To show that the observed oxygen isotope shifts are intrinsic, we have performed isotope back-exchange experiments ($^{16}O \to {}^{18}O$; $^{18}O \to {}^{16}O$). In Fig. 14, we show the normalized magnetization for the ^{16}O and ^{18}O samples before and after isotope back-exchange. The symbol ($\times$) denotes the ^{16}O sample which has been back-exchanged from the original ^{18}O sample (denoted by open square). The symbol (+) represents the ^{18}O sample which has been back-exchanged from the original ^{16}O sample (denoted by open circle). It is evident that the T_{CO} of the ^{16}O (^{18}O) sample goes back completely to that of the original ^{18}O (^{16}O) sample after the isotope back-exchange. This clearly indicates that the shift of T_{CO} is caused only by changing the oxygen isotope mass.

In Fig. 15, we show the charge-ordering temperatures of the ^{16}O and ^{18}O samples of $Nd_{0.5}Sr_{0.5}MnO_3$ as a function of the external magnetic field. It is clear that, with an increase of the external magnetic field, the charge-ordering temperatures of both ^{16}O and ^{18}O samples decrease, but the decreasing rate for the ^{18}O sample is much slower than for the ^{16}O sample. This leads to an dramatic increase in the oxygen isotope shift with increasing magnetic field.

The observed large oxygen isotope shift of the charge-ordering temperature and its

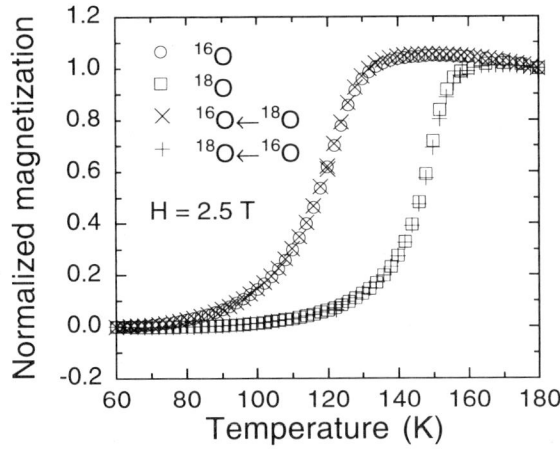

Figure 14: Oxygen isotope back-exchange result for $Nd_{0.5}Sr_{0.5}MnO_3$.

strong dependence on applied magnetic field are difficult to understand on the basis of the existing theories. In most theoretical models, the charge-ordering arises from a long-range Coulomb repulsive interaction between carriers [63, 64, 65]. So these models predict no isotope effect, and cannot explain our present results.

If the charge-ordering in the manganites is related to small polaron ordering, one may be able to explain the oxygen isotope shift of the charge-ordering temperature in zero magnetic field. This is because the effective mass of polaronic carriers is larger

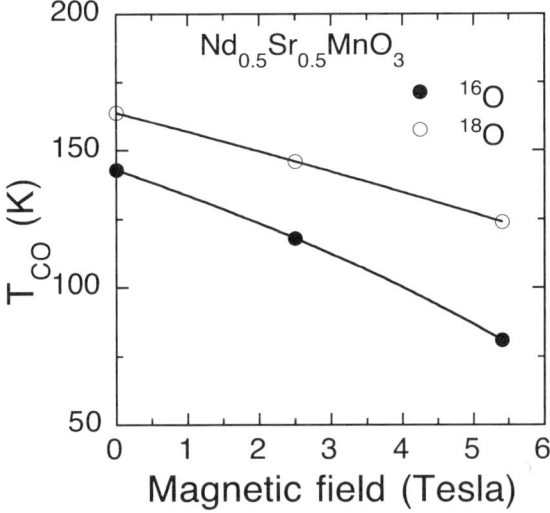

Figure 15: Magnetic field dependence of the charge-ordering temperatures of the ^{16}O and ^{18}O samples of $Nd_{0.5}Sr_{0.5}MnO_3$.

for a heavier isotope mass, and the larger the carrier mass, the more stable the charge-ordering state. As a result, one would expect a higher T_{CO} in samples with a heavier isotope mass. Thus this model can qualitatively explain the observed isotope effect in zero magnetic field. However, it cannot explain why a small magnetic field (< 10

Tesla corresponding to an energy of about 10 K) can destroy the charge-ordering state in $Nd_{0.5}Sr_{0.5}MnO_3$ where a large energy gap (> 1000 K) was observed [61, 62]. The observed large energy gap implies that the charge-ordering state, if it is a polaronic ordering state, must be stable and cannot be destroyed by a small magnetic field.

Now we consider a theory of bipolaronic charge ordering [66, 12]. On the basis of this theory [66, 12], the bipolaron charge ordering temperature is a function of both the repulsive interaction v among bipolarons and the bipolaron hopping integral t^*,

$$T_{CO} \propto zv\sqrt{1-(t^*/v)^2}\sqrt{1+(1-2n)^2 - \frac{2(1-2n)}{\sqrt{1-(t^*/v)^2}}}, \quad (5)$$

where z is the number of the nearest neighbour ions, n is the average number of bipolarons per unit cell.

In the present case, $z = 6$ and $n = 0.25$, so

$$T_{CO} \propto 6v\sqrt{1-(t^*/v)^2}\sqrt{1.25 - 1/\sqrt{1-(t^*/v)^2}}. \quad (6)$$

Since t^* could strongly depend on the isotope mass, one might expect a large oxygen isotope effect on the charge-ordering temperature as t^*/v is close to 0.6 (when seen from Eq. 6). A small magnetic field could melt the bipolaron ordered state by reducing the density of the bipolaronic carriers. This is because the charge-ordering temperature is very sensitive to n when t^*/v is close to 0.6 and n is near 0.25 (see Eq. 5). Meanwhile there is no difficulty in explaining the observed large energy gap which can be related to the binding energy of the bipolarons.

5 CONCLUDING REMARKS

We have observed various oxygen isotope effects in doped manganite systems. Our results clearly indicate that lattice vibrations play an essential role in various magnetic and electrical properties of manganites. The observed colossal oxygen isotope effect on the charg-ordering temperature, and the unusual temperature dependence of the EPR linewidth suggest that bipolaronic charge carriers may exist in these compounds. This is consistent with other experiments such as tunneling [49], photoemission [29], high-resolution electron diffration [67], and optical measurements [68]. It is likely that the charge carriers in manganites are heavy onsite bipolarons [67]. This is in contrast with the intersite bipolarons in cuprates, which is not heavy according to a model [12], and recent experiments [8]. Any viable theory which is proposed to describe the physics of manganites must be able to explain all the isotope effects observed.

ACKNOWLEGEMENT: We would like to thank A. S. Alexandrov, and A. J. Millis for useful discussions. The work is partially supported by the Swiss National Science Foundation and NSF of the United States.

References

[1] M. K. Crawford, M. N. Kunchur, W. E. Farneth, E. M. McCarron III, and S. J. Poon, Phys. Rev. B **41**, 282 (1990)

[2] J. P. Franck, S. Harker and J. H. Brewer, Phys. Rev. Lett. **71**, 283 (1993)

[3] H. J. Bornemann, and D. E. Morris, Phys. Rev. B **44**, 5322 (1991)

[4] G. M. Zhao, M. B. Hunt, H. Keller, and K. A. Müller, Nature (London) **385**, 236 (1997)

[5] G. M. Zhao and D. E. Morris, Phys. Rev. B **51**, 16487 (1995)

[6] G. M. Zhao, K. K. Sinha, A. P. B. Sinha, and D. E. Morris, Phys. Rev. B **52**, 6840 (1995)

[7] G. M. Zhao, J. W. Ager III, and D. E. Morris, Phys. Rev. B **54**, 14982 (1996)

[8] G. M. Zhao, K. Conder, H. Keller, and K. A. Müller, J. Phys.:Condens. Matter (in press)

[9] G. M. Zhao, V. Kirtikar, K. K. Singh, A. P. B. Sinha, and D. E. Morris, Phys. Rev. B **54**, 14956 (1996)

[10] D. Zech, H. Keller, K. Conder, E. Kaldis, E. Liarokapis, N. Poulakis, and K. A. Müller, Nature (London) **371**, 681 (1994)

[11] D. Zech, K. Conder, H. Keller, E. Kaldis, and K. A. Müller, Physica B **219-220**, 136 (1996)

[12] A. S. Alexandrov and N. F. Mott, *Polarons and bipolarons* (World Scientific, Singapore, 1995)

[13] G. M. Zhao, K. Conder, H. Keller, and K. A. Müller, Nature **381**, 676 (1996)

[14] E. O. Wollan, and W. C. Koeler, Phys. Rev. **100**, 545 (1955)

[15] G. H. Jonker and J. H. van Santen, Physica **16**, 337 (1950)

[16] S. Jin, T. H. Tiefel, M. McCormack, R. A. Fastnacht, R. Ramesh, and L. H. Chen, Science **264**, 413 (1994)

[17] R. von Helmolt, J. Wecker, B. Holzapfel, L. Schltz, and K. Samwer, Phys. Rev. Lett. **71**, 2331 (1993)

[18] C. Zener, Phys. Rev. **82**, 403 (1951)

[19] P. W. Anderson and H. Hasegawa, Phys. Rev. **100**, 675 (1955)

[20] A. J. Millis, P. B. Littlewood, and B. I. Shraiman, Phys. Rev. Lett. **74**, 5144 (1995).

[21] A. J. Millis, B. I. Shraiman, and R. Müller, Phys. Rev. Lett. **77**, 175 (1996)

[22] G. M. Zhao, M. B. Hunt and H. Keller, Phys. Rev. Lett. **78**, 955 (1997)

[23] G. M. Zhao, H. Keller, J. Hofer, A. Shengelaya, and K. A. Müller, Solid State Commun., **104**, 57 (1997)

[24] M. Jaime, M. B. Salamon, M. Rubinstein, R. E. Treece, J. S. Horwitz, and D. B. Chrisey, Phys. Rev. B **54**, 11914 (1996)

[25] S. J. L. Billinge, R. G. DiFrancesco, G. H. Kwei, J. J. Neumeier, and J. D. Thompson, Phys. Rev. Lett. **77**, 715 (1996)

[26] J. M. De Teresa, M. R. Ibarra, P. A. Algarabel, C. Ritter, C. Marquina, J. Blasco, J. Garcia, A. del Moral, and Z. Arnold, Nature (London), **386**, 256 (1997)

[27] Y. Okimoto, T. Katsufuji, T. Ishikawa, A. Urushibara, T. Arima, and Y. Tokura, Phys. Rev. Lett. **75**, 109 (1995)

[28] H. L. Ju, H.-C. Sohn, and K. M. Krishnan, Phys. Rev. Lett. **79**, 3230 (1997)

[29] D. S. Dessau, T. Saitoh, C.-H. Park, Z.-X. Shen, P. Villella, N. Hamada, Y. Moritomo, and Y. Tokura, Preprint (1998)

[30] H. Röder, J. Zang, and A. R. Bishop, Phys. Rev. Lett. **76**, 1356 (1996)

[31] T. Saitoh, A. E. Bocquet, T. Mizokawa, H. Namatame, A. Fujimori, M. Abbate, Y. Takeda, and M. Takano, Phys. Rev. B **51**, 13942 (1995)

[32] T. Holstein, Ann. Phys. (N. Y.), **8**, 325 (1959)

[33] D. Emin and T. Holstein, Ann. Phys. (N. Y.), **53**, 439 (1969)

[34] K.-H. Höck, H. Nickisch, and H. Thomas, Helv. Phys. Acta. **56**, 237 (1983)

[35] H. Kamimura, Int. J. Mod. Phys. B **1**, 873 (1987)

[36] P. W. Anderson, Phys. Rev. Lett. **34**, 953 (1975)

[37] R. A. Street and N. F. Mott, Phys. Rev. Lett. **35**, 1293 (1975)

[38] S. Lakkis, C. Schlenker, B. K. Chakraverty, R. Buder, and M. Marezio, Phys. Rev. B **14**, 1429 (1976)

[39] K. A. Müller, G. M. Zhao, K. Conder, and H. Keller, J. Phys.:Condens. Matter **10**, L291 (1998)

[40] G. M. Zhao and D. E. Morris, 1995 (unpublished)

[41] Y. Moritomo, A. Asamitsu, and Y. Tokura, Phys. Rev. B **51**, 16491 (1995)

[42] G. M. Zhao, K. Ghosh, and R. L. Greene, 1998 (to be published)

[43] J.-S. Zhou and J. B. Goodenough, Phys. Rev. Lett. **80**, 2665 (1998)

[44] J.-S. Zhou, W. Archibald, and J. B. Goodenough, Nature (London) **381**, 770 (1996)

[45] J. J. Neumeier, M. F. Hundley, J. D. Thompson, and R. H. Heffner, Phys. Rev. B **52**, R7006 (1995)

[46] J. P. Franck, I. Isaac, W. Chen, J. Chrzanowski, and J. C. Irwin, Preprint (1998)

[47] A. Shengelaya, G. M. Zhao, H. Keller, and K. A. Müller, Phys. Rev. Lett. **77**, 5296 (1996).

[48] S. E. Lofland, P. Kim, P. Dahiroc, S. M. Bhagat, S. D. Tyagi, S. G. Karabashev, D. A. Shulyatev, A. A. Arsenov, and Y. Mukovskii, Phys. Lett. A **233**, 476 (1997)

[49] A. Biswas, S. Elizabeth, A. K. Raychaudhuri, and H. L. Bhat, cond-matt/9806084

[50] A. P. Ramirez, P. Schiffer, S-W. Cheong, C. H. Chen, W. Bao, T. T. M. Palstra, P. L. Gammel, D. J. Bishop, and B. Zegarski, Phys. Rev. Lett. **76**, 3188 (1996)

[51] J. M. D. Coey, M. Viret, L. Ranno, and K. Ounadjela, Phys. Rev. Lett. **75**, 3910 (1995)

[52] N. A. Babushkina, L. M. Belova, O. Yu. Gorbenko, A. R. Kaul, A. A. Bosak, V. I. Ozhogin and K. I. Kugel, Nature (London) **391**, 159 (1998)

[53] M. R. Ibarra, G. M. Zhao, J. M. De Teresa, B. Garcia-Landa, Z. Arnold, C. Marquina, P. A. Algarabel, and H. Keller, Phys. Rev. B **57**, 7446 (1998)

[54] M. Marezio, D. B. Mcwhan, P. D. Dernier, and J. P. Remeika, Phys. Rev. Lett. **28**, 1390 (1972)

[55] C. H. Chen, S.-W. Cheong, and A. S. Cooper, Phys. Rev. Lett. **71**, 2461 (1993)

[56] C. H. Chen, and S.-W. Cheong, Phys. Rev. Lett. **76**, 4042 (1996)

[57] G. M. Zhao, K. Ghosh, H. Keller, and R. L. Greene, 1998 (to be published)

[58] M. Tokunaga, N. Miura, Y. Tomioka, and Y. Tokura, Phys. Rev. B **57**, 5259 (1998)

[59] H. Kuwahara, Y. Tomioka, A. Asamitsu, Y. Moritomo, and Y. Tokura, Science **270**, 961 (1995)

[60] Y. Tomioka, A. Asamitsu, Y. Moritomo, H. Kuwahara, and Y. Tokura, Phys. Rev. Lett. **74**, 5108 (1995)

[61] A. Chainani, H. Kumigashira, T. Takahashi, Y. Tomioka, H. Kuwahara, and Y. Tokura, Phys. Rev. B **56**, R15513 (1997)

[62] A. Biswas, A. K. Raychaudhuri, R. Mahendiran, A. Guha, R. Mahesh, and C. N. R. Rao, J. Phys.:Condens. Matter **9**, L355 (1997)

[63] V. I. Anisimov, I. S. Elfimov, M. A. Korotin, and K. Terakura, Phys. Rev. B **55**, 15494 (1997)

[64] S. K. Mishra, R. Pandit, and S. Satpathy, Phys. Rev. B **56**, 2316 (1997)

[65] L. Sheng and C. S. Ting, Phys. Rev. B **57**, 5265 (1997)

[66] A. S. Alexandrov and J. Ranninger, Phys. Rev. B **23**, 1796 (1981)

[67] W. Bao, S. A. Carter, C. H. Chen, S.-W. Cheong, B. Batlogg, and Z. Fisk, Solid State Commun. **98**, 55 (1996)

[68] P. Calvani, P. Dore, S. Lupi, A. Paolone, P. Maselli, P. Giura, B. Ruzicka, S.-W. Cheong, and W. Sadowski, J. Supercond. **10**, 293 (1997)

ELECTRONIC TRANSPORT IN La-Ca MANGANITES

Marcelo Jaime and Myron B. Salamon[†]

Los Alamos National Laboratory
Los Alamos, NM 87545

[†]Department of Physics and Materials Research Laboratory.
University of Illinois at Urbana-Champaign.
Urbana, IL 61801

INTRODUCTION

So called colossal magnetoresistance (CMR) manganites of composition $A^{3+}_{1-x}B^{2+}_{x}MnO_3$ (A: La, Nd; B: Ca, Sr, Pb) have been subject of intense research during the last years, especially after their re-discovery[1], in part due to their potential use as device applications in the magnetic storage industry, and in part because of the complexity of the mechanisms responsible for unusual, interdependent electric, magnetic and structural properties. The materials and some of their transport properties, however, have been known since the early experiments by Jonker and van Santen[2], and Volger[3] in the early fifties. Somewhat later, resistivity, magnetoresistance, specific heat and magnetization were studied as functions of temperature in $La_{1-x}Sr_xMnO_3$ polycrystalline samples. The transport data show clearly most of the relevant physics that we know today, including a surprising disagreement between the thermoelectric and Hall effects regarding the sign of the charge carriers, accompanied by a quite small Hall mobility. Volger interpreted the simultaneous metal-insulator and ferromagnetic phase transitions observed in his samples in terms of the double exchange theory (DE) for metallic ferromagnets first suggested by Zener[4] and, perhaps worried about specimen quality issues, intergrain-barrier effects, neglected the clues that his data provided in support of some lattice involvement. This trend persisted over the ensuing years, inspiring both theoretical models [5-13] and the interpretation of experimental results [14-17].

We revisit the transport properties of CMR in detail below, since they have been confirmed in a number of CMR-exhibiting materials, and present new data considered the hallmark of an unusually high electron-phonon coupling responsible for charge localization and small polaron transport. The discussion is organized in four different sections. In the first place we briefly review materials and phase diagrams, as well as general properties, with remarks on differences between intrinsic and extrinsic properties. Second, the high temperature properties are analyzed, emphasizing those which more clearly show the role of the coupling to the lattice. Next, the very low temperature limit is considered, were the double exchange physics is believed to rule. Finally, in the fourth section, an statistical model for the phase transitions of manganites is presented, in an attempt to bridge the far less well understood region of intermediate temperatures with a reasonable extrapolation of what we know and understand in the extremes. We will see that transport experiments give us again an important insight into the dominant physics, that of localized charge carriers that reinforce the tendency toward magnetic order when they gain mobility through the ferromagnetic transition. Most of the discussion is restricted to the doping region $x \sim 1/3$, where the low temperature metallic properties are optimized.

GENERAL PROPERTIES

LaMnO$_3$ is a cubic antiferromagnetic (AFM) semiconductor perovskite, with Neel temperature $T_N \approx 140$ K, where magnetic moments at Mn sites are ferromagnetically (FM) coupled in planes that alternate spin orientations in what is known as A-phase[14], as displayed in Fig. 1a. In this structure Mn^{3+}, surrounded by six oxygen atoms, is a Jahn-Teller (JT) atom. The d-shell electronic energy levels t_{2g} (triplet) and e_g (doublet), in consequence, split under a distortion of the octahedrally coordinated Mn-O bonds. The JT splitting reduces the electronic energy as schematically shown in Fig. 1c. Three strongly coupled and localized (t_{2g}) electrons, occupy the bottom-most levels and form the core spin $S = 3/2$. The fourth electron, occupying the first e_g level, is coupled to the core spin through the intra-atomic Hund's coupling constant $J > 0$, estimated on the scale of ~ 1 eV.

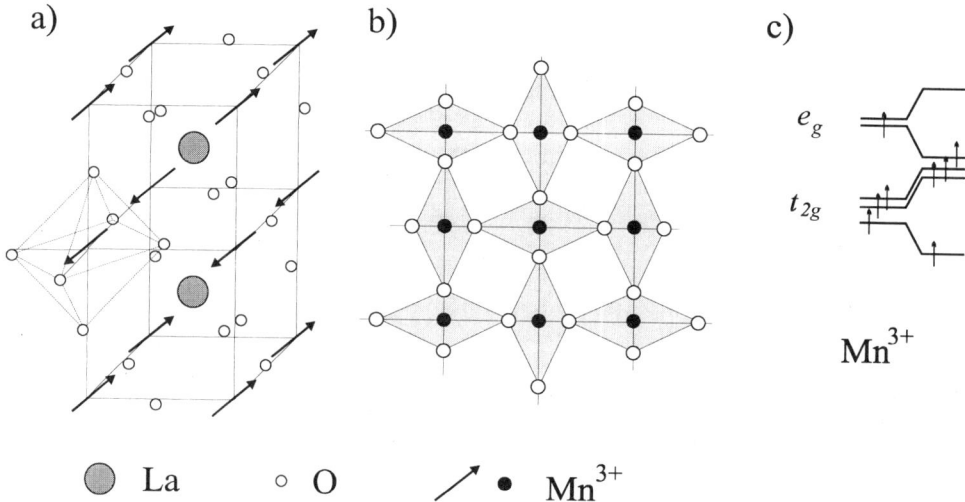

Figure 1. a) Lattice structure of LaMnO$_3$ b) MnO planes showing the characteristic periodic distortion c) Mn^{3+} electronic levels t_{2g} and e_g split as a result of the Mn-O bond length distortion qualitatively displayed in b).

This apparently simple system is by itself a materials-science challenge. In first place pure and ideal LaMnO$_3$ crystals are in principle semiconductors, while real laboratory samples are often good conductors that show a variety of magnetic structure and ordering. This behavior originates in unavoidable cationic (La^{3+}) vacancies that naturally occur during sample preparation[18–20] and introduce charge carriers, magnetic frustration and lattice stress in the system. Cationic vacancies in particular and deviations from ideal stoichiometry in general are the most common causes for discrepancy among experimental results reported in the literature. Associated with cationic vacancies it is not unusual to observe also oxygen deficiency.

Band calculations using density-functional methods (LDA) predicted a metallic ground state for an hypothetical cubic/undistorted version LaMnO$_3$, a finding that is at odds with the experimental results. Satpathy et al.[21] have investigated and identified the physical reason for this behavior by introducing different distortions into the oxygen octahedrons. They have studied three different distortions, *i.e.* the breathing mode (Q1), the basal-plane distortion mode (Q2) displayed in Fig. 1b, and a stretching mode (Q3) in addition to a small rotation of the octahedron. They claim that *"for LaMnO$_3$ a Jahn-Teller distortion of the Q2 type, with the basal-plane oxygen atoms displaced by at least the amount ≈ 0.1 Å from their ideal positions, is necessary for an insulating band structure within the LDA and that the Q1 or the Q3 distortions are not effective in opening up the gap."* They also argue that Jahn-Teller (JT) like distortions favor antiferromagnetic rather than ferromagnetic order. The implications of these findings are quite relevant. Even though they only discuss static distortions, its clear that the electronic band structure is extremely sensitive to particular phonon modes and that a large enough distortion of the Mn-O bonds can drive the system through a metal-insulator phase transition.

Another surprising, and has turned out to be, important characteristic of LaMnO$_3$ is that hole doping by means of chemical substitution of La^{3+} with Ca^{2+} or Sr^{2+}, while increasing the electrical conductivity, does not always produce metallic samples. Again, a rich spectrum of magnetic ordering and/or charge localization has been found experimentally. In particular, as much as 30% content of B^{2+} is required to observe a metallic behavior at room temperature, as it can be seen in the phase diagram in Fig 2.

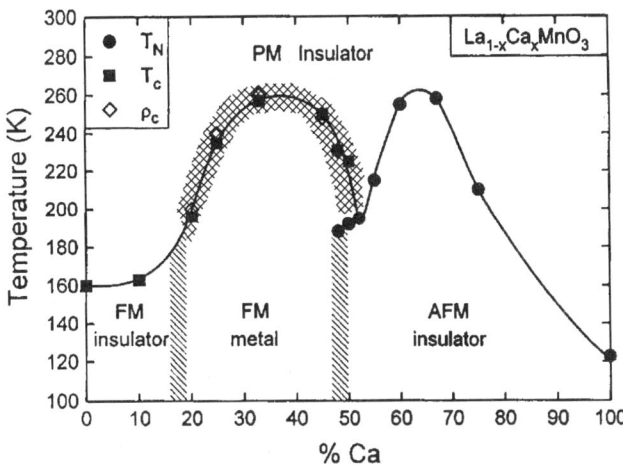

Figure 2. Phase diagram for A$_x$B$_{1-x}$MnO$_3$ manganites. Modified from Schiffer et al.[76], mesh is coexistance region discussed below.

When a divalent atom replaces La in the structure, electrical neutrality is granted by the mixed valence nature of Mn. Indeed, Mn^{3+} gives up one electron per dopant atom in order to keep oxygen happy, resulting in x Mn^{4+} atoms per formula. As a result, a random elastic strain field is introduced in the lattice. Because e_g levels are empty in Mn^{4+} ions, they cannot profit from the JT effect; there is no energy gain obtained in the e_g level and the distortion is no longer favored. In consequence the lattice long-range periodic distortion is now frustrated by a non-JT atom, as schematically displayed in Fig. 3. Holes are likely to stay localized at those Mn^{4+} sites, since some elastic energy must be paid to move them into a Mn^{3+} site, and eventually they reach some kind of periodic distribution, generating charge ordering in the system as the density of holes increases. This would likely be the scenario in the case of spin-less charge carriers. In the case of the manganites, however, the spin-1/2 holes move in a spin-2 environment resulting in a remarkable enrichment of the physics and phenomenology. Indeed, for a critical concentration of holes experimentally found to be close to $x = 1/3$, and temperatures low enough to keep the spin fluctuations small, holes improve their jump probability, reducing their kinetic energy, while moving between Mn sites with core t_{2g} moments that point toward the same direction. This is direct consequence of a very strong Hund's rule at Mn atoms, and is the essence of the double exchange mechanism. When these clusters of magnetically aligned Mn are large enough to overlap the same holes that benefited form local magnetic order, delocalize acting as the driving force for global ferromagnetism and a phase transition into a FM metallic state.

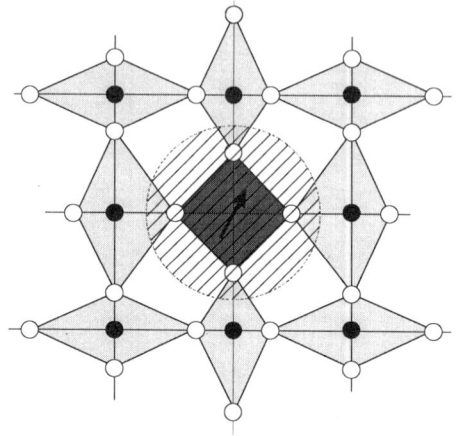

Figure 3. Strain field induced in the structure by a non-JT, s = 3/2 spin, Mn^{4+} atom.

A large electron-phonon coupling is evidenced by an overwhelming amount of experimental data. Outstanding unambiguous evidence of coupling to Jahn-Teller lattice distortions include large pressure effects [22–24], magnetostriction effects associated with the metal-insulator transition [25–28], a discrepancy between the chemical potential estimation by means of thermoelectric effects and thermally activated behavior of the electrical conductivity [29–35], the sign anomaly of the Hall effect, and the Arrhenius behavior of the drift and Hall mobilities [36]. Further, optical properties [37], charge ordering observed in the low doping limit [38], local atomic structure studies [39], neutron diffraction studies [40,41], isotope effect [42–44], X-ray absorption fine-structure measurements that indicate delocalization of charge carriers at the Curie temperature (T_C) as well as coupling between distortions, charge distribution and magnetism [45,46], electron paramagnetic resonance [47], thermal transport [48], and muon spin relaxation (μSR) [49] add to the list.

The transport properties can be discussed qualitatively with the help of a resistivity vs. temperature curve. We will use the simple diagram in Fig. 4 to illustrate them. The temperature range is divided in three regions, e.g. much lower than T_C (I), much higher than T_C (II), and around to T_C (III). The most important energy scale in this diagram is obviously T_C, which is determined by i) bandwidth, ii) band filling, iii) local disorder, and iv) effective electron-phonon coupling. The bandwidth is fundamentally fixed once the structure of the system is fixed, e.g. atomic radii determine the structure and an average Mn-O-Mn bond angle for a particular composition (A, B, x). A clever way to characterize the structure is to use

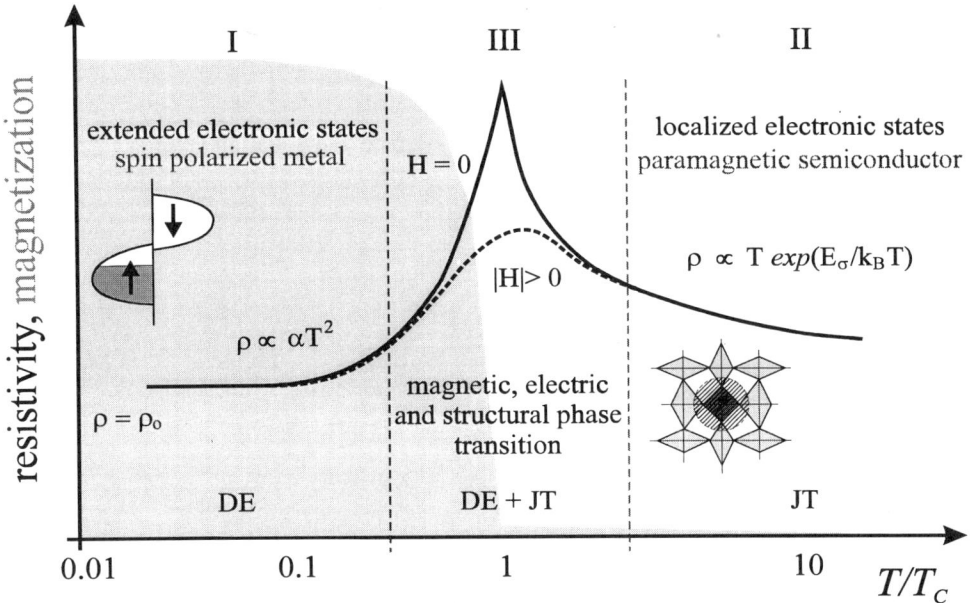

Figure 4. Different relevant temperature ranges for transport properties in CMR manganites for doping levels $x \approx 1/3$. At low temperatures DE effects are dominant, while at high temperature dynamic structural effects control the transport properties.

the so called "tolerance factor" $tf = \overline{\text{A-O}}/\sqrt{2}\overline{\text{Mn-O}}$ where $\overline{\text{A-O}}$ and $\overline{\text{Mn-O}}$ are equilibrium metal-oxygen bond lengths for, respectively, twelve-fold and sixfold oxygen coordination and its physical meaning is quite clear. The more colinear the three atoms are, the larger the transfer integral for charge carriers between them. This is a static, average property of the system. Temperature-tf phase diagrams as well as bandwidth effects have been discussed with detail in the bibliography [50–52], although Hwang et al. seem to have used incorrect (ninefold) coordination numbers. Band filling is determined by the doping level in the system (x), also quite relevant. Because of the Coulomb repulsion in the paramagnetic state, there is a strong tendency of the system towards charge order as x increases. There is always in the system an underlying competition between charge order and metallicity, and band filling is what inclines the scale towards one or the other. Finally, local disorder effects produced mainly by difference in atomic radii between A^{3+} and B^{2+} ions but also by cationic vacancies and oxygen defects play a very important role. These random defects introduce elastic stress in the lattice that interferes with the relaxation of the JT effect described in Fig. 3 affecting the lattice dynamics as well as the hopping process at high temperatures. The disorder is quantified by means of the variance of the A-cation radius distribution (σ^2) defined as $\sum y_i r_i^2 - \langle r_A \rangle^2$, where y_i are the fractional occupancies of the species[53,54]. Rodriguez-Martinez and Attfield find that the temperature at which the resistivity peaks (T_p) and T_C are monotonically decreasing functions of σ^2. At constant $\langle r_A \rangle$ (corresponding to the maximum T_C in Fig. 5) they report $d(Tp)/d(\sigma^2) = 20600$ K/Å^2 for a series of six samples.

The physical properties of interest for this review are T_C, T_p, and lattice transition temperature T_{latt}, all determined by the bandwidth, band filling and local disorder but with different intensity. As a consequence, while the physical properties are coupled to each other, in general $T_C \neq Tp \neq T_{latt}$ [55]. It has been reported, on the other hand, that A-Mn transference is also important, but the experimental situation is still far from clear[56]. Regarding the extrinsic transport properties of CMR manganites, they are most likely dominated by grainsize

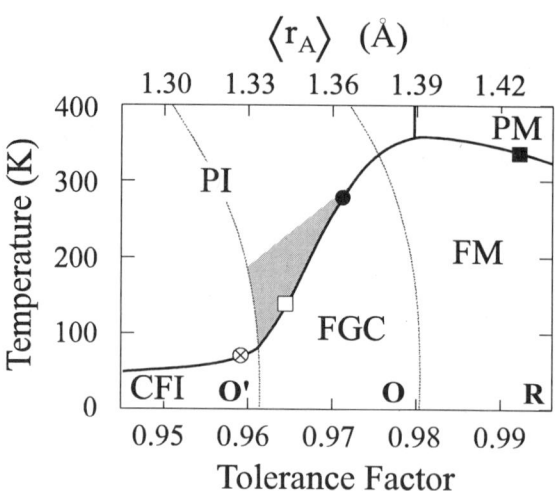

Figure 5. Transition temperature vs. tolerance factor for the $A_{0.7}B_{0.3}MnO_3$ family of compounds, modified from ref. 53. PI: paramagnetic insulator, PM: paramagnetic metal, CFI: canted ferromagnetic insulator, FGC: ferromagnetic glass conductor, FM: ferromagnetic metal, O and O: orthorrombic, R: rombohedric phases. The shadowed area indicates coexistence of extended and localized electronic states. Some samples discussed bellow were included: ($\otimes$) $(La_{0.33}Gd_{0.33})Ca_{0.33}MnO_3$, ($\square$) $(La_{0.5}Gd_{0.17})Ca_{0.33}MnO_3$, ($\bullet$) $La_{0.67}Ca_{0.33}MnO_3$, and ($\blacksquare$) $La_{0.67}(Ca_{0.11}Pb_{0.22})MnO_3$

effects, grain boundary scattering and/or spin-flip, irreversible disorder produced by partial annealing of polycrystalline and film samples, mechanical strain induced by the substrates[57], and very importantly deviations from nominal stoichiometry. Extrinsic effects are quite important, they are evidently responsible for the largest magnetoresistance values reported in the literature[1] and are critical for the many prospective technological applications of the materials. They have been identified as the cause of large low-temperature magnetoresistance by spin polarized tunneling through intergrain barriers, as well as anisotropic magnetic effects.

HIGH TEMPERATURE TRANSPORT

The transport properties have been, for more that forty years, the easier and more straightforward characterization and study method for CMR manganites.[3] However, not until very recently have the clues in favor of lattice involvement in the electronic properties been discussed[29,30,32]. Fig. 6a and 6b display typical results obtained in polycrystalline samples of composition $La_{2/3}Ca_{1/3}MnO_3$ prepared by standard solid state reaction techniques. The resistance in zero field peaks at $T_p \approx 267$ K, somewhat above $T_C(H <1Oe) \approx 261$ K. The magnetoresistance peaks at T_p, but does not vanish at low temperatures as a consequence of the granular nature of the specimen. The granular behavior is not relevant in the high temperature region because the mean free path for charge carriers much smaller that the grain size; an external magnetic field then has little effect on the transport properties. A large (intragrain) mean free path in the metallic phase below T_C, on the other hand, makes the transport extremely sensitive to intergrain barriers caused by magnetic misalignment between weakly coupled grains. An external magnetic field reduces those barriers by aligning neighboring grains. At high temperatures, the thermoelectric power or Seebeck effect $S(T)$ is also sensitive to the metal-insulator phase transition. A sharp change from semiconductor-like absolute values $|S| \approx 10$ μV/K toward metallic values $|S| \approx 1$ μV/K is found coincident with T_C, in Fig 6b). That the Seebeck coefficient approaches a negative value at high temperatures has been attributed in part to the reduction in spin entropy produced when a hole converts a Mn^{3+} ion to Mn^{4+} and is not in disagreement with hole doping in the system. The Seebeck effect, a zero current experiment, shows no indication of grain boundary effects in the proximity or above T_C. These effects are present at much lower temperatures, where the phonon mean free path approaches the grain size, as a large spike centered at 30-50 K[20,30].

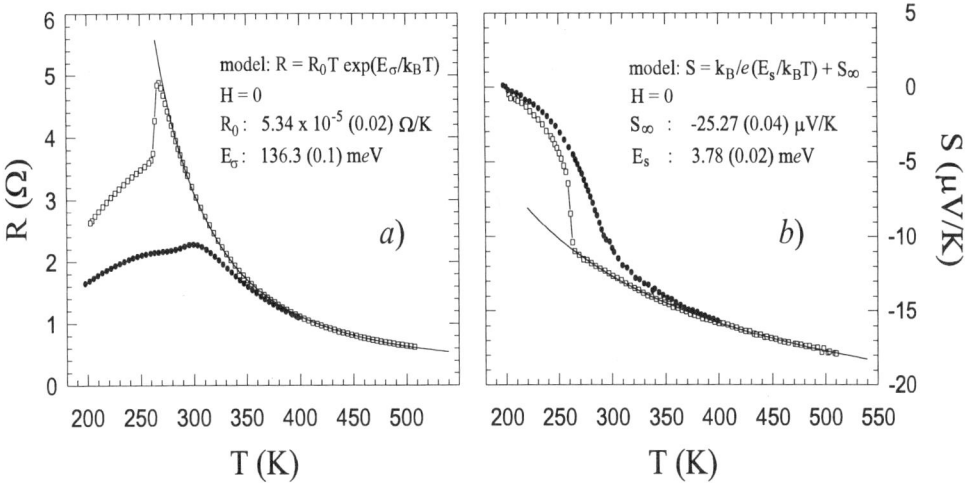

Figure 6. a) Resistance vs temperature for a polycrystalline sample in zero field (□) and H=5T (●) and exponential fit. The data follows very well the model for adiabatic small polarons. b) The thermoelectric power vs temperature and fit to a function of the form A/T + B

The thermopower of semiconductors differ from that of metals also in its temperature dependence, since is governed by thermal activation of carriers thus increasing with decreasing temperature:

$$S = \frac{k_B}{e}\left(\frac{E_S}{k_B T} + B\right) \quad (1)$$

In the case of band semiconductors $E_S = E\sigma$ is the semiconductor gap defined from the temperature dependence of the conductivity by $\sigma(T) \propto exp(-E\sigma/k_B T)$. This is clearly not the case in manganites, where it has been found that $E\sigma$ (~100 meV) $>> E_S$ (~4 meV). The relatively large activation energy in the electrical conductivity has to be interpreted in a different way. $E\sigma$ is, then, not just the semiconducting gap but the gap added to the "hopping energy" W_H, a consequence of a thermally activated mobility of localized carriers jumping between neighboring sites.

The formation and transport properties of small lattice polarons in strong electron-phonon coupled systems, in which charge carriers are susceptible to self-localization in energetically favorable lattice distortions, were first discussed in disordered materials [58] and later extended to crystals [60]. Emin [61] considered the nature of lattice polarons in magnetic semiconductors, where magnetic polarons are carriers self-localized by lattice distortions but also dressed with a magnetic cloud. A transition from large to small polaron occurs as the ferromagnet disorders, successfully explaining the metal-insulator transition observed experimentally in EuO. If the carrier together with its associated crystalline distortion is comparable in size to the cell parameter, the object is called a small, or Holstein, polaron (HP). Because a number of sites in the crystal lattice can be energetically equivalent, a band of localized states can form. These energy bands are extremely narrow, and

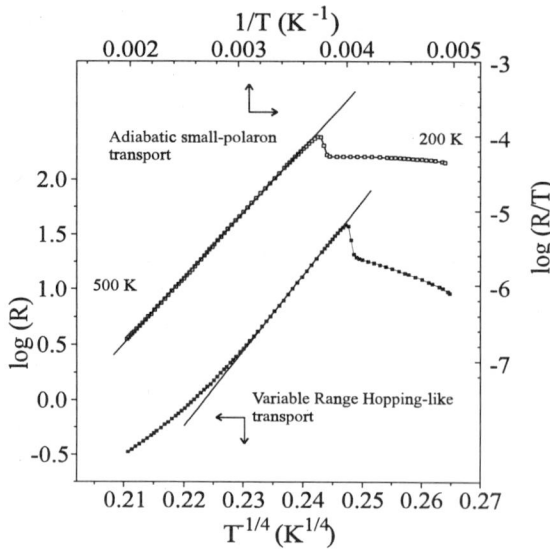

Figure 7. The resistance for a polycrystalline sample of composition La$_{0.67}$Ca$_{0.33}$MnO$_3$ ploted in two different scales, the one expected for adiabatic small polaron hopping and for Variable Range Hopping.

the carrier mobility associated with them is predominant only at very low temperatures. It is important to note that these are not extended states even at the highest temperatures, where the dominant mechanism is thermally activated hopping, with an activated mobility $\mu_p = [x(x-1)ea^2/h](T_0/T)^s exp[-(W_H - J^{3-2s})/k_BT]$ where a is the hopping distance, J the transfer integral, x the polaron concentration, and W_H one-half of the polaron formation energy. In the non-adiabatic limit, we have $s = 3/2$ and $k_BT_0 = (pJ^4/4W_H)^{1/3}$ and, in the adiabatic limit, $s = 1$ and $k_BT_0 = \hbar\omega_0$, where ω_0 is the optical phonon frequency. The criterion for non-adiabatic behavior is that the experimental $k_BT_0 << \hbar\omega_0$. Using experimental values for σ, $E\sigma$, S and cell parameter we find that $k_BT_0/\hbar \approx 10^{14} s^{-1}$, comparable to optical phonon frequencies, although it could be considered a marginal case. We will assume the adiabatic limit to hold, in which case the electrical conductivity, $\sigma = eN\mu_p$, where N is the equilibrium polaron number at a given temperature, can be expressed as

$$\sigma = \frac{x(1-x)e^2T_0}{\hbar aT} \exp\left(-\frac{\epsilon_0 + W_H - J}{k_BT}\right) \quad (2)$$

Figure 7 shows the same resistance data of figure 6a displayed in two different charge localization scenarios, i.e. adiabatic small polaron-like: $\log(R/T)$ vs T^{-1} and Variable Range Hopping-like: $\log(R)$ vs $T^{-1/4}$. While the data mimics VRH behavior at temperatures close enough to T_C, the adiabatic small polaron (Eq. 2) describes the system well in the entire temperature range. $R(T)$ data obtained up to $T \approx 5T_C$ show excellent agreement with this model[71,89], and we can assume that the localization of holes persists up to the material's melting point.

The field dependence of the activation energies E_S and $E\sigma$ has been discussed for film samples of similar nominal composition [32]. Within experimental resolution, changes in activation energies are different but of the same order of magnitude. An estimation of average experimental values is $\Delta W_H/\Delta H = -2.9 \times 10^{-5}$ meV/Oe or 2.8×10^{-5} %/Oe and $\Delta E_S/\Delta H = \Delta\varepsilon_0/\Delta H = -1.4 \times 10^{-5}$ meV/Oe or -1.4×10^{-4} %/Oe. While ε_0 reflects changes in the Fermi energy that can be related to the reported magnetostriction of

CMR materials [25], changes in W_H imply an increase of the radius of the small polaron with field and consequently some magnetic character of the quasiparticle. Because of this double character of the localized holes, elastic as well as magnetic at temperatures up to $2T_C$, they are named *magnetoelastic polarons* in an attempt to emphasize differences with purely magnetic polarons[7,60] and HP.

In Eq. 1, the temperature independent term $B < 0$ is given by two contributions, namely $-(k_B/e)\ln\{[2x(3/2)+1]/[2x^2+1]\} = -(k_B/e)\ln(4/5) = -19\,\mu\text{V/K}$ associated with the spin entropy appropriate for a spin-1/2 hole moving in a spin-2 background[62]; and a mixing entropy term that counts in how many different ways x holes can be distributed between n Mn sites. In the case of correlated hopping with weak near-neighbor repulsion[63] this term is $\ln[x(1-x)/(1-2x)^2]$ and at the nominal doping level $x = 1/3$ contributes $-60\,\mu\text{V/K}$; without the repulsive interaction, the mixing term contributes $+60$ mV/K at the same hole concentration. In either case the prediction is unable to reproduce the experimental value $S_\infty \approx -25$ μV/K, see Fig. 6b. Attempts to understand the high temperature limit of the thermopower $B = S_\infty$ following its changes with hole concentration via Ca concentration changes have been frustrating so far. An alternative way to modify the doping level is via control of the concentration of oxygen vacancies, which can be accomplished with thermogravimetric methods. As part of a cooperative program with the group at the Centro Atómico Bariloche, Argentina[64] polycrystalline samples were placed in an oven equipped with atmosphere control capabilities where the isotherm displayed in Fig. 8 was obtained[65]. The maximum concentration of vacancies observed without mass instability effects characteristic of phase segregation was $d = 0.051$, enough to depress T_C from 265 K to 221 K. Figure 9a shows $S(T)$ vs $1000/T$ for three polycrystalline samples of composition $La_{0.67}Ca_{0.33}MnO_3$, $La_{0.67}Ca_{0.33}MnO_{2.49}$, and $La_{0.75}Ca_{0.25}MnO_3$. Indeed, the variation in S_∞ confirms changes in the doping level. The overall changes however, are not as large as they would be expected. Fig. 9b show data by Mahendiran et al.[20] as well as our samples and different model predictions. Besides the Chaikin-Beni (Ch-B) model discussed above, alternative models considered in the bibliography are the correlated and uncorrelated limits by Heikes[66]:

$$S_{ME}^{H_{corr}} = \ln\left(\frac{1+x}{1-x}\right) \; ; \; S_{ME}^{H_{uncorr}} = \ln\left(\frac{x}{1-x}\right) \qquad (3)$$

and the D-dimensional extension of Ch-B formula[67]:

$$S_{ME}^D = \ln\left\{\frac{x(1-Dx)^D}{2[1-x(D+1)]^{D+1}}\right\} \qquad (4)$$

These predictions are in general unable to reproduce results, except perhaps in the case of Heikes uncorrelated limit. There are three possible explanations for this behavior. A disproportionation theory, where two Mn^{3+} atoms generate Mn^{2+} and Mn^{4+} sites with the transference of one electron. The disproportionation density is related to oxygen non-stoichiometry[35]. In an alternative explanation by Emin[36], the small polarons are proposed to be correlated with divalent atoms in real space due to the elastic stress introduced in the lattice by atomic size mismatch. In this kind of "impurity" conduction, the number of available sites for localized states increases with doping and as a consequence the mixing entropy remains unchanged. Finally, the Heikes uncorrelated limit suggest that multiple occupancy or collective behavior could be possible for small polarons. This possibility, unlikely in principle, finds some support in recent neutron diffraction data where spin clusters (charge droplets) of a few holes localized in regions of ~20Å were identified[68].

Perhaps the most distinctive property of steady-state small-polaronic transport is its Hall mobility μ_H. The activation energy of the Hall mobility is calculated to be always less

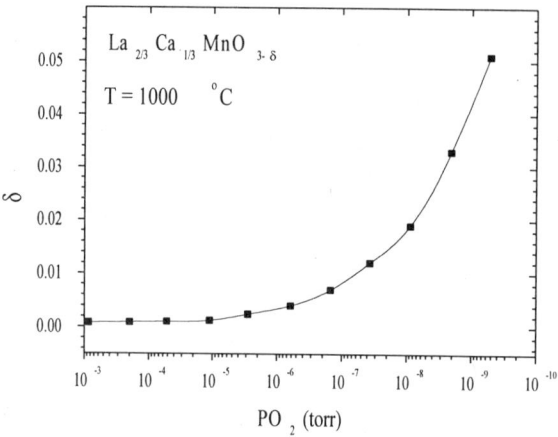

Figure 8. The 1000 °C isoterm for $La_{0.67}Ca_{0.33}MnO_3$.

than that for drift mobility E_d. The simplest model predicts $\approx E_d/3$, and this has been observed [69] before in, for example, oxygen-deficient $LiNbO_3$. The sign of the Hall effect for small polaron hopping can be "*anomalous.*" A small polaron based on an electron can be deflected in a magnetic field as if it were positively charged and, conversely, a hole-based polaron can be deflected in the sense of a free electron. As first pointed out by Friedman and Holstein, the Hall effect in hopping conduction arises from interference effects of nearest neighbor hops along paths that define an Aharonov-Bohm loop. Sign anomalies arise when the loops involve an odd number of sites. The first successful measurement of the high-temperature Hall coefficient in manganite samples was reported by Jaime et al.[36], finding that it exhibits Arrhenius behavior and a sign anomaly relative to both the nominal doping and the thermoelectric power. The results are discussed in terms of an extension of the Emin-Holstein (EH) theory of the Hall mobility in the adiabatic limit.

The authors exploit the sensitivity of CMR materials to rare-earth substitutions to lower the transition temperature in $(La_{1-y}R_y)_{0.67}Ca_{0.33}MnO_3$ from ~260 K at $y = 0$ to ~ 130 K, thereby extending the accessible temperature range to $\approx 4T_C$. The samples used in this study were laser ablated from ceramic targets and deposited on $LaAlO_3$ substrates, as described previously[30], showing a temperature dependence in resistivity and thermopower data qualitatively similar to Fig. 6.

Sections of these specimens were patterned by conventional lithographic methods into a five-terminal Hall geometry. Hall experiments were carried out in a high-temperature insert constructed for use at the 20 Tesla superconducting magnet at the National High Magnetic Field Laboratory (Los Alamos, NM). The transverse voltage data taken while sweeping the field from -16 Tesla to +16 Tesla, and that taken while sweeping back to -16 Tesla, were each fit to a second-order polynomial with the term linear in field attributed to the Hall effect. We verified in each case that the longitudinal magnetoresistance is completely symmetric in field. Figure 10 shows the Hall coefficient derived from the linear term. Several points for the $y = 0$ film are included. Due to the much higher T_C of that sample, extraction of the Hall contribution leads to greater uncertainty. The data are, however, consistent with the Gd-substituted film. The line through the data points is an Arrhenius fit, giving the expression $R_H = -(3.8 \times 10^{-11} \text{ m}^3/\text{C}) \exp(91 \text{ meV}/k_B T)$. Note that the sign is negative, even though

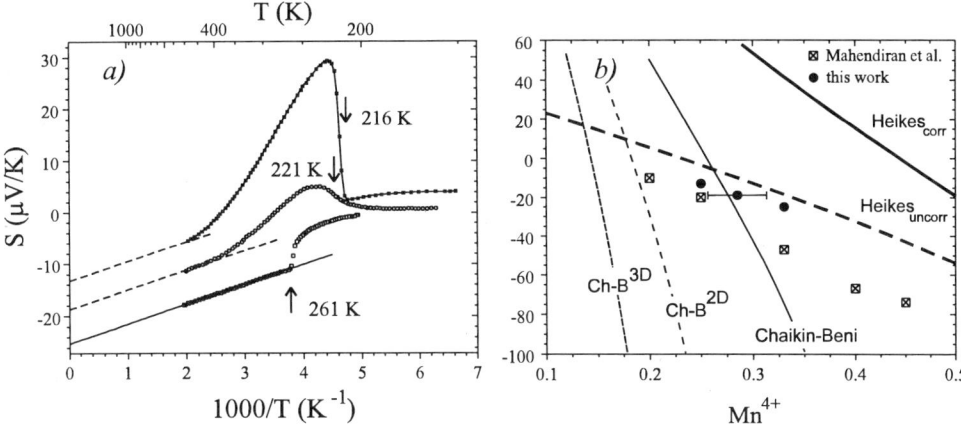

Figure 9. a) Thermopower vs. 1000/T for polycristalline samples of composition La$_{0.67}$Ca$_{0.33}$MnO$_3$, La$_{0.67}$Ca$_{0.33}$MnO$_{2.49}$ and La$_{0.75}$Ca$_{0.25}$MnO$_3$. b) S_∞ vs Mn^{4+} content for our samples and theoretical predictions discussed in the text.

divalent dopants should introduce holes. This is shown more clearly in an Arrhenius-like plot (inset b). Inset a, displays $1/R_H$ vs T. If the observed linear term in the Hall data is due to the well known *skew* scattering, then R_H is expected to be proportional to the magnetization and in consequence the data should extrapolate to T_C. Our data extrapolates to 245 K, more than a hundred degrees above T_C. *Skew* scattering is then unlikely to explain the negative sign of the Hall data in these samples.

Detailed expressions for the Hall effect in the adiabatic limit have been calculated by EH[70] for the hopping of electrons with positive transfer integral J_H on a triangular lattice, and results in a normal (electron-like) Hall coefficient. However, the sign of both the carrier and the transfer integral changes for hole conduction, leaving the sign of the Hall coefficient electron-like, and therefore anomalous. However, no anomaly would arise if the hopping involves 4-sided loops with vertices on nearest-neighbor Mn atoms. A sign anomaly, then, implies that hopping involves odd-membered Aharonov-Bohm loops. Such processes arise when next-nearest neighbor (*nnn*) transfer processes across cell face diagonals are permitted. If the Mn-O-Mn bonds were strictly colinear, the former processes would be disallowed by symmetry. However, the bond angles are substantially less than 180°, implying the presence of p-bond admixture, and opening a channel for diagonal hops. The triangular-lattice calculation of EH is extended to the situation in which a hole on a Mn ion can hop to any of its four nearest neighbors in the plane normal to the applied field with transfer matrix element $J < 0$ and to its four *nnn* with a reduced transfer energy γJ. The effect of these diagonal hops (plus those in the plane containing both electric and magnetic fields) has also to be considered on the conductivity prefactor. The Hall coefficient can be written as $R_H = R_H^0(T) \exp(2E\sigma/3k_BT)$, with

$$R_H^0 = -\frac{g_H}{g_d} \frac{F(|J_H|/k_BT)}{ne} \exp\left\{-[\varepsilon_0 + (4|JH| - E_S)/3]/k_BT\right\} \qquad (5)$$

EH found that the factor $g_H = 1/2$ for three-site hopping on a triangular lattice. In Eq. 7 the carrier-density is included as $n\exp(-E_S/k_BT)$, where E_S is estimated to be 8 meV from the thermopower data. The quantity ε_0 is the J_H-dependent portion of a carrier's energy achieved when the local electronic energies of the three sites involved in an Aharonov-Bohm loop are equal. For the problem considered by EH, an electron hopping within a lattice composed of equilateral triangles, $\varepsilon_0 = -2|J_H|$, and $g_H/g_d = 1/3$. Within the domain of validity of EH, the temperature dependence of R_H arises primarily from the factor $\exp(2E_\sigma/3k_BT)$ when $E_\sigma \gg E_S$. For holes hopping within a cubic lattice in which three-legged Aharonov-Bohm loops include ε_0 varies from $-\sqrt{2}|J_H|$ to $-|J_H|$ as γ increases from zero to unity, and the temperature dependence of R_H remains dominated by the factor $\exp(2E_\sigma/3k_BT)$. Indeed, the energy characterizing the exponential rise of the Hall coefficient that we observe, $E_H = 91 \pm 5$ meV, is about 2/3 the measured conductivity activation energy, $E_H/E_\sigma = 0.64 \pm 0.03$, in excellent agreement with theory.

The geometrical factor g_d depends on the ratio of the probability Pnnn of nnn hops to Pnn, that of nn hops, through $g_d = (1 + 4Pnnn/Pnn)$. If these probabilities are comparable ($\gamma \approx 1$) $g_d = 5$, $g_H = 2/5$ and the exponential factor in Eq. 7 becomes $\exp[(E_S - |J_H|)/3k_BT] \approx 1$. In the regime $|J_H| \geq k_BT$, the function $F(|J_H|/k_BT)$ is relatively constant with a value ≈ 0.2, and we find $R_H^0 \approx -0.02/ne = -3.8x10^{-11}$ m^3/C. This yields an estimated carrier density $n = 3.3 \times 10^{27}$ m^{-3}, quite close to the nominal level of 5.6×10^{27} m^{-3}.

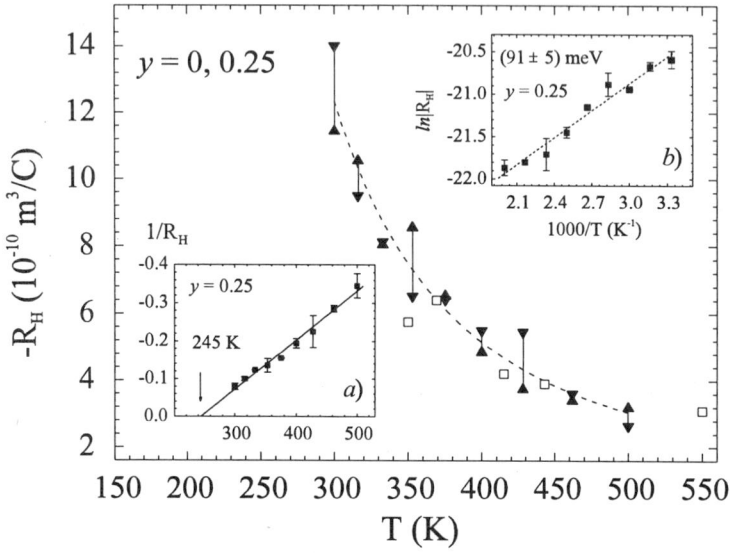

Figure 10. The Hall coefficient vs T for film samples of composition (La$_{1-y}$Gd$_y$)$_{0.67}$ Ca$_{0.33}$MnO$_3$, y = 0 ($\square$) and y = 0.25 ($\blacktriangle$). The dashed line correspond to a fit of the form $R_H = R_H^0 \exp[E_H/k_BT]$. Inset (a): $1/R_H$ vs T, for y = 0.25 showing an extrapolation to 245 K $\gg T_C$ = 142 K. Inset (b) $\ln|RH|$ vs 1000/T, for y = 0.25, showing an activation energy $E_H = 91 \pm 5$ meV.

Before moving over to the low temperature transport properties the following conclusions can be reached. The high-temperature Hall coefficient in manganite films is consistent with small-polaron charge carriers that move by hopping. The magnitude of the conductivity prefactor indicates that the carrier motion is adiabatic. The sign anomaly in the Hall effect implies that small polarons hop not only among near-neighbor sites (making Aharonov-Bohm loops with an even number of legs) but must have a significant probability of traversing Hall-effect loops with odd numbers of legs. As such, the results indicate the occurrence of significant nnn transfer across face diagonals, and therefore a crucial role for deviations of the Mn-O-Mn bond angle from 180°. In other words, the sign anomaly its a simple consequence of the geometry of the sublattice where the small polarons move and the fact that it is triangular and not square indicate an interesting possibility, that may also relate to unusual high-temperature values observed for the Seebeck coefficient [30,32]. That is that transport is a type of impurity conduction in which carriers remain adjacent to divalent cation dopants (*i.e.*

Ca ions). The local distortions associated with the presence of the impurity may also increase the admixture of π-bonds, and enhance diagonal hopping.

In a recent paper, Worledge et al.[71] discuss the temperature and doping dependence of the resistivity in $La_{1-x}Ca_xMnO_3$ laser ablated films measured up to $T = 1200$ K for $0 > x > 1$. They conclude that the results can be unambiguously explained by adiabatic small polaron hopping, which is limited by on-site Coulomb repulsion. The magnitude of the conductivity prefactor, however, is too large to be accounted by the classical theory by Emin and Holstein[70] and the authors claim that a proper description should consider hopping beyond nearest neighbors, in good agreement with high temperature Hall effect results. A few other reports on the Hall effect of manganites are now available[72,73], they are restricted however to the relatively low temperature side of the diagram in Fig.4, i.e. regions I and III. The low temperature Hall effect is not less intriguing than the high temperature counterpart, and is not yet understood. The Hall resistance in the metallic regime imply carrier concentrations up to $3\times$ the nominal values suggesting some compensation effects and/or two-band conduction, the spontaneous Hall contribution is opposite in sign from the normal Hall effect with the overall effect exhibiting a sign change around T_C from hole-like in the ferromagnetic phase to electron-like in the paramagnetic phase. In order to clarify the subject, more experiments in the very high temperature limit ($T \geq 2T_C$) are desirable..

LOW TEMPERATURE TRANSPORT

The low temperature region I, in Fig. 4 in optimally doped cubic manganites is perhaps the most interesting one since it corresponds to a ground state that is the closest to half metallic systems ever synthesized. Unfortunately, the transport properties of polycrystalline samples in this regime are dominated, or at best highly influenced, by grain boundary scattering, and close attention has to be paid to sample quality issues. These problems have led in the past to misunderstanding about the most basic transport properties, like resistivity and Seebeck effect for example. One common problem in the resistivity of polycrystalline specimens is the presence of a minimum at $T \approx$ 10-20 K that has been attributed to localization effects. In the same temperature range, the Seebeck effect shows large anomalies (as big as -40 μV/K) by no means compatible with a metallic state. None of these features have been reproduced in carefully prepared ($x \sim 1/3$) single crystals and are thus considered non-intrinsic. One of the simplest methods of characterization seems to be the magnetoresistance (MR), since granular samples show large ratios in small applied magnetic fields down to the lowest temperatures. On the other hand, MR rapidly vanishes below T_C in long time annealed films[30] and single crystals.

The double exchange mechanism is generally agreed to provide a good description of the ferromagnetic ground state. In that model, strong Hund's Rule coupling enhances the hopping of e_g electrons between neighboring Mn^{3+} and Mn^{4+} ions by a factor $\cos(\theta/2)$, where θ is the angle between the spin of their respective t_{2g} cores, thereby producing a ferromagnetic interaction. In KO's treatment of the problem[9], occupied sites are assumed to have total spin $S = 2$, the combination of the spin-3/2 t_{2g} core and spin-1/2 e_g electron demanded by strong Hund's rule exchange. Holes are then assumed to couple antiparallel to each localized spin. In the ferromagnetic ground state, with all local spins aligned, only spin-down holes can move to form a band. However, once the system begins to disorder, a locally down-spin hole is the appropriate combination of majority and minority carriers, referred to the global magnetization axis. Therefore, the minority-spin hole band reappears as the system disorders, even though the local moments retain their Hund's rule value. Furukawa has treated this explicitly in a many-body context, demonstrating that both minority and majority bands are split by Hund's rule exchange in the paramagnetic state. As the system magnetizes, the lower majority-spin band gradually gains spectral weight at the expense of the lower minority-spin band. Both treatments predict that the ground state is half-metallic; that is, that the carriers are fully spin-polarized. We can leave aside here electron-phonon coupling which dominates near and above T_C.

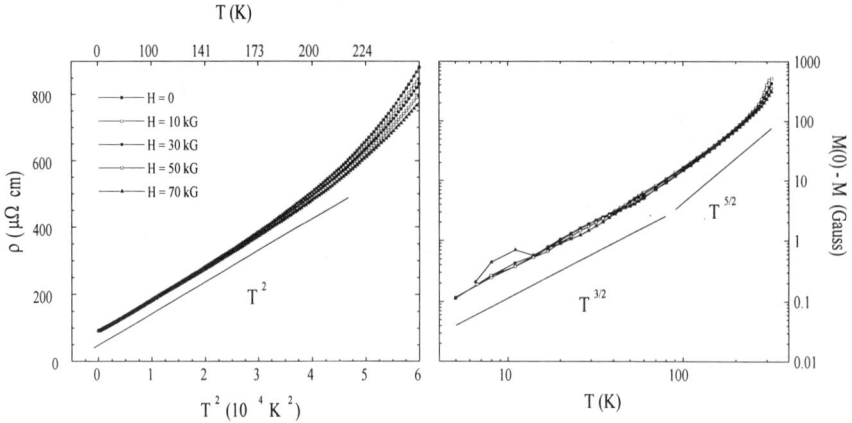

Figure 11. Left: the resistivity vs T^2 for magnetic fields up to 70 kOe in a single crystal sample of composition La$_{0.67}$(Ca,Pb)$_{0.33}$MnO$_3$. Right: the magnetization deviation from saturation vs temperature on a log-log plot, with $T^{3/2}$ and $T^{5/2}$ contributions.

At low temperature, there are no propagating minority-spin hole states in the $S = 2$ manifold; they only exist on sites at which the t_{2g} core is not ferromagnetically aligned. As a consequence, single-magnon scattering processes, which cause the resistivity of conventional ferromagnets to vary as T^2, are suppressed. KO extended the standard perturbation calculation of Mannari[6] to consider two-magnon processes, predicting a leading $T^{9/2}$ temperature dependence of the resistivity. However, a dominant T^2 contribution is universally observed in the manganites, and has usually been ascribed to electron-electron scattering.[30,75,76] New resistivity data on single crystals is discussed below, demonstrating that the quadratic temperature dependence is strongly suppressed as the temperature is reduced. The constancy of the low-temperature resistivity has been noted elsewhere, but not explained.[77] It is argued that the observed T^2 contribution reflects the reappearance of minority spin states that are accessible to thermally excited magnons. Quite recently, spin-polarized photoemission data, taken on films exhibiting square hysteresis loops, indicate 100% spin polarization at low temperatures, decreasing as the temperature is increased[78]. Single crystals, which have essentially no hysteresis, would be expected to depolarize more rapidly. To explore the consequences, Mannari's calculation is extended to the situation in which a minimum magnon energy is required to induce spin-flip transitions. At temperatures well below that energy, single magnon scattering is suppressed exponentially as predicted by KO. The treatment discussed by Jaime *et al.*[74] is in the context of the relaxation time approximation while a proper theory would consider lifetime effects from magnon scattering using Furukawa's many-body approach. Nonetheless, the results are in qualitative agreement with the data. Band structure calculations also indicate that minority spin states persist at E_F, even at $T = 0$ K.[79]

High quality single crystals of nominal composition La$_{0.66}$(Pb$_{0.67}$Ca$_{0.33}$)$_{0.34}$MnO$_3$, determined by inductively coupled plasma spectroscopy on samples from the same batch, were grown from a 50/50 PbF$_2$/PbO flux and used to study the low temperature properties. X-ray diffractometry shows a single pseudo-orthorhombic structure with lattice parameters $a = 5.472(4)$ Å, $b = 5.526(6)$ Å, and $c = 7.794(8)$ Å. Gold pads were evaporated onto both oriented and unoriented crystals using both standard four-terminal and Montgomery eight-corner contact arrangements as described elsewhere[74]. Fig. 11 shows the resistivity of sample sc3, a single crystal of dimensions $1.04 \times 1.24 \times 0.3$ mm^3 with $T_C = 300$ K, vs the square of the temperature in fields up to 70 kOe. The data show a dominant T^2 temperature dependence with evidence of a small T^5 contribution (10 $\mu\Omega$cm at 100 K). A calculation of the $T^{9/2}$ contribution predicted by KO for two-magnon processes predicts only 0.5 $\mu\Omega$cm at 100 K with appropriate parameters. It is likely, then, that this is the usual T^5 contribution from electron-phonon processes. Within the spin-wave approximation, the low temperature magnetization is given by $M(T) = M(0) - BT^{3/2} - ...$, where $B = 0.0587g\mu_B(k_B/D)^{3/2}$. The stiffness constant D has been determined by neutron scattering [80,81] and muon spin resonance [49] to be $D \approx 135 - 170$ meV Å^2. The right side of Fig. 11 shows the magnetization for this sample, from which we extract $B(10$ kOe$)$ and the value $D = 165$ meV Å^2, in good agreement with other results.

The plot in the right on Fig. 12 shows that the data do not follow a T^2 dependence to the lowest temperatures. Rather, they deviate gradually from the curve $\rho_0 + \alpha(H)T^2$, fit over the range $60 \leq T \leq 160$ K, saturating at an experimental residual resistivity $\rho_0^{exp} = 91.4$ $\mu\Omega$cm, comparable to values observed by Urushibara et al.[75], but ~7% larger than ρ_0. This conclusion is not changed by including the T^5 contribution. Fits to data taken in various fields show that $\alpha(H)$ decreases with increasing field and is the source of the small negative magnetoresistance at low temperatures. To quantify the disappearance of the T^2 contribution, the authors calculate $(\rho - \rho_0^{exp})/\alpha(H)T^2$ and display it in Fig. 13a. The T^5 contribution which gives a slight upward curvature to the data at higher temperatures has not been substracted. Should the T^2 description be valid in the low temperature range, we should have $(\rho - \rho_0^{exp})/\alpha(H)T^2 \equiv 1$. Note that this description of the data is extremely sensitive to the value of ρ_0^{exp}, and it must be determined very carefully. An alternative description is possible by means of a numerical derivative, as discussed elsewhere[74], similar conclusions are arrived at.

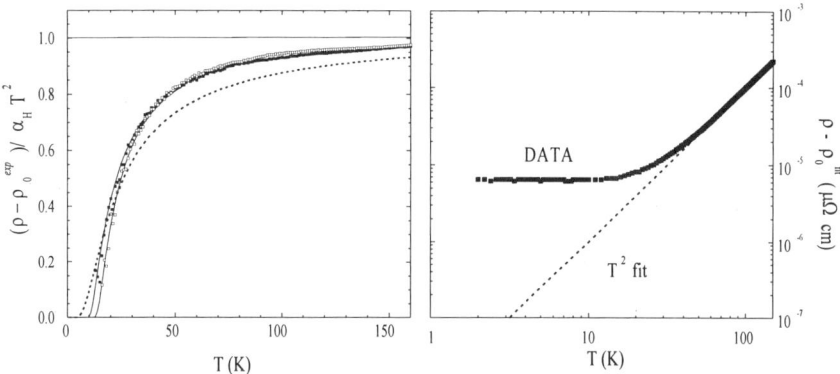

Figure 12. Left: the experimental resistivity $\rho(T)$ after substraction of the residulal value ρ_0^{exp}, divited by $\alpha_H T^2$ vs temperature. The dashed line is the result of the one-magnon calculation described in the text. Right::$\rho(T)$ after substraction of the fitted value ρ_0^{fit} vs temperature in a log-log plot. In both plots significant deviations from the T^2 behavior displayed in Fig. 12 are observed.

Previous investigators have attributed the T^2 term in the resistivity to electron-electron scattering. An empirical relationship has been found between the coefficient α and the coefficient γ_{hc} of the electronic specific heat by Kadowaki and Woods[82] : $\alpha/\gamma^2 = 1 \times 10^{-5}$ $\mu\Omega$cm(mole K/mJ)^2. Using our experimental value and $\gamma \sim 4$ mJ/mol K^2 from ref. [84] we find a value $\sim 60\times$ the Kadowaki-Woods parameter which argues against e-e scattering. With an electron density at the nominal doping level $n = 5.7 \times 10^{27}$ m^{-3}, the effective mass that follows from γ_{hc} is $m^*/m = 2.5$ and the Fermi energy is $E_F = 0.5$ eV and the $e-e$ relaxation rate of the order of 2×10^{11} s^{-1} at 100 K. The experimentally observed T^2 contribution at that temperature is 100 $\mu\Omega$ which with the same parameters correspond to a relaxation rate of 6×10^{13} s^{-1} more that two orders of magnitude larger. This disagreement is not fixed by using low temperature Hall-deduced effective concentration of carriers[73] . Rather than vanishing, what is more, $e-e$ scattering should become more apparent as the temperature is reduced. We conclude that $e-e$ scattering is an unlikely explanation for the observed quadratic dependence on temperature.

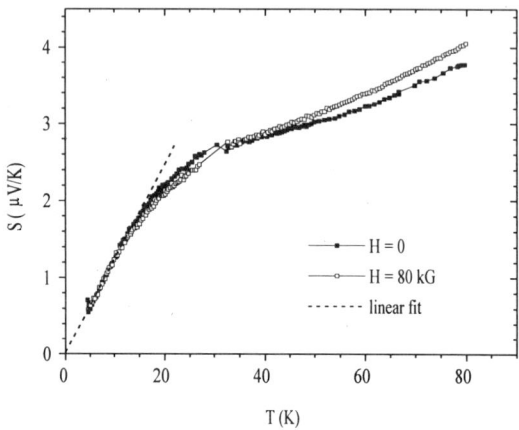

Figure 13. The Seebeck coeficient S vs temperature. It is positive and metal-like at low temperature, has an anomalous kink near 30 K, and develops a positive field dependence above 40 K. The dashed line is a linear fit in the low temperature regime.

When the usual calculation of the electron-magnon resistivity[6] is extended to allow the minority-spin sub-band to be shifted upward in energy such that its Fermi momentum differs by an amount q_{min} from that of the majority sub-band, the one-magnon contribution can be written as $\rho\epsilon = \alpha\epsilon T^2$, where $\alpha\epsilon = (9\pi^3 N^2 J^2 \hbar^5/8e^2 E_F^4 k_F)(k_B/m^*D)^2 I(\epsilon)$. Here NJ is the electron-magnon coupling energy which is large and equal to $\mu = W - E_F$ in the DE Hamiltonian of KO; $2W$ is the bandwidth.. The magnon energy is given by Dq^2, and

$$I(\epsilon) = \int_{\epsilon}^{\infty} \frac{x^2}{sinh^2 x} dx \qquad (6)$$

The lower limit is $\epsilon = Dq_{min}^2/2k_B T$, where Dq_{min}^2 is the minimum magnon energy that connects up- and down-spin bands; result that reproduces Mannari's calculation in the limit $\epsilon \to 0$, and KO's exponential cut-off for large ϵ. At high temperatures, the lower limit of the integral in Eq. 6 can be set equal to zero, leaving only the coupling energy $NJ = W - E_F$

as a parameter. Equating the calculated value to the experimental α fixes the coupling to be $W - E_F \approx 1.0$ eV or $W \simeq 1.5$ eV, in good agreement with a virtual crystal estimate of the band width. [79] In Fig. 12 we have plotted $I(\epsilon, T)$ assuming $D(0)q_{min}^2 = 4$ meV and including the temperature dependence observed experimentally, $D(T)/D(0) = (1 - T/T_C)^{0.38}$ [81] which is important only at higher temperatures. While the curve follows the data qualitatively, it is clear that the minimum magnon energy is substantially larger than 4 meV at low temperatures, and decreases rapidly with increasing temperature.

Fig. 13 shows the Seebeck coefficient $S(T)$, measured on the same unoriented sample. Below 20 K, $S(T)$ is positive as expected for hole conduction, linear in temperature, and extrapolates to zero as $T \to 0$. If we take the scattering to be independent of energy, which is the case below 20 K, Seebeck coefficient can be expressed as $S(T) = (\pi^2/2e)(k_B^2 T/E_F)$.[83] Using the simplistic approximation of parabolic band $E_F = \hbar^2 k_F^2/2m^*$, and spherical Fermi surface $k_F^3 = 3n\pi^2$, the effective mass results to be $m^*/m \approx 3.7$, comparable to the value obtained from specific heat measurements. The sharp deviation from linear behavior in the temperature range $20 - 40$ K correlates with the onset of electron-magnon scattering which, being a spin flip process, must involve the minority spin band, and which therefore has a different dependence on energy near E_F.

In conclusion, the low temperature transport data cannot be explained by electron-electron scattering as proposed before[30,75–77] and, while oversimplified, the extension of the standard calculation of one-magnon resistivity to describe spin-split bands gives a qualitative account of the half metallic suppression of the spin-wave scattering at very low temperatures.

INTERMEDIATE TEMPERATURES, $T \simeq T_C$

As Millis and coworkers[85–88] have emphasized, the Jahn-Teller effect in Mn^{3+}, if strong enough, can lead to polaron formation and the possibility of self-trapping. The effective JT coupling constant λ_{eff}, in this picture, must be determined self-consistently, both because it depends inversely on the bandwidth and because the effective transition temperature increases with decreasing λ_{eff}. If λ_{eff} is larger than a critical value λ_c, the system consists of polarons in the paramagnetic phase. As the temperature is lowered to the Curie temperature T_C, the onset of ferromagnetism increases the effective bandwidth, which reduces λ_{eff}, thereby increasing the effective transition temperature. As a result, the polarons may dissolve into band electrons if λ_{eff} drops below λ_c and the material reverts to a half-metallic, double exchange ferromagnet at low temperatures. The tendency toward polaron formation is monitored by a local *displacement* coordinate r, which is zero for $\lambda_{eff} < \lambda_c$, and grows continuously as λ_{eff} increases beyond that value. However, polarons are typically [70] bimodal–large or small–so that we should consider r to be a measure of the relative proportion of large polarons (band electrons for which $r \approx 0$) and small polarons (for which r is an atomic scale length).

Indeed, there is growing experimental evidence[39,41,92,93] that polaronic distortions, evident in the paramagnetic state, persist over some temperature range in the ferromagnetic phase, as displayed in Fig 2 as a coexistence zone. This possibility can be explored by considering the observed electrical resistivity to arise from the parallel conduction of a field- and temperature-dependent polaronic fraction (with activated electrical conductivity) and band-electron fraction (with metallic conductivity). The validity of this model is tested by applying it to the thermoelectric coefficient using an extension of the well-known Nordheim-Gorter rule for parallel conducting channels. The $La_{2/3}Ca_{1/3}MnO_3$ film samples used in this study were prepared by pulsed laser deposition onto $LaAlO_3$ substrates to a thickness of 0.6 μm. As described previously [32], they were annealed at 1000 °C for 48 hr. in flowing oxygen. Measurements were carried out in a 7T Quantum Design Magnetic Property Measurement System with and without an oven option provided by the manufacturer. A modified sample rod brought electrical leads and type-E thermocouples to the sample stage. A bifilar coil of

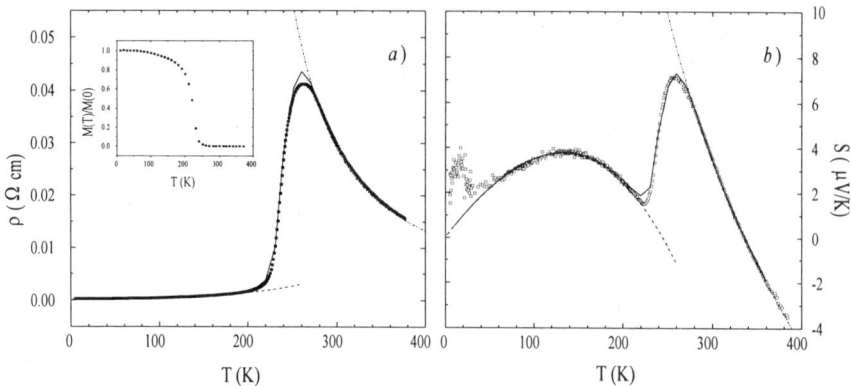

Figure 14. *a)* Resistivity vs temperature in zero field. The broken lines indicate extrapolations of the fits to the low and high temperature regions of the curve. The solid line is the parallel combination of the two conductivities using the magnetization (inset) as a mixing factor. *b)* Similar results for the Seebeck coefficient.

12 μm Pt wire was calibrated to serve both as a thermometer and to provide a small heat input for the thermopower measurements. Measurements in fields up to 70 kOe could be carried out over the temperature range 4 K $\leq T \leq$ 500 K. Following the transport measurements, magnetization data $M(H,T)$ were acquired up to 380 K by conventional methods.

Figure 14a shows the resistivity data in zero field over the full temperature range. The data below 200 K exhibit metallic behavior, and are well fit by a power law, $\rho_{lt}(T) = [0.22 + 2 \times 10^{-5} \text{ K}^{-2}T^2 + 1.2 \times 10^{-12} \text{ K}^{-5}T^5]$ mΩ cm. Above 260 K, the resistivity is exponential, given[30] by the form expected for the adiabatic hopping of small polarons, $\rho_{ht} = (1.4\mu\Omega$ cm K$^{-1})T \exp(1276 \text{ K}/T)$. These are shown as broken lines. The assumption is that these represent the resistivity of band electrons and polarons, irrespectively, and that the transition region can be represented by a parallel combination characterized by a mixing factor $c(H,T)$ which is envisaged to be the fraction of the carriers that are in the metallic state; that is

$$\rho(H,T) = \left[\frac{c(H,T)}{\rho_{lt}(T)} + \frac{1-c(H,T)}{\rho_{ht}(T)}\right]^{-1} \quad (7)$$

As a first approximation $c(0,T) = M(0,T)/M_{sat}$ is chosen, using the data in the inset of Fig. 14a. The solid curve through the data shows the result of this process with no further adjustable parameters. As a second test of this approach, the Seebeck coefficient $S(H,T)$ is considered, measured over the same temperature range and plotted in Fig. 14b. We fit the low temperature thermopower arbitrarily to a power law, $S_{lt}(T) = [(0.051 \text{ K}^{-1})T - (1.3 \times 10^{-4} \text{ K}^{-2})T^2 - (3.2 \times 10^{-7} \text{ K}^{-3})T^3]$ μV/K, and the high temperature data [32] to the form expected for small polarons, $S_{ht}(T) = [(9730 \text{ K})T^{-1} - 29]$ μV/K. Broken lines in Fig. 14b show the extrapolation of these fits into the transition region. The Nordheim-Gorter rule[94] can now be applied to compute the thermopower for parallel conduction,

$$S(H,T) = \rho_{exp}(H,T)\left[\frac{c(H,T)S_{lt}(T)}{\rho_{lt}(T)} + \frac{(1-c(H,T))S_{ht}(T)}{\rho_{ht}(T)}\right] \quad (8)$$

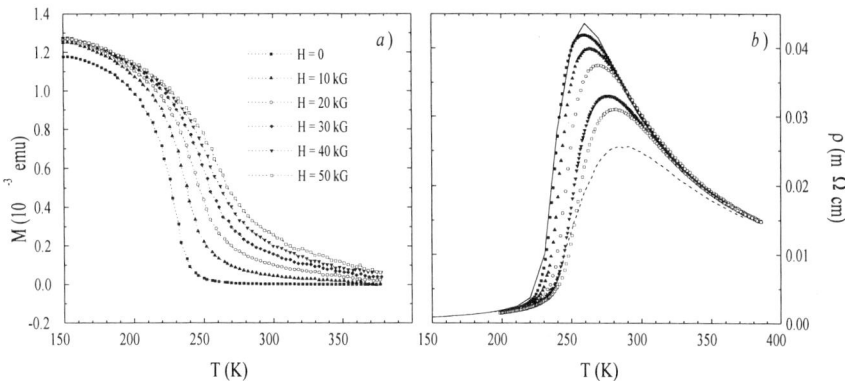

Figure 15. *a)* Magnetization data for this sample. *b)* Resistivity data as functions of field and temperature. The dashed curve is a parallel admixture using the reduced magnetization measured at 10 kOe.

The result is shown as a solid line in Fig. 14*b*, again using the reduced magnetization as a measure of the relative concentration of band electrons and polarons.

The association of $c(H,T)$ with $m(H,T) \equiv M(H,T)/M_{sat}$ does not hold in applied fields. Fig 15*a* shows the magnetization in fields up to 50 kOe. In Fig. 15*b*, the dashed curve shows the calculated $\rho(10 \text{ kOe},T)$ along with the experimental data. Clearly, $m(H,T)$ significantly overestimates the mixing factor $c(H,T)$. In order to explore this two-fluid approach further, the mixing coefficient is *computed* from the *field-independent* low and high temperature resistivities, and its validity tested by calculating from it the field-dependent Seebeck coefficient. Explicitly, $c(H,T)$ is defined through the expression

$$c(H,T) = \frac{\rho_{ht}(T)/\rho_{exp}(H,T) - 1}{\rho_{ht}(T)/\rho_{lt}(T) - 1} \qquad (9)$$

which clearly approaches zero and unity in the high and low temperature limits respectively. Fig. 16*a* shows the mixing factor at various applied fields extracted from the data of Fig. 15*b*. In Fig. 16*b*, these experimental mixing factors are used in Eq. 8 to generate curves for the field dependent Seebeck coefficient. These give an excellent account of the data, providing an independent check on the validity of this two-fluid approach. The main effect of the magnetic field is to shift the onset of the band-electron phase without broadening the transition. However, as we shall see, the vanishing of $c(H,T)$ does not represent a shifted critical point for the material.

The essential feature of the Millis *et al.* model is that the effective Jahn-Teller coupling constant is very near its critical value in the paramagnetic phase. In this case, coupling to the magnetization via the associated band-broadening of the double exchange model, reduces λ_{eff} through its critical value λ_c, inducing the expansion of small polarons into band electrons. A simple mean-field model is proposed here, that reproduces the essential features of the microscopic model and provides a comparison with experiment. The assumed ferromagnetic free-energy functional is of conventional form

$$F_{mag} = \frac{1}{2}(T/T_C - 1)m^2 + \frac{1}{4}bm^4 - mh \qquad (10)$$

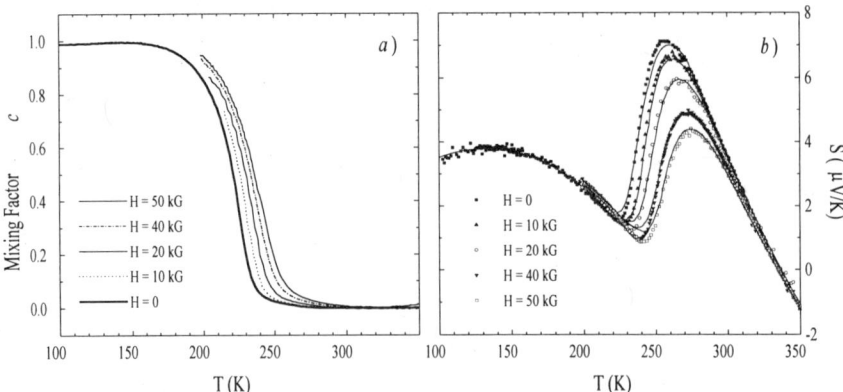

Figure 16. *a)* The mixing factor $c(H,T)$ extracted from the resistivity as described in the text. *b)* Seebeck coefficient data and results of a computation using $c(H,T)$ from *(a)* in Eq. 8.

where the free energy is written in units of[95] $3Sk_BT_C/(S+1) = 1.94k_BT_C$ for $S = 2(1-x) + 3x/2 = 1.83$ and $x = 1/3$, and $h = g\mu_B(S+1)H/3k_BT_C = H/2360$ kOe. The dependence of λ_{eff} on the magnetization can be approximated by writing $\lambda_{eff} - \lambda_c \propto \alpha - \gamma m^2 + ...$, where α is small and positive. The electronic free energy can then be written, in the same dimensionless units as Eq.10, as

$$F_{el} = \frac{1}{2}(\alpha - \gamma m^2)c^2 + \frac{1}{4}\beta c^4 \qquad (11)$$

Here $c(H,T)$ is a nearly-critical secondary order parameter, driven by the difference $\lambda_{eff} - \lambda_c$. Minimizing the total free energy two coupled equations are obtained, $(T/T_C - 1 - \gamma c^2)m + bm^3 - h = 0$ and $(\alpha - \gamma m^2)c + \beta c^3 = 0$. From the later it is obvious that the concentration of metallic electrons is zero until the magnetization reaches the value $m = \sqrt{\alpha/\gamma}$, beyond which point c increases. In the limit $\alpha \to 0$, c is proportional to m, with the result that $b \to b - \gamma^2/\beta$, signalling a tendency for the system to approach a tricritical point and first order transitions as the coupling constant is increased. Note that the existence of a non-zero concentration $\bar{c}$ can be considered to increase the critical point to $(1 + \gamma\bar{c}^2)T_C$, causing the magnetization to increase more rapidly than would be the case without coupling to the metallic electron concentration. Solutions to the coupled equations are,

$$m = B_S\left(\frac{3ST_C}{(S+1)T}[(1+\gamma c^2)m + h]\right) \qquad (12)$$

and

$$c = \tanh\left[(1 - \alpha + \gamma m^2)c\right] \qquad (13)$$

In Fig. 17 the simultaneous solutions of Eqs. 12 & 13 for $\alpha = 0.02$ and $\gamma = 0.3$ at $H = 0$, 24 kOe, and 48 kOe is found. Application of the magnetic field increases the temperature at which c becomes non-zero by 7% or 20 K, consistent with the experimental data in Fig. 16a, but does not produce a high-temperature tail. As no thermal factors are included in the definition of c, the concentration of free carriers does not approach unity, and therefore differs slightly from the experimentally defined $c(H,T)$ in Eq. 9. The abrupt appearance of band electrons in this model produces a kink in the zero field magnetization

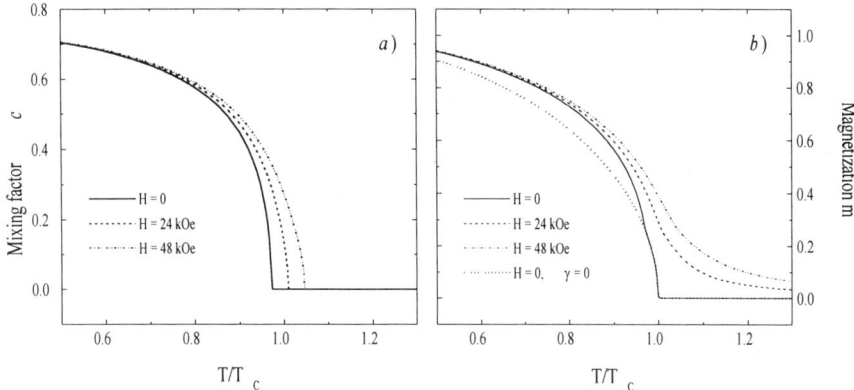

Figure 17. *a*) the mixing factor $c(H,T)$ calculated in the mean-field model with $\alpha = 0.02$ and $\gamma = 0.3$. *b*) The magnetization calculated with the same parameters. the dotted line shows the non-interactive case for comparison.

curve at the onset temperature T_D, seen as a deviation from the $H = 0$, $\gamma = 0$ curve.

In non-zero field, the kink persists as seen in Fig. 18*a* where we plot $\chi^{-1} \equiv H/m$ at several magnetic fields. These results show clearly how the delocalization of charge carriers produces a rise in the effective T_C, in good agreement with experimental data[96] for $La_{0.79}Ca_{0.21}MnO_3$. The magnitude of the kink present in the experimental $\chi^{-1}(T)$ is larger and more evident in samples that show broader ferromagnetic transitions at constant doping,[97] and are consequently considered of "lower quality". This deserve a further analysis. Fig. 18*b* shows the resistivity curves determined using the mean-field $c(H,T)$. Clearly, the model must be extended to include critical fluctuations and associated rounding.

The proposed model differs from a percolation-like picture in which more or less static regions of high conductivity are weakly connected by surrounding insulating material. If that were the case, the standard Nordheim-Gorter rule for series connection would emphasize the increasing Seebeck coefficient of the resistive polaronic contribution, rather than the small thermopower of the more conductive component. There is ample experimental evidence, from studies of spin waves for example, that the ordered phase emerges with its full three-dimensional properties —albeit with strong evidence of slow, diffusive contributions— in materials in the composition regime discussed here. This mean-field model ignores a number of features that should be included in a complete treatment. In particular a term m^2c is missing, because it leads to a first-order transition for all values of the parameters; it cannot be ruled out on symmetry grounds. Similarly, there should be a mixing entropy in the electronic free energy which, at sufficiently high temperatures, will lead to thermal dissociation of the polarons. Finally, no gradient terms were included and therefore ignore inhomogeneous thermal fluctuations that are certain to be significant in a system such as this where there are competing order parameters. Nonetheless, this phenomenological approach provides a qualitative understanding of the field and temperature dependence of the transport properties while correctly predicting the existence of kinks in the magnetization curves.

In summary, we have discussed the transport properties of optimally doped manganite materials and showed how they play a key role in the understanding of their ground state, as well as their different magnetic phases. Transport properties allow us to distinguish different temperature regimes and also to identify the relevant physics ruling them. As indicated in

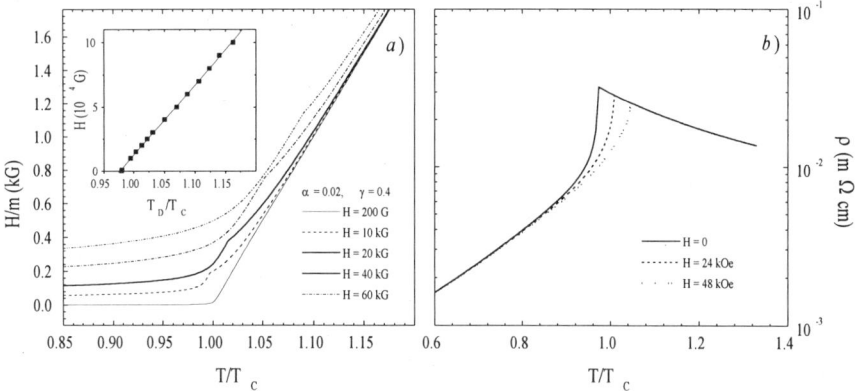

Figure 18. *a)* Inverse magnetic susceptibility H/m vs temperature near the M-I transition. The appearance of free carriers induce the rise of the effective T_C, in qualitative agreement with data by Goodwing et al. *b)* Calculated resistivity using $c(H,T)$ from Fig. 17 *a)* in Eq. 7.

Fig. 4, DE physics dominates the very low temperature region. Additional theoretical work is needed however to describe the details of the gradual changes in the band structure with temperature, from 100% spin-polarized to partial polarization just bellow T_C. Such a model should allow us to properly calculate the temperature dependence of the electrical resistivity at very low temperature. JT and localization of charge dominates at high temperatures, but more experimental work is needed to improve the understanding of the Seebeck and Hall effects. One of the hardest experimental problems is related to sample quality issues. As discussed before, the physical properties of manganites are strongly dependent on bandwidth, doping, and local defects. Most of the experimental work done until now has concentrated on samples where these three parameters are changed simultaneously. For example, studies of T_C vs doping do not usually take into account the concentration of local defects nor the tolerance factor. It would be useful for the understanding of the high temperature Seebeck effect to be able to prepare samples with different doping levels and different concentration of local defects, keeping the tolerance factor tf a constant. Samples like these were prepared at Urbana, to test the impurity conduction model proposed by Emin, however discrepancy between nominal and measured chemical compositions make the experimental results hard to analyze[98]. At intermediate temperatures, both DE and JT mechanisms are required to understand the details of the phase transition between a paramagnetic insulator and a highly polarized ferromagnetic metal. The gradual delocalization of charge carriers is driven by a temperature dependent effective coupling between charge and lattice, which at the same time is determined by a DE controlled bandwidth.. While the coexistence of itinerant and localized charges explain some experimental properties the situation is still unclear as respects the origins and mechanisms of very slow spin dynamics and cluster formation just above T_C. More careful measurements in this regime as well as a theoretical description that includes both double exchange, strong electron-phonon coupling, and spin fluctuations, should help.

We would like to thank A. Caneiro and F. Prado, Centro Atómico Bariloche; M. Rubinstein, U.S. Naval Research Laboratory; D. Emin, University of New Mexico, Albuquerque, N.M.; P. Han, P. Lin, and S.-H. Chun, University of Illinois at Urbana. This work was possible thanks to the support from U.S. Department of Energy at Los Alamos National Laboratory, NM, USA. MBS acknowledges support by the Department of Energy, Office of Basic Energy Sciences through Grant No. DEFG0291ER45439 at the University of Illinois and by National Science Foundation Grant No. DMR-9120000 through the Science and Technology Center for Superconductivity.

REFERENCES

1. K. Chahara, T. Ohno, M. Kasai, and Y. Kozono, Appl. Phys. Lett. 63: 1990 (1993). R. von Helmolt, J. Wecker, B. Holzapfel, L. Shultz, and K. Samwer, Phys. Rev. Lett. 71: 2331 (1993). S. Jin, T.H. Tiefel, M. McCormack, R.A. Fastnacht, R. Ramesh, and L.H. Chen, Science 64: 413 (1994). M. McCormack, S. Jin, T.H. Tiefel, R.M. Fleming, J. M. Phillips, and R Ramesh, Appl. Phys. Lett. 64: 3045 (1994). S. Jin, M. McCormack, T.H. Tiefel, and R. Ramesh, J. Appl. Phys. 76: 6929 (1994).
2. G.H. Jonker and J.H. Van Santen, Physica 16: 337 (1950). J. H. Van Santen and G.H. Jonker, *ibid.* 16: 599 (1950).
3. J. Volger, Physica 20: 49 (1954).
4. C. Zener, Phys. Rev. 2: 403 (1951).
5. P.W. Anderson and H. Hasegawa, Phys. Rev. 100: 675 (1955).
6. I. Mannari, Prog. Theor. Phys. 22: 325 (1959).
7. T. Kasuya, Prog. Theor. Phys. 22: 227 (1959).
8. P.G. de Gennes, Phys. Rev. 118: 141 (1960).
9. K. Kubo and N. Ohata, Jou. Phys. Soc. Japan 33: 21 (1972).
10. J. Mazaferro, C. A. Balseiro, and B. Alascio, J. Phys. Chem. Solids 46: 1339 (1985).
11. N. Furukawa, Jou. Phys. Soc. Japan 63: 3214 (1994).
12. C.M. Varma, Phys. Rev. B54: 7328 (1996).
13. R. Allub and B. Alascio, Phys. Rev. B 55: 14113 (1997).
14. E.O Wollan, K.C. Kochler, Phys. Rev. 100: 545 (1955).
15. R.M. Kusters, J. Singleton, D.A. Keen, R. McGreevy, and W. Hayes, Physica B 155: 362 (1989).
16. K. N. Clausen, W. Hayes, D.A. Keen, R.M. Kusters, R.L. McGreevy, and J. Singleton, J. Phys.: Condens. Matter 1: 2721 (1989).
17. R. von Helmolt, L. Haupt, K. Bärner, and U. Sonderman, Sol. St. Comm. 82: 693 (1992).
18. A. Maigan, F. Damay, A. Barnabe, C. Martin, M. Herview, and B. Raveau, Phil. Trans. A 356: 1635 (1998).
19. A. Arulraj, R. Mahesh, G.N. Subbanna, R. Mahendiran, A.K. Raychaudhuri, and C.N.R. Rao, Jou. Sol. St. Chem. 127: 87 (1996).
20. R. Mahendiran, S.K. Tiwary, A.K. Raychaudhuri, T.V. Ramakrishnan, R. Mahesh, N. Rangavittal, and C.N.R. Rao, Phys. Rev. B 53: 3348 (1996). Also R. Mahendiran, S.K. Tiwary, and A.K. Raychaudhuri, Sol. St. Comm. 98: 701 (1996).
21. S. Satpathy, Z.S. Popovic, and F.R. Vukajlovic, Phys. Rev. Lett. 76: 960 (1996).
22. K. Khazeni, Y.X. Jia, L. Lu, C.H. Crespi, M. Cohen, and A. Zettl, Phys. Rev. Lett. 76: 295 (1996).
23. J.J. Neumeier, M.F. Hundley, J.D. Thompson, and R.H. Heffner, Phys. Rev. B 52: R7006 (1995).
24. H.Y. Hwang, T.T.M. Palstra, S-W. Cheong, and B. Batlogg, Phys. Rev. B 52: 15046 (1995).
25. M.R. Ibarra, P.A. Algarabel, C. Marquina, J. Blasco, and J. García, Phys. Rev. Lett. 75: 3541 (1995).
26. P.G. Radaelli, D.E. Cox, M. Marezio, S-W. Cheong, P.E. Schiffer, and A.P. Ramirez, Phys. Rev. Lett. 75: 4488 (1995).
27. P. Dai, J. Zhang, H.A. Mook, S.-H. Liou, P.A. Dowben, and E.W. Plummer, Phys. Rev. B 54: R3694 (1996).
28. J.M. De Teresa, M.R. Ibarra, C. Marquina, P.A. Algarabel, and S. Oseroff, Phys. Rev. B 54: R12689 (1996).
29. J. Tanaka, M. Umehara, S. Tamura, M. Tsukioka, and S. Ehara, Jou. Phys. Soc. Japan 51: 1236 (1982).
30. M. Jaime, M. B. Salamon, K. Pettit, M. Rubinstein, R.E. Treece, J.S. Horwitz, and D.B. Chrisey, Appl. Phys. Lett. 68: 1576 (1996).
31. V.H. Crespi, L. Lu, Y.X. Jia, K. Khazeni, A. Zettl, and M. Cohen, Phys. Rev. B 53: 14303 (1996).
32. M. Jaime, M.B. Salamon, M. Rubinstein, R.E. Trece, J.S. Horwitz, and D.B. Chrisey, Phys. Rev. B 54: 11914 (1996).
33. J.-S. Zhou, W. Archibald, and J.B. Goodenough, Nature 381: 770 (1996).
34. T.T. Palstra, A.P. Ramirez, S-W. Cheong, B.R. Zegarski, P. Schiffer, and J. Zaanen, Phys. Rev. B 56: 5104 (1997).
35. M.F. Hundley and J.J. Neumeier, Phys. Rev. B 55: 11511 (1997).
36. M. Jaime, H.T. Hardner, M.B. Salamon, M. Rubinstein, P. Dorsey, and D. Emin, Phys. Rev. Lett. 78: 951 (1997). Also M. Jaime, H. Hardner, M.B. Salamon, M. Rubinstein, P. Dorsey, and D. Emin, J. Appl. Phys. 81: 4958 (1997).
37. S.G. Kaplan, M. Quijada, H.D. Drew, D.B. Tanner, C.G. Xiong, R. Ramesh, C. Kwon, and T. Venkatesan, Phys. Rev. Lett. 76: 2081 (1996).

38. Y. Yamada, O. Hino, S. Nohdo, R. Kanao, T. Inami, and S. Katano, Phys. Rev. Lett. 77: 904 (1996).
39. S.J.L. Billinge, R.G. DiFrancesco, G.H. Kwei, J.J. Neumeier, and J.D. Thompson, Phys. Rev. Lett. 77: 715 (1996).
40. P.G. Radaelli, M. Marezio, H.Y. Hwang, S-W. Cheong, and B. Batlogg, Phys. Rev. B54: 8992 (1996).
41. D. Louca, T. Egami, E.L. Brosha, H. Röder, and A.R. Bishop, Phys. Rev. B 56: R8475 (1997).
42. G. Zhao, K. Conder, H. Keller, and K.A. Müller, Nature 381: 676 (1996).
43. G. Zhao, M.B. Hunt, and H. Keller, Phys. Rev. Lett. 78: 955 (1997).
44. I. Isaac and J. P. Franck, Phys. Rev. B 57: R5602 (1998).
45. C.H. Booth, F. Bridges, G.J. Snyder, and G.T. Geballe, Phys. Rev. B 54: R15606 (1996).
46. C.H. Booth, F. Bridges, G.H. Kwei, J.M. Lawrence, A.L. Cornelius, and J.J. Neumeier, Phys. Rev. B 57: 10440 (1998).
47. A. Shengelaya, G. Zhao, K. Heller, and K.A. Müller, Phys. Rev. Lett. 77: 5296 (1996).
48. J.L. Cohn, J.J. Neumeier, C.P. Popoviciu, K.J. McClellan, and Th. Leventouri, Phys. Rev. B 56: R8495 (1997).
49. R.H. Heffner, L.P. Le, M.F. Hundley, J.J. Neumeier, G.M. Luke, K. Kojima, B. Nachumi, Y.J. Uemura, D.E. MacLaughlin, and S-W. Cheong, Phys. Rev. Lett. 77:1869 (1996).
50. H.Y. Hwang, S-W. Cheong, P.G. Radaelli, M. Marezio, and B. Batlogg, Phys. Rev. Lett. 75: 914 (1995).
51. J.B. Goodenough, J.-S. Zhou, Mat. Res. Soc. Symp. Proc. 494: 335 (1998).
52. H. Yoshizawa, R. Kajimoto, H. Kawano, Y. Tomioka, and Y. Tokura, Phys. Rev. B 55: 2729 (1997).
53. L.M. Rodriguez-Martinez, J.P. Attfield, Phys. Rev. B 54: R15622 (1996).
54. A. Sundaresan, A. Maigan, and B. Raveau, Phys. Rev. B 56: 5092 (1997).
55. G.H. Rao, J.R. Sun, J.K Liang, W.Y. Zhou, Phys. Rev. B 55: 3742 (1997). Also F. Damay, C. Martin, A Maignan, and B. Raveau, J. Appl. Phys. 82: 6181 (1997).
56. J.L. García-Muñoz, J. Fontcuberta, B. Martínez, A. Seffar, S. Piñol, and X. Obradors, Phys. Rev. B 55: R668 (1997).
57. J.N. Eckstein, I. Bozovic, J. ODonnell, M ONellion, M.S. Rzchowski, Appl. Phys. Lett. 69: 1312 (1996).
58. T. Holstein, Ann. Phys. 8: 343 (1959).
59. L. Friedman and T. Holstein, Ann.Phys. (N.Y.) 21: 494 (1963).
60. N.F. Mott and E.A. Davis, in Electronic Processes in Non-Crystalline Materials (Clarendon Press, Oxford, 1971).
61. D. Emin, in Electronic Structure Properties of Amorphous Semiconductors, edited by P.G. Le Comber and N.F. Mott (Academic Press, New York, 1973). Also M.S. Hillery, D. Emin and N.H. Liu, Phys. Rev B 38: 9771 (1988).
62. R. Raffaelle, H.U. Anderson, D.M. Sparin, and P.E. Parris, Phys. Rev. B 43: 7991 (1991).
63. P.M. Chaikin, and G. Beni, Phys. Rev. B 13: 647 (1976).
64. F. Prado and A. Caneiro (unpublished).
65. A. Caneiro, P. Bavdaz, J. Fouletier, and J. P. Abriata, Rev. of Sci. Instr. 53: 1072 (1982).
66. R.R. Heikes, in Thermoelectricity, edited by P.H. Egli (Wiley, New York, 1965). Also R.R. Heikes, in Transition Metal Compounds, edited by E.R. Schatz (Gordon and Breach, New York, 1963).
67. A. Rojo, private communication.
68. J.W. Lynn, R.W. Erwin, J.A. Borchers, A. Santoro, Q. Huang, J.L. Peng, and R.L. Greene, Jou. Appl. Phys. 81: 5488 (1997).
69. P. Nagels in The Hall Effect and its Applications, edited by C.L. Chien and C.R. Westgate (Plenum, New York, 1980), p. 253.
70. D. Emin and T. Holstein, Ann. Phys. (N.Y.) 53: 439 (1969).
71. D.C. Worledge, L. Mieville, and T.H. Geballe, Phys. Rev. B 57: 15267 (1998).
72. J.E. Nunez-Regueiro, D. Gupta, and A.M. Kadin, J. Appl. Phys. 79: 5179 (1996). G.J. Snyder, M.R. Beasley, T.H. Geballe, R. Hiskes, and S. diCarolis, Appl. Phys. Lett. 69, 4254 (1996).
73. P. Matl, N.P.Ong, Y.F. Yan, Y.Q. Li, D. Studebaker, T. Baum, and G. Doubinira, Phys. Rev. B 57: 10248 (1998). G. Jakob, F. Martin, W. Westerburg, and H. Adrian, ibid. B 57: 10252. P. Mandal, K. Barner, L. Haupt, A. Poddar, R. vonHelmolt, A.G.M. Jansen, and P. Wyner, ibid. B 57: 10256. K. Li, R. Cheng, Z. Chen, J. Fang, X. Cao, and Y. Zhang, Jou. Appl. Phys 84: 1467 (1998).
74. M. Jaime, P. Lin, M.B. Salamon, and P.D. Han, Phys. Rev. B 58: R5901 (1998).
75. A. Urushibara, Y. Moritomo, T. Arima, A. Asamitsu, G. Kido, and Y. Tokura, Phys. Rev. B 51: 14103 (1995).
76. P. Schiffer, A.P. Ramirez, W. Bao, and S-W. Cheong, Phys. Rev. Lett. 75: 3336 (1995).
77. G.J. Snyder, R. Hiskes, S. Dicarolis, M.R. Beasley, T.H. Geballe, Phys. Rev. B 53: 14434 (1996).
78. J-H. Park, et al. Nature 392: 794 (1998).
79. W.E. Pickett and D.J. Singh, Phys. Rev. B 53: 1146 (1996). Also D.J. Singh and W.E. Pickett Phys. Rev. B 57: 88 (1998).
80. J.W. Lynn et al., Phys. Rev. Lett.76: 4046 (1996).

81. J.A. Fernandez-Baca, P. Dai, H.Y. Hwang, C. Kloc, and S.W. Cheong, Phys. Rev. Lett. 80: 4012 (1998).
82. K. Kadowaki, and S.B. Woods, Sol. St. Comm. 58: 507 (1986).
83. N.W. Ashcroft and N.D. Mermin, *Solid State Physics* (Sauders College Publishing, 1976) p.33
84. J.J. Hamilton, E.L. Keatley, H.L. Ju, A.K. Raychaundhuri, V.N. Smolyaninova, and R.L. Greene, Phys. Rev. B 54: 14926 (1996).
85. A.J. Millis, P.B. Littlewood, and B.I. Shraiman, Phys. Rev. Lett. 74: 5144 (1995).
86. A.J. Millis, B.I. Shraiman, and R. Mueller, Phys. Rev. Lett. 77: 175 (1996).
87. A.J. Millis, Phys. Rev. B 53: 8434 (1996). A.J. Millis, Phys. Rev. B 55: 6405 (1997).
88. A. J. Millis, R. Mueller, and B. I. Shraiman, Phys. Rev. B 54: 5389 (1996), *ibid.* B 54: 5405 (1996).
89. D.C. Worledge, G.J. Snyder, M.R. Beasley, and T. Geballe, J. Appl. Phys. 80: 5158 (1996).
90. D. Emin, Philos. Mag. 35: 1189 (1977).
91. C. Wood and D. Emin, Phys. Rev. B 29: 4582 (1984).
92. S. Yoon, H.L. Liu, G. Schollere, S.L. Cooper, P.D. Han, D.A. Payne, S.W. Cheong, and Z. Fisk, Phys. Rev. B 58: 2795 (1998).
93. C. H. Booth, F. Bridges, G.H. Kwei, J.M. Lawrence, A.L. Cornelius, and J.J. Neumeier, Phys. Rev. Lett. 80: 853 (1998). Similar results are discussed by A. Lanzara *et al.*, Phys. Rev. Lett. (to be published).
94. R.D. Bernard, *Thermoelectricity in Metals and Alloys* (Taylor and Francis, London, 1972) p.140.
95. D. C. Mattis, *The Theory of Magnetism II*, (Springer-Verlag, Berlin, 1985) p. 22.
96. D.H. Goodwing, J.J. Neumeier, A.H. Lacerda, and M.S. Torikachvili, Mat. Res. Soc. Symp. Proc. 494: 95 (1998).
97. J. Gardner, J. Thompson, and J. Sarrao, private communication.
98. P. Lin, M.B. Salamon, and M. Jaime, Bull. Am. Phys. Soc. 43: 293 (1998).

SPIN INJECTION IN MANGANITE–CUPRATE HETEROSTRUCTURES

V. A. Vas'ko, K. R. Nikolaev, V. A. Larkin, P. A. Kraus, and A. M. Goldman

School of Physics and Astronomy
University of Minnesota
Minneapolis, MN 55455

ABSTRACT

The growth of oxide manganite–cuprate heterostructures and the results of experiments on the injection of spin-polarized carriers into a high-T_c superconductor are discussed. The investigation of spin injection and spin-polarized tunneling into superconductors can provide valuable information on spin-dependent properties, such as the spin relaxation length. It may also result in the development of new superconducting devices, such as switches and transistors.

INTRODUCTION

The first experiments on spin-polarized transport and tunneling in superconductors were carried out in the seventies by Tedrow and Meservey.[1] A theoretical description of these phenomena was provided by Aronov.[2] Experimental and theoretical studies of the injection of spin-polarized carriers into normal metals were carried out by Johnson and Silsbee.[3,4] The subject of spin-polarized transport and tunneling in metallic systems has recently been reviewed by Prinz.[5] Interest in these phenomena has been heightened by the rediscovery that the doped lanthanum manganite compounds exhibit large magnetoresistance which has been referred to as "colossal" magnetoresistance (CMR).[6,7]

Insulating and antiferromagnetic in its undoped state, $LaMnO_3$ becomes ferromagnetic and metallic when doped with divalent metal ions of Ba, Ca, Sr, or Pb. The Curie temperature is in the range 100-400 K depending on the dopant ion and its level. The doped compound also exhibits strong magnetoresistance in the vicinity of its magnetic transition and this has been attributed to the double exchange interaction of the manganese ions.[8,9] The double exchange picture implies that in these compounds electrical transport involves spin-polarized carriers, and that the degree of polarization of these carriers is close to 100%. Recent experiments involving the study of magnetic tunnel junctions (MTJ) have shown that the degree of the polarization of the carriers in these half-metallic compounds is indeed much higher than that of conventional ferromagnetic metals and alloys.[10] For this reason, CMR materials would appear to be useful sources of spin-polarized carriers which can be used to study the influence of such carriers on the properties of other materials.

The CMR materials are perovskites and therefore may be epitaxially compatible with other perovskites. This would appear to facilitate the formation of perovskite heterostructures using thin film growth techniques. There exists a variety of perovskites possessing a wide range of different physical properties, such as high-T_c superconductivity, ferromagnetism, ferrimagnetism, paramagnetism, and ferroelectricity. Here we will discuss the growth and properties of heterostructures consisting of ferromagnetic and superconducting perovskites. The similar values of the lattice parameters of the ferromagnet, $La_{2/3}M_{1/3}MnO_3$ (M = Ca, Sr, Ba), and the high-T_c superconductor $DyBa_2Cu_3O_7$, in two directions (*a* and *b*), can be exploited to facilitate the growth of epitaxial heterostructures of these compounds. Recently it was demonstrated that these materials can be incorporated into a single structure.[11, 12]

Molecular beam epitaxy (MBE) is an effective technique for the fabrication of high-quality multicomponent oxide heterostructures because of the possibility of precise control of composition and the use of an *in-situ* characterization technique to monitor film growth, reflection high-energy electron diffraction (RHEED).

Tri- and bilayers of MBE-grown perovskite ferromagnets and superconductors have been used to demonstrate the suppression of superconductivity by injection of spin-polarized carriers.[13, 14, 15] To develop an improved physical understanding of the phenomena taking place at the ferromagnet–superconductor interface during the injection of spin-polarized carriers into a superconductor we also studied the I-V characteristics of the interface between the compounds. Our results suggest that there is a modification of Andreev reflection at the ferromagnet–superconductor interface as a consequence of the spin-polarization of the carriers in the ferromagnet.[16] We also suggest that spin injection into superconductors opens the possibility of a new class of devices.

GROWTH OF HETEROSTRUCTURES

The $DyBa_2Cu_3O_7/La_{2/3}M_{1/3}MnO_3$ heterostructures (M = Ba, Sr, or Ca) were grown on $SrTiO_3$ (001) substrates using ozone-assisted MBE.[17] In the present work, the substrates were radiatively heated to 650-700 °C. The evaporation of Dy, Ba, Ca, Sr, Cu, and Mn was carried out using conventional effusion cells (T < 1350 °C), whereas a high-temperature effusion cell (T < 2200 °C) was used for La. Deposition rates were measured using a quartz crystal monitor. The amounts of each element deposited were controlled by computerized shutters. The O_3 partial pressure in the growth chamber during deposition was maintained at 2×10^{-5} Torr. O_3 is much more effective oxidizer than oxygen and permits full oxygenation of films at relatively low pressures that are characteristic of an MBE process.

We employed the "block-by-block" deposition mode for the growth of these heterostructures. This approach has been shown to be effective in reducing the number of defects during the deposition of complex, multicomponent oxide films.[18] It has been used to produce films with ultra-smooth surfaces which are essential for the synthesis of heterostructures.[19] In this approach to film deposition, atoms of different elements required

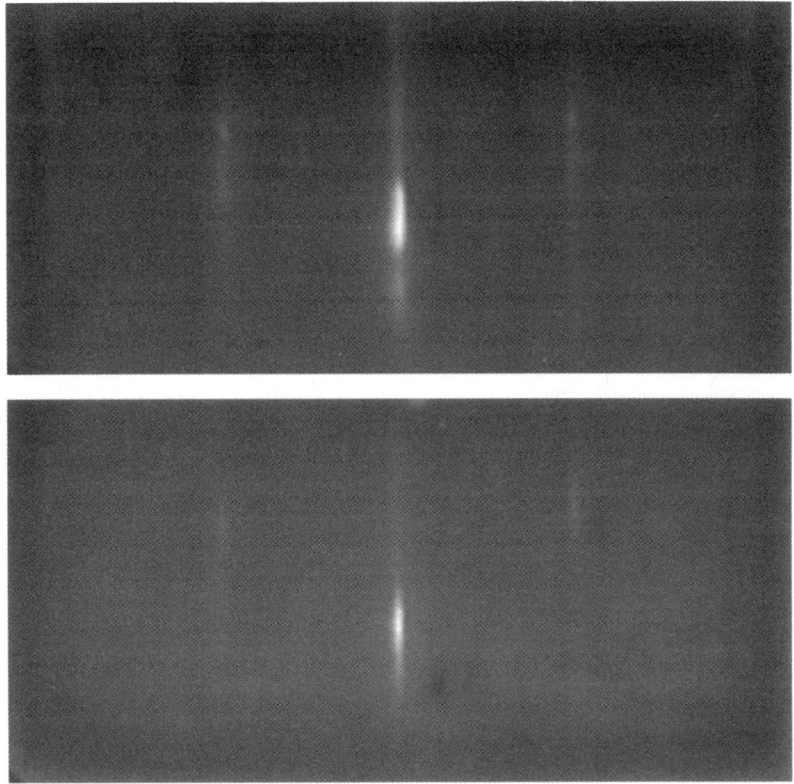

Figure 1. RHEED patterns acquired during deposition of $La_{2/3}Sr_{1/3}MnO_3$ (top) and $DyBa_2Cu_3O_7$ (bottom) layers of heterostructure.

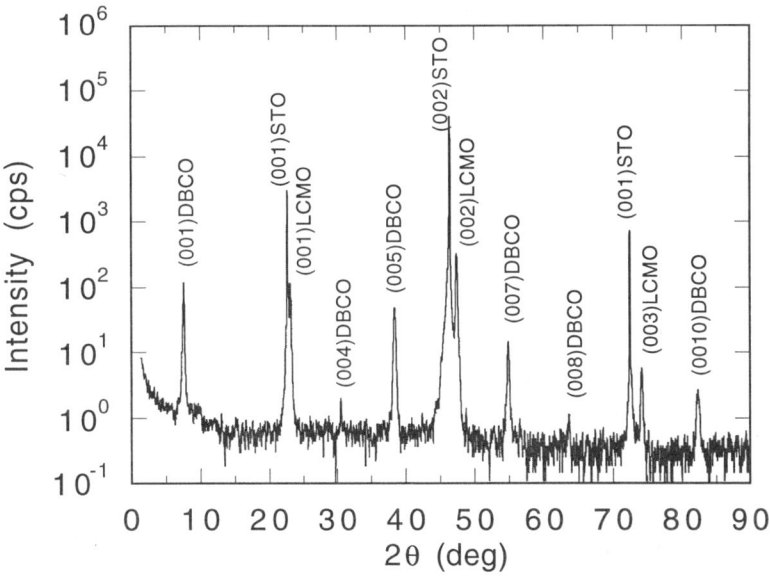

Figure 2. $2\theta-\Omega$ scan of $DyBa_2Cu_3O_7/La_{2/3}Ca_{1/3}MnO_3$ heterostructure grown on $SrTiO_3$ (100) substrate. All peaks are indexed to $(00l)$ peaks of the substrate and the heterostructure.

for the formation of each layer of unit cell thickness are deposited sequentially as blocks. The sequences are chosen to prevent the formation of the stable compounds before completion of a unit cell thick layer. For a $La_{2/3}M_{1/3}MnO_3$ layer, the sequence M, La, Mn, in which all the M (M= Ca, Sr, or Ba) atoms required for the growth of an unit cell are deposited first, followed by all the La atoms, and then by the Mn atoms, resulting in the growth of a layer of $La_{2/3}M_{1/3}MnO_3$. For the $DyBa_2Cu_3O_7$ layer, a sequence of successive depositions of Ba, Dy, Cu was used.

The heterostructures were prepared *in-situ* by first growing a 400 Å thick layer of $La_{2/3}M_{1/3}MnO_3$, followed by a 600 Å thick layer of $DyBa_2Cu_3O_7$. After growth, the structure was cooled down to 100 °C in an O_3 pressure of 3.5 x 10^{-5} Torr. Subsequent cooling to room temperature was carried out without O_3 in the chamber. The RHEED diffraction patterns and their variation during the formation of both layers were characteristic of highly oriented epitaxial growth. Typical patterns observed during the growth of the all layers of the heterostructure are shown in Fig. 1. Note that these are streaky without the presence of diffraction spots. The *c*-axes of the crystal lattices of both the $La_{2/3}Sr_{1/3}MnO_3$ and $DyBa_2Cu_3O_7$ layers were normal to the (001) plane of the substrate. In-plane orientation was such that the [100] and [010] crystallographic directions of all of the layers of the heterostructures were aligned with the same directions of the substrate. Intensity oscillations in the RHEED pattern were observed with streaks becoming sharper and more intense upon the completion of the deposition of each unit cell. The absence of spots in the RHEED patterns at this point of the growth of a film is indicative of the smoothness of the surface and stoichiometric composition of the layers. Similar RHEED patterns were obtained during the growth of heterostructures with Ca and Ba doped lanthanum manganite films.

The crystal structures of the heterostructures were studied using high resolution X-ray diffraction (XRD) (Philips MRD), Rutherford back scattering (RBS), and 2.3 and 3.0 MeV He^+ ion channeling. A high-resolution symmetric $2\theta-\Omega$ scan of a $DyBa_2Cu_3O_7/La_{2/3}Ca_{1/3}MnO_3$ heterostructure on a $SrTiO_3$ (001) substrate is shown in Fig. 2. This was obtained using a four-crystal Bartells monochromator. The diffraction peaks are indexed in the figure. It should be noted that only the peaks corresponding to the substrate, $La_{2/3}Ca_{1/3}MnO_3$, and $DyBa_2Cu_3O_7$ are visible. No other peaks were detected, consistent with epitaxial growth and the absence of the second-phase precipitates. The orientation of the layers is with (00l) planes parallel to the (001) plane of the substrate. The c-axis lattice parameters were calculated to be 3.84 Å for $La_{2/3}Ca_{1/3}MnO_3$ and 11.69 Å for $DyBa_2Cu_3O_7$. Rocking curve scans also confirm a high degree of perfection of the both layers. Their width was 0.09° for the $La_{2/3}Ca_{1/3}MnO_3$ layer, and 0.08° for the $DyBa_2Cu_3O_7$ layer. These values are typical of MBE grown oxide films on $SrTiO_3$ substrates. The fact that the value of the rocking curve width of the second layer ($DyBa_2Cu_3O_7$) is not higher than the corresponding value for first layer ($La_{2/3}Ca_{1/3}MnO_3$) indicates that interface chemistry does not result in the degradation of the quality of the growth and the two compounds are fully compatible in their epitaxial growth.

Channeling data also confirmed a high degree of epitaxy. The channeling yield for the heterostructure peaks is about 12%. The analysis of the individual inputs of the layers is hindered by overlapping of the corresponding peaks. Polar scans of the Rutherford Backscattering Spectroscopy confirmed RHEED data on the in-plane epitaxy of the heterostructures.

Four-probe measurements of electrical resistance, R(T), were carried out. Figure 3 shows R(T) of representative examples of the three types of heterostructures grown (with M = Ca, Sr, and Ba). The conduction at high temperatures is dominated by the

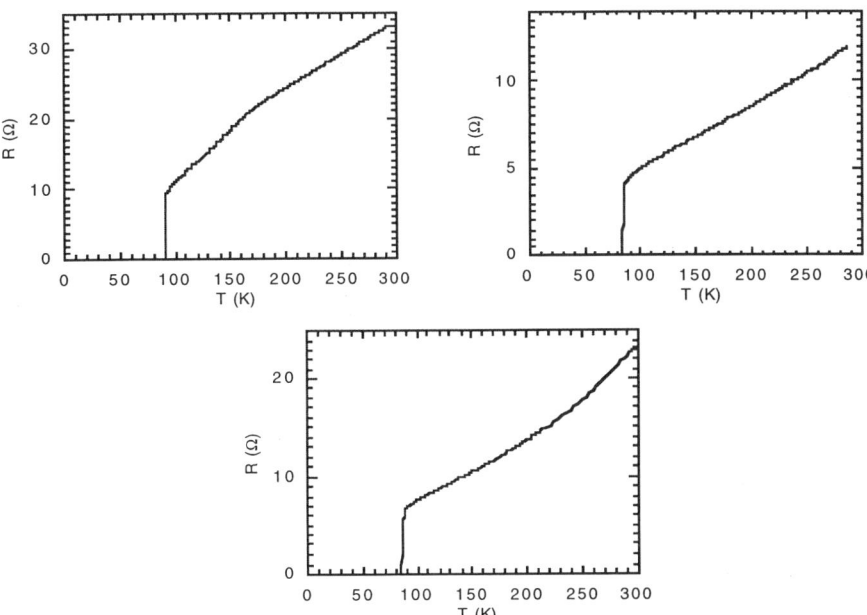

Figure 3. Temperature dependence of the resistance of heterostructures with three different compositions: $DyBa_2Cu_3O_7/La_{2/3}Ca_{1/3}MnO_3$ (top left), $DyBa_2Cu_3O_7/La_{2/3}Sr_{1/3}MnO_3$ (top right), and $DyBa_2Cu_3O_7/La_{2/3}Ba_{1/3}MnO_3$ (bottom center).

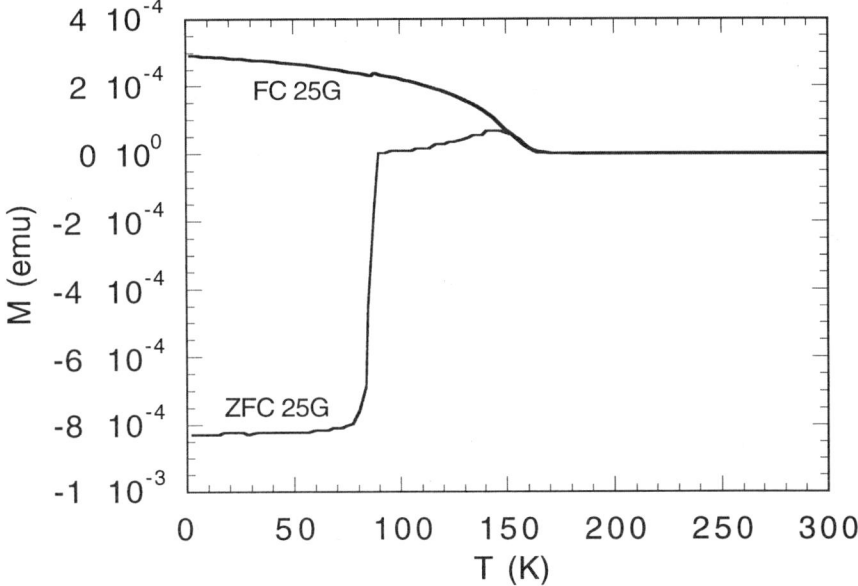

Figure 4. Temperature dependence of the magnetization of $DyBa_2Cu_3O_7/La_{2/3}Ca_{1/3}MnO_3$ heterostructure. Both field cooled (FC) and zero-field-cooled (ZFC) data are presented.

superconducting layer because its resistance in the normal state is much lower than that of the ferromagnetic layer. The observed temperature dependences are characteristic of high quality superconductors with sharp ($\_T_c \sim 1$ K) superconducting transitions in the range of 85 - 90 K. Note that the plots of R(T) deviate from linearity near the Curie temperatures of the corresponding ferromagnetic layers. This is a consequence of the peak in the resistance associated with the metal-insulator transitions in the manganites near their ferromagnetic transitions. The peak in R(T) associated with the metal-insulator transition is found at $\sim$ 170 K for Ca doped, $\sim$ 330 K for Sr doped, and $\sim$ 300 K for Ba doped lanthanum manganites, respectively.

Magnetic measurements were made using a superconducting susceptometer. The data for a $DyBa_2Cu_3O_7/La_{2/3}Ca_{1/3}MnO_3$ heterostructure is presented in Fig. 4. Magnetization vs. temperature was measured with the structure both field-cooled (FC) and zero-field-cooled (ZFC). Zero field in this instance was a magnetic field of less than 0.5 G. An applied field of 25 G was used in measurements, and data was taken on warming. Both ferromagnetic and superconducting (Meissner effect) behavior can be seen clearly from this data, unlike the case of resistance measurements while the contribution of the manganite to the conductivity was subtle. The transition temperature of the superconductor was measured to be 85 K with a width of 2 K. The ferromagnetic transition temperature of the $La_{2/3}Ca_{1/3}MnO_3$ layer was 165 K, decreased relative to the bulk value of 250 K possibly as a consequence of strain effects. This value of the Curie temperature is in agreement with the value deduced from R(T). These measurements establish the coexistence of ferromagnetism and superconductivity in the heterostructure. For heterostructures containing Sr and Ba doped lanthanum manganites, Curie temperatures of the ferromagnetic layer were found to be in agreement with those deduced from measurements of R(T).

We have also grown superlattices consisting of ferromagnetic and superconducting layers to demonstrate the feasibility of using MBE for complex oxide multilayer growth. Figure 5 shows R(T) for a superlattice consisting of $La_{2/3}Sr_{1/3}MnO_3$ and $DyBa_2Cu_3O_7$ layers with a period of 200 Å, and with individual layer thicknesses of 150 Å and 50 Å for the superconductor and ferromagnet respectively. A total of 4 periods were deposited. The as-prepared superlattice was not superconducting. Superconductivity appeared only after annealing the multilayer at 450 °C in one atmosphere of oxygen. The need for this additional step could be a consequence of manganite layers preventing oxygen from diffusing into the cuprate layers during the usual post-deposition cooling of the heterostructure at low O_3 pressure. The superconducting transition temperature of the annealed multilayer was only 62 K. The Curie temperature of the ferromagnet was also reduced to 270 K, as determined from the temperature dependence of the magnetization. These reduced values are possibly due to the relatively small thicknesses of the individual layers and associated strain effects. However, the fact of the coexistence of ferromagnetism and superconductivity supports the assertion that these phenomena are compatible in heterostructures.

The high quality bilayer heterostructures of ferromagnetic $La_{2/3}M_{1/3}MnO_3$ (M = Ca, Sr, Ba) and superconducting $DyBa_2Cu_3O_7$ that we have produced demonstrate that these compounds are compatible in a sense of the epitaxial growth with a high degree of in-plane and out-of-plane orientation. The heterostructures possess both ferromagnetic and superconducting properties with well-defined transitions. The heterostructures are of such a quality that they can be used to study superconductor–ferromagnet interfaces and spin injection into superconductors.

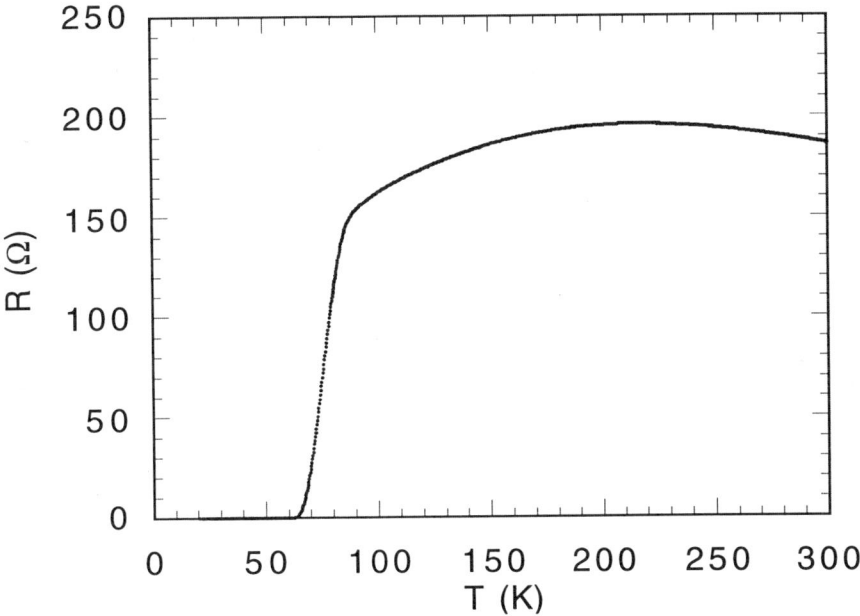

Figure 5. Temperature dependence of the resistance of (150 Å $DyBa_2Cu_3O_7$ / 50 Å $La_{2/3}Sr_{1/3}MnO_3)_4$ superlattice.

THE SUPPRESSION OF SUPERCONDUCTIVITY BY SPIN INJECTION

Here we discuss results on the suppression of superconductivity by spin injection[13] as well as published results of other groups.[14, 15] In our investigations we studied epitaxially grown heterostructures consisting of layers of $La_{2/3}Sr_{1/3}MnO_3$, La_2CuO_4, and $DyBa_2Cu_3O_7$ which are ferromagnetic, insulating, and superconducting respectively. We found that current injected from a ferromagnet suppresses the critical current and critical temperature of the superconducting film. We attributed this suppression of superconductivity to pair-breaking by spin-polarized carriers diffusing across the boundary between the ferromagnet and the superconductor.

The heterostructures were prepared by consecutive growth of the layers of $La_{2/3}Sr_{1/3}MnO_3$, La_2CuO_4 and $DyBa_2Cu_3O_7$ with thicknesses of 400Å, 24Å, and 600Å, respectively. The buffer layer of the insulator La_2CuO_4 was used in the heterostructure to improve the crystal structure of the $DyBa_2Cu_3O_7$ layer. It was noticed from the *in-situ* RHEED patterns that a barrier layer of this type reduces the roughness of the surface of the $La_{2/3}Sr_{1/3}MnO_3$ layer. As a result, the nucleation and growth of the $DyBa_2Cu_3O_7$ layer proceeds with a smaller density of defects. Only two unit cells of the buffer layer material were deposited so as to avoid modifying the electric transport between the ferromagnet and the superconductor in a significant manner.

The heterostructures were patterned using conventional photolithography to achieve a configuration, which would allow the measurement of the superconducting properties while injecting the current from the ferromagnet. The difference between the etching rates for the cuprate and manganite films when exposed to 0.1% hydrochloric acid was exploited to selectively remove the cuprate layer. A simple geometry was chosen to reduce the number of

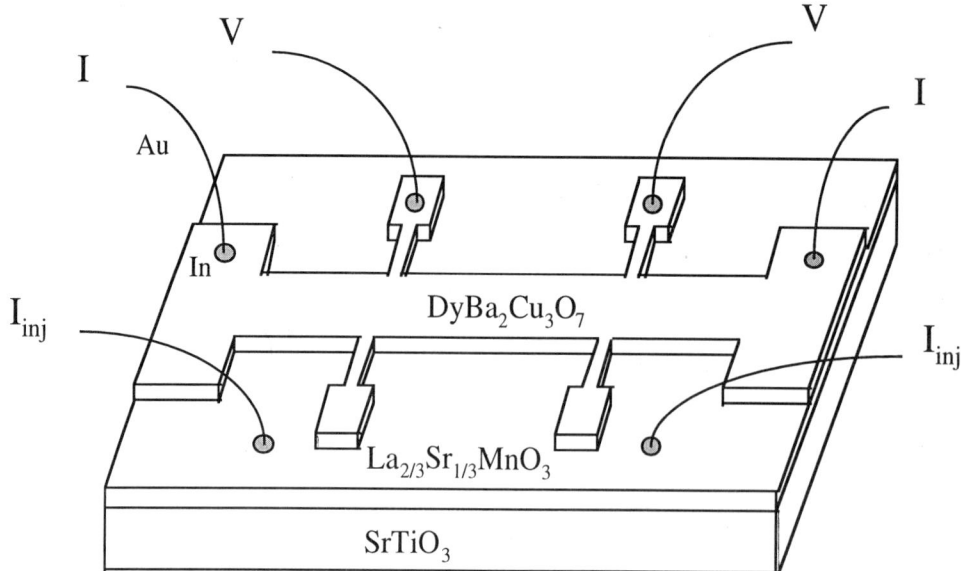

Figure 6. Geometry of the ferromagnet–superconductor sample used for measurements. The width of the $DyBa_2Cu_3O_7$ bridge was 300 μm. The distance between the voltage leads was 3mm. The substrate was 6mm x 6mm square.

lithographic processing steps that could lower quality of the sample. The schematic geometry of the sample and the measurement setup is shown in Fig. 6. Running current in the ferromagnet parallel to the superconducting bridge results in the injection of carriers from the ferromagnet into the superconductor across the interface. This happens because superconductor shorts out part of the ferromagnetic film.

To study the suppression of the superconductivity by injection of the carriers from the ferromagnet we measured the I-V characteristics of the superconductor while injecting additional current from the ferromagnet. A series of I-V characteristics taken at 50 K at several different injection currents is shown in Fig. 7. All but one I-V curves are shifted along the current axis due to the current injected from the ferromagnet. The injected current simply adds to the measuring current supplied to the superconductor for the purpose of determining its I-V characteristic. This shift from zero current of the symmetry point of the I-V characteristics is a measure of the injection current. Thus, the geometry of the experiment allows a direct measurement of the magnitude of the injection current. In addition to the shift of the I-V characteristics, there is a reduction of the width of the current plateau with zero voltage drop. This is a measure of the critical current of the superconductor, which is the half-width of the plateau. The conventionally used figure of merit of a 1 μV voltage drop was used to determine the critical current.

The dependence of the critical current of the superconducting bridge on the injection current for three different temperatures is shown in Fig. 8. The plot also displays the dependence of the effective transition temperature of the superconductor on the value of the injected current. In this case, reduction of the critical current to zero is taken to be equivalent to the full suppression of superconductivity at temperature of the measurement. The transition temperature of the particular superconducting film in the absence of current

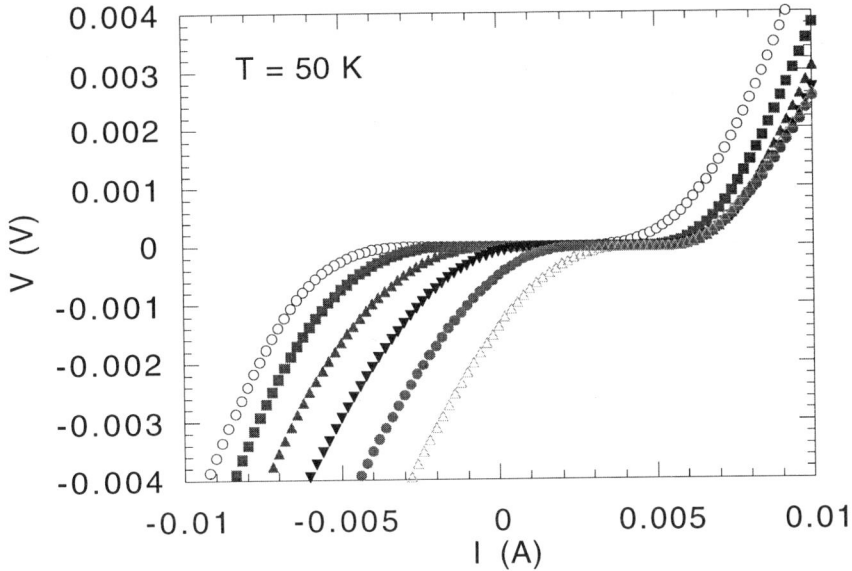

Figure 7. Voltage–current characteristics of the $DyBa_2Cu_3O_7$ layer of the $La_{2/3}Sr_{1/3}MnO_3/La_2CuO_4/DyBa_2Cu_3O_7$ heterostructure when the parallel current in the ferromagnet equals 0 mA (open circles), 2 mA (squares), 4 mA (solid triangles), 6 mA (inverted triangles), 8 mA (solid circles), and 10 mA (open triangles).

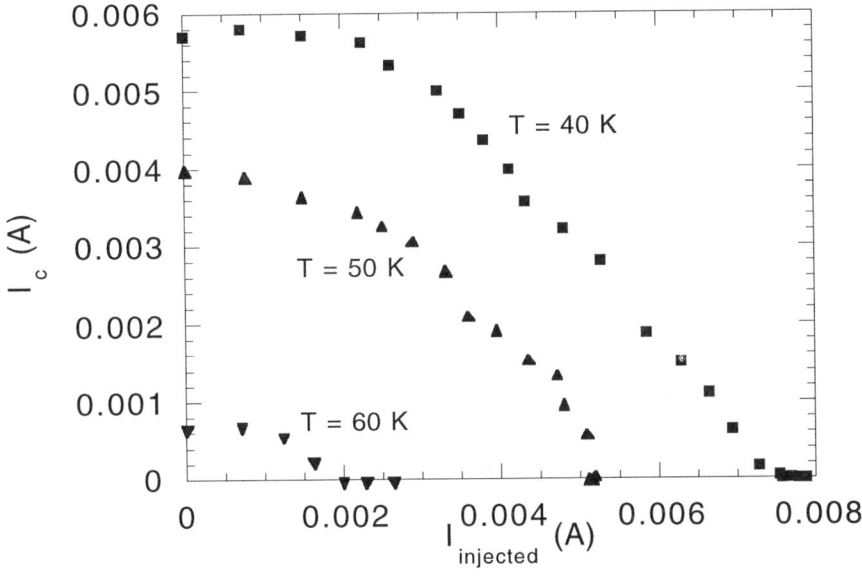

Figure 8. Dependence of the critical current of the $DyBa_2Cu_3O_7$ layer of the $La_{2/3}Sr_{1/3}MnO_3/La_2CuO_4/DyBa_2Cu_3O_7$ heterostructure on the current injected from the $La_{2/3}Sr_{1/3}MnO_3$ layer. A critical current of 1 mA corresponds to current density of 5.6×10^4 A/cm^2.

injection from the ferromagnet was 71 K. It should be noted that the reduction of the value of the critical current is comparable to the value of the injected current. This feature is most likely a geometrical property of the particular sample as it will become clear from the discussion below.

We will now argue that the observed effect is a consequence of spin polarization of the injected carriers. That is why the possible domain structure of the ferromagnetic layer might be important. The effect of the injected current on the superconducting properties was found to be independent of the magnetic field in the range 10-500 G. This leads us to conclude that the magnetic domains were either larger than the spin relaxation length of the superconductor or the domains were substantially oriented, but not randomly polarized. We also found that the magnetization of the $La_{2/3}Sr_{1/3}MnO_3$ underlayer in a 0.5 G field applied in the plane of the sample was about 1/3 of the saturation magnetization.

The observed effects are most likely due to the spin polarization of the injected current. Unpolarized current injected into superconductor does not reduce the value of the critical current significantly. This was demonstrated in a series of control experiments. In place of the ferromagnetic layer we deposited an Au film on top of a $DyBa_2Cu_3O_7$ film patterned into a bridge for four-terminal measurements. It was observed that current injected from the Au film produced the changes in the critical current of the superconductor 20 times smaller than those produced by the injection of current from the ferromagnet. The small effect observed could be due to suppression of superconducting properties by charge injection[20] or is a consequence of heating of the Au layer by the current. The resistances per square of the ferromagnetic and the Au film were almost identical. Heating was also excluded in the case of the ferromagnet–superconductor heterostructure because we could estimate the change of the sample temperature by analyzing the I-V characteristics of the $La_{2/3}Sr_{1/3}MnO_3$ film. We concluded that heating did not exceed 1 K for the maximum current, which is far less than the shift in the temperature needed to explain the suppression of the superconductivity by the current injected from the ferromagnet.

The effect of spin injection on superconductivity can be qualitatively related to the phenomena already discussed in the literature. It well known that doping of a superconductor with magnetic impurities results in pair-breaking because of spin-flip scattering.[21] Aronov[2] treated the out-of-equilibrium situation arising when spin-polarized carriers are injected from a ferromagnetic metal into a non-ferromagnetic normal metal. The chemical potentials of carriers with opposite spins are different, which produces a net out-of-equilibrium polarization of the carriers of the non-ferromagnetic metal. A similar asymmetry in the population of the quasiparticles in the superconductor will result in a reduction of the superconducting order parameter due to the impossibility of the spin-polarized quasiparticle recombination into pairs. In addition, the presence of the spin-polarized quasiparticles may produce additional pair-breaking due to the destruction of pairs scattered by spin-polarized quasiparticles.

These effects would be limited to the distances shorter than the spin diffusion length, which may be much greater than the scattering length of unpolarized quasiparticles. The spin diffusion length in light metals like Al at low temperatures can be as high as few hundred microns[22], and the "average" atomic mass of the conducting CuO_2 planes of $DyBa_2Cu_3O_7$ is comparable to that of Al. Large spin diffusion lengths and the associated large spin relaxation times could result in a substantial buildup of spin polarization in the superconductor and hence a substantial suppression of the superconductivity.

A more profound analysis of the problem is certainly needed. Such an analysis should include other phenomena associated with the boundary between the ferromagnet and

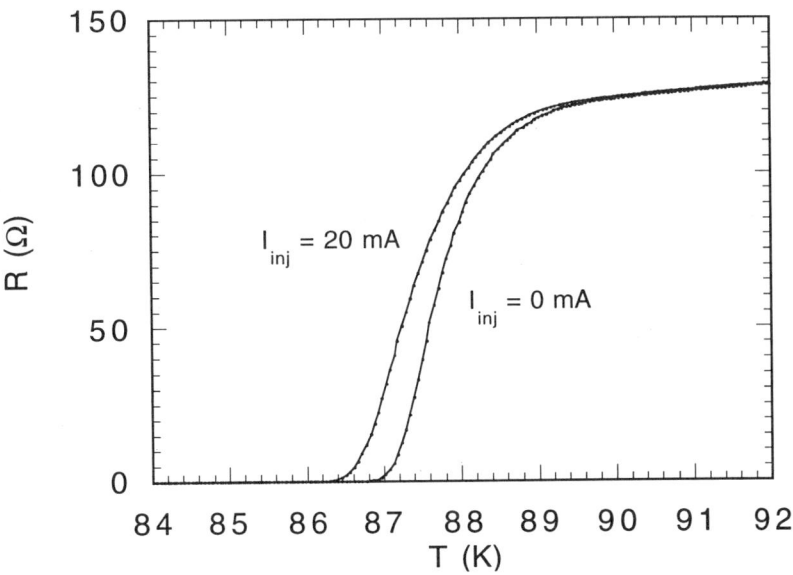

Figure 9. Dependence of the resistance of the superconducting $DyBa_2Cu_3O_7$ bridge on temperature without and with applied injection current from the ferromagnet $La_{2/3}Sr_{1/3}MnO_3$. The plot for the case of applied injection current is shifted by 4 K up in temperature to compensate for heating of the sample.

the superconductor. The extent of spin-flip scattering by the barrier is not known. The injection process itself may be modified by the spin polarization of the carriers in the ferromagnet. For example, the injection of the spin polarized carriers with energies below the gap of the superconductor will result in a modification of Andreev scattering. This will prevent many carriers from entering the superconductor, as discussed in the next section. On the other hand, carriers, which are spin-polarized, should be readily injected when their energies are greater than the gap. It is possible that the superconducting order parameter at the interface with the ferromagnet may be reduced by the pair-breaking effect of carriers scattering off of the ferromagnetic boundary. In this case spin polarized carriers might easily penetrate into the superconductor.[23] The ferromagnet–superconductor proximity effect may weaken the order parameter of the superconductor. Finally, possible asymmetry of the gap in k-space as a consequence of the d-wave character of the order parameter may be important.

Similar experiments carried out by other groups which have employed different ferromagnetic and superconducting materials and different geometries, have confirmed the existence of spin injection effects for both high-T_c[14] and conventional[15] superconductors. These studies have also shown that magnitude of the suppressed critical current can be a multiple of the current injected from the ferromagnet, thus demonstrating significant gain of a possible novel superconducting device.

Finally, we would like to mention that heating accompanying the injection of the spin-polarized carriers can pose a serious problem for device applications. In the simple structures that we studied heating was a problem if a robust superconductor ($T_c \sim 90$ K) was used. In this case, to suppress the superconductivity significantly, a large injection current was needed, resulting in substantial Joule heating. This heating was proportional to the square of the injection current. The measurement of the resistance of the superconducting layer in the vicinity of the superconducting transition temperature can be used to distinguish between the spin injection effect and heating as shown in Fig. 9. Heating results in a simple shift of the R(T) curves along the temperature axis, whereas the spin injection results in the widening of the transition. The magnitude of the heating problem can be mitigated through the use of sample geometries that minimize the voltage drop in the manganite layer, which is naturally resistive.

THE MODIFICATION OF ANDREEV REFLECTION AT FERROMAGNETIC METAL-SUPERCONDUCTOR INTERFACES

We now discuss the study of the differential conductance of ferromagnet-superconductor interfaces in heterostructures consisting of layers of the ferromagnet $La_{2/3}Ba_{1/3}MnO_3$ and the high-T_c superconductor $DyBa_2Cu_3O_7$. We have found that the differential conductance is a function of the voltage bias across the interface and has a dip at zero bias. The heterostructures were bilayers consisting of a 400 Å thick layer of the $La_{2/3}Ba_{1/3}MnO_3$ and a 600 Å thick layer of $DyBa_2Cu_3O_7$. To ensure good contact between the films no buffer layer was used between the ferromagnet and the superconductor.

Again, the heterostructures were patterned using conventional photolithography. They were first etched with concentrated HCl to define the area of the ferromagnetic and superconducting layers on the substrate. Then a dilute HCl solution was used to differentially pattern the $DyBa_2Cu_3O_7$ layer without damaging the $La_{2/3}Ba_{1/3}MnO_3$ layer. Figure 10 shows the geometry of the sample and the corresponding measurement configuration. Five such superconductor-ferromagnet-superconductor junctions with different lengths of ferromagnetic spacers (from 3 to 102 µm) were defined on the same chip. The measurement of the conductance of each of the junctions and the subsequent extrapolation of the results to zero spacer length allowed for the determination of the conductance of an individual ferromagnet–superconductor interface. We assumed that each of the interfaces of the sample had identical properties.

We studied the variation of the differential conductance of the interface with temperature, voltage bias across the interface, and applied magnetic field. The dependence of the differential conductance of the ferromgnet–superconductor interface on temperature and

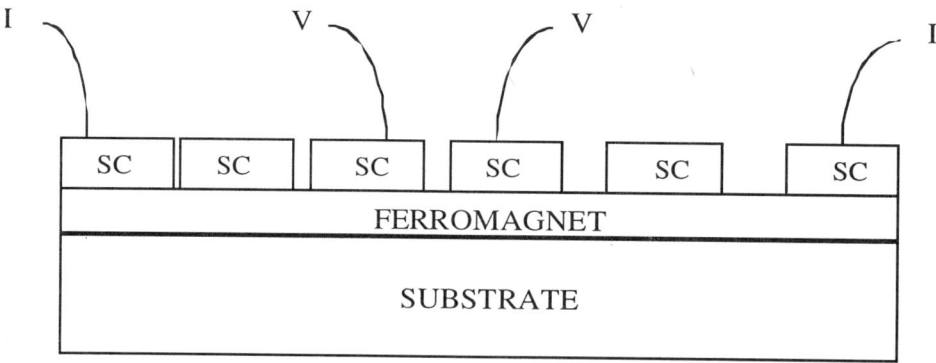

Figure 10. Geometry of the investigated ferromagnet/superconductor sample. The width of the $DyBa_2Cu_3O_7$ bridge was 400 µm, the lengths of the $La_{2/3}Ba_{1/3}MnO_3$ spacers were 3, 5, 12, 32, and 102 µm.

voltage bias is shown in Fig. 11. A well defined zero bias deep can be seen below 60 K. Both the shape and the amplitude of the dip are temperature-dependent. The amplitude decreases rapidly with increasing temperature, whereas the width of the dip is almost independent of temperature. In addition to the dip, a weak parabolic dependence of the differential conductance on the voltage bias is present. This may be a consequence of heating of the sample by the current flowing through the ferromagnetic spacer. On the other hand, the dip in the conductance is not a consequence of heating and is most probably a result of the injection of spin-polarized carriers into superconductor, as discussed below.

If magnetic field is applied to the sample the magnitude of the dip decreases. The dependence of the differential conductance on magnetic field is linear as it follows from Fig. 12. The magnetoconductance of the interface is also three times higher at zero voltage bias than at the maximum voltage bias applied in our experiment (40 mV). The magnetoconductance increases as the temperature of the sample increases. It was also determined that the magnetoconductance does not depend on the direction of the field with respect to the crystallographic axes of the sample and direction of the current in the sample. It is worth noting that magnetoconductance of the interface is much higher than the corresponding value for the ferromagnetic layer itself (less than 1% in the 12 T field over the investigated temperature range 1.5 -105 K).

Just as in the case of the spin injection experiment discussed in the previous section, we have performed a series of the control experiments to determine if the effect described (the conductance dip) is due to spin polarization. Replacement of the ferromagnetic layer with Au layer did not produce a differential conductance dip at zero voltage bias. Only a weak parabolic background was observed at all temperatures, similar to the one observed for the ferromagnet–superconductor sample above 60 K. A second control experiment in which

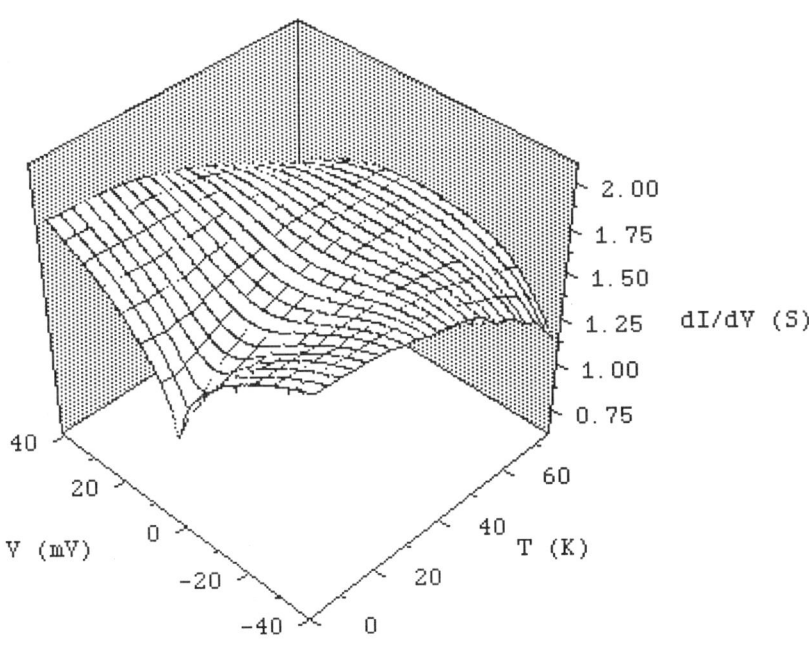

Figure 11. A surface representing temperature and voltage bias dependence of the differential conductance of the interface between the ferromagnet $La_{2/3}Ba_{1/3}MnO_3$ and the superconductor $DyBa_2Cu_3O_7$.

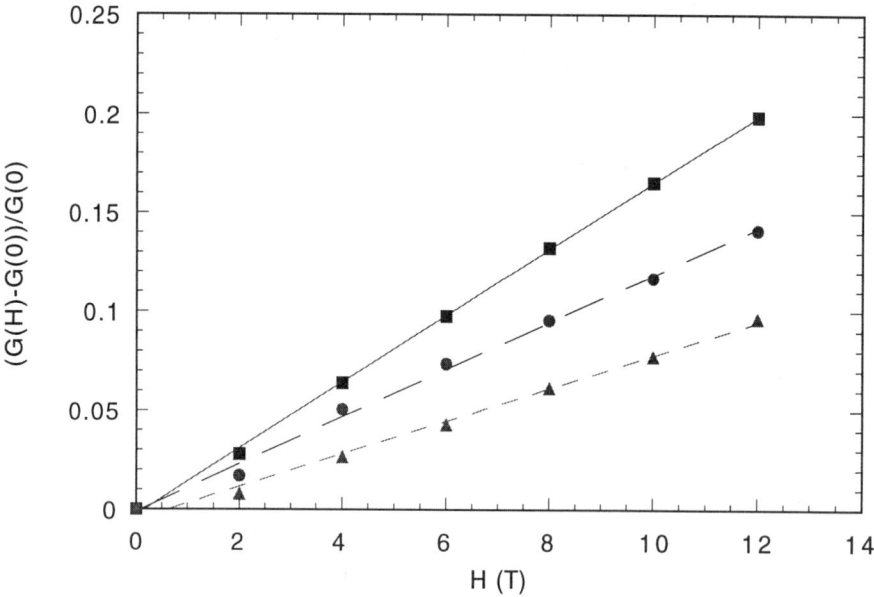

Figure 12. Magnetoconductance of the $La_{2/3}Ba_{1/3}MnO_3/DyBa_2Cu_3O_7$ interface at 5 K for three different voltage biases: 0 mV (squares), 18 mV (circles), and 36 mV (triangles).

ferromagnetic $La_{2/3}Ba_{1/3}MnO_3$ was replaced by the isostructural paramagnetic metallic perovskite $LaNiO_3$ also resulted in the disappearance of the zero bias conductance dip. The study of the $Au/La_{2/3}Ba_{1/3}MnO_3$ interface produced no sign of the conductance dip. Taking all the measurements together, we conclude that the zero bias conductance dip is the property of the ferromagnet–superconductor boundary.

To understand the above phenomena it is necessary to consider certain aspects of electrical transport across normal metal-superconductor boundaries. It is well established that there are two limiting behaviors associated with clean and dirty interfaces. Andreev reflection[24] is the property of the clean interface, whereas tunneling occurs at a disordered interface, which can be described by a potential barrier. Both cases, and the case of barriers of intermediate nature, have been treated thoroughly in the literature.[25] Andreev reflection for a conventional superconductor results in an increase of the conductance of the interface for transport at voltage biases below the superconducting gap. This is due to retroreflection of a hole in the process. In the ideal case for an *s*-wave superconductor, the differential conductance at biases below the gap is twice the value of the differential conductance at biases above the gap. In the case of the ideal tunneling in the limit of zero-temperature, the conductance vanishes below the gap.

We do not believe that transport across the ferromagnet–superconductor interface of our heterostructures involves tunneling for the following reasons: we did not intentionally put any barrier between the two layers. Monitoring of the growth process with the RHEED suggests that there was a clean and abrupt transition between the layers, which took place over one unit cell. Finally, a transmission electron microscopy (TEM) study of similar heterostructures did not reveal disorder at the interface, which appeared to be abrupt.

It was suggested recently that Andreev reflection is modified significantly when a normal metal is replaced by a ferromagnet.[26] Andreev reflection in this case must be strongly affected by the exchange interaction in the ferromagnet. The retroreflected hole, found in the conventional Andreev process when carriers are injected with energies less than the gap, is precluded when the normal metal is replaced by a half-metallic ferromagnet. This is the case because the retroreflected hole must have its spin opposite to that of the incident electron. The propagating states in the minority carrier band of the ferromagnet are too high in energy to be available for such a hole. Thus, transport across the ferromagnet–superconductor interface will not occur at voltage biases below the superconducting gap. As a result, Andreev reflection is suppressed and the conductance at voltage biases lower than the gap value could be lowered significantly or become zero. Indeed, Byers has shown that the ratio of the zero-bias conductances of ferromagnet–superconductor and normal metal-s-wave superconductor is given by

$$G_F(0)/G_N(0) = 2(1-P), \quad (1)$$

where P is the degree of spin polarization.[27]

The results of our experiment are in qualitative agreement with a picture of suppressed Andreev reflection. However the conductance dip is smeared out and does not have the sharp characteristic of the conventional s-wave gap. It is consistent with an anisotropic d-wave gap. A directional average of the current at the interface could be the reason for the smearing of the observed conductance dip. A recent calculation of the differential conductance of similar structures[28] is in qualitative agreement with our experiment. The comparison of quasiparticle transport across ferromagnet–superconductor interfaces with normal metal-superconductor interfaces demonstrated that the subgap conductance behavior in the case of ferromagnet is different from the nonmagnetic case. The results observed using a high-T_c superconductor are different from what is expected for an s-wave superconductor. For a ballistic ferromagnet–d-wave superconductor interface a zero subgap conductance might be achieved if the current flowed in an antinodal direction.

The temperature and magnetic field dependence of the magnitude of the observed conductance dip is consistent with the picture of the suppressed Andreev reflection as well. The decrease in the magnitude of the dip with increasing temperature could be due to the decrease in the amplitude of the order parameter with increasing temperature as well as to a decrease of the degree of spin polarization of the carriers in the ferromagnet with increasing temperature. The temperature at which the dip disappears (60 K) suggests that the superconducting material near the interface may be different from that of the bulk of the film. Suppression of the superconductivity near the interface could be due to spin injection[13] or the proximity effect.[29] An alternative explanation is that the superconductor in the vicinity of the interface with ferromagnet is in the orthorhombic II phase which has a superconducting transition temperature of 60 K.[30, 31] This is plausible because the temperature of the conductance dip disappearance was the same for all measured samples, which may have differed somewhat in their processing. The magnetic field dependence of the amplitude of the dip is most likely due to the suppression of the order parameter of the superconductor in the cores of the introduced vortices. Because the energy gap is zero in the cores of vortices, transport of spin polarized carriers from the ferromagnet into the normal cores would not require an Andreev reflection channel.

CONCLUSIONS

By means of ozone-assisted "block-by-block" MBE we have produced high quality heterostructures of CMR ferromagnets and high-T_c superconductors and demonstrated that these materials are compatible in a sense of epitaxial growth with a high degree of in-plane and out-of-plane orientation. Rocking curve widths of less than 0.1° were measured for all layers. The heterostructures posses both ferromagnetic and superconducting properties with sharp superconducting transitions. The heterostructures were used successfully to study the effects of the spin injection into a high-T_c superconductor and properties of the ferromagnet–superconductor interface.

It was suggested that the current injected from a ferromagnet into a superconductor reduces both the critical current and transition temperature as a consequence of the pair-breaking by spin-polarized carriers. It was also shown that the differential conductance of the ferromagnet–superconductor boundary has a zero bias dip at temperatures below the superconducting transition. The latter is most likely due to the suppression of Andreev reflection as a consequence of the spin polarization of the carriers in the ferromagnet. Possible applications of both effects could result in fabrication of fast switching devices. A new configuration that we propose would be a three-terminal ferromagnet–superconductor–ferromagnet device in which suppressed Andreev reflection at one interface would detect the change of the superconducting order parameter due to the spin injection at the other interface. The creation of such a device would need further investigation of the discussed effects, especially their temporal response. The problem of heating would also require thorough investigation.

ACKNOWLEDGEMENTS

This work was supported in part by AFOSR under Grants No. F49620-96-1-0043 and No. F49620-95-1-0397, the ONR under Grants No. N00014-92-1368 and No. N00014-98-1-0098, and by NSF Grant No. NSF/DMR-9421941.

REFERENCES

1. P. M. Tedrow and R. Meservey, Phys. Rev. Lett. **26**, 192 (1971); Phys. Rev. B **7**, 318 (1973).
2. A. G. Aronov, JETP Lett. **24**, 32 (1976); Sov. Phys. JETP **44**, 193 (1976).
3. M. Johnson and R. H. Silsbee, Phys. Rev. B **37**, 5326 (1988).
4. M. Johnson, Appl. Phys. Lett. **65**, 1460 (1994).
5. G. A. Prinz, Physics Today **48**, #4, 58 (1995).
6. K. Chahara, T. Ohno, M.Kasai, and Y. Kozono, Appl. Phys. Lett. **63**, 1990 (1993).
7. R. Von Helmolt, J. Weckerg, B. Holzapfel, L. Schultz, and K. Samwer, Phys. Rev. Lett. **71**, 2331 (1993).
8. C. Zener, Phys. Rev. **82**, 403 (1951).
9. P. G. de Gennes, Phys. Rev. **118**, 141 (1960).
10. J. Z. Sun, L. Krusin-Elbaum, P. R. Duncombe, A. Gupta, and R. B. Laibowitz, Appl. Phys. Lett. **70**, 1769 (1997).
11. M. Kasai, T. Ohno, Y. Kanke, Y. Kozono, M. Hanazono, and Y. Sugita, Jpn. J. Appl. Phys. **29**, L2219 (1990).

12. G. Jakob, V. V. Moschalkov, and Y. Bruynseraede, Appl. Phys. Lett. **66**, 2564 (1995).
13. V. A. Vas'ko, V. A. Larkin, P. A. Kraus, K. R. Nikolaev, D. E. Grupp, C. A. Nordman, and A. M. Goldman, Phys. Rev. Lett. **78**, 1134 (1997).
14. Z. W. Dong, R. Ramesh, T. Venkatesan, M. Johnson, Z. Y. Chen, S. P. Pai, V. Talyansky, R. P. Sharma, R. Shreekala, C. J. Lobb, and R. L. Greene, Appl. Phys. Lett. **71**, 1720 (1997).
15. D. B. Chrisey, M. S. Osofsky, J. S. Horwitz, R. J. Soulen, B. Woodfield, J. Byers, G. M. Daly, P. C. Dorsey, J. M. Pond, M. Johnson, R. C. Y. Auyeung, IEEE Trans. Appl. Supercond., **7**, 2067 (1997).
16. V. A. Vas'ko, K. R. Nikolaev, V. A. Larkin, P. A. Kraus, and A. M. Goldman, Appl.Phys. Lett. **73**, (1998).
17. V. S. Achutharaman, K. M. Beauchamp, N. Chandrasekhar, G. C. Spalding, B. R. Johnson, and A. M. Goldman, Thin Solid Films **216**, 14 (1992).
18. J.-P. Locquet, A. Catana, E. Mächler, C. Gerber, and J. G. Bednorz, Appl. Phys. Lett. **64**, 372 (1994).
19. V. A. Vas'ko, C. A. Nordman, P. A. Kraus, V. S. Achutharaman, A. R. Ruosi, and A. M. Goldman, Appl. Phys. Lett. **68**, 2571 (1996).
20. T.-W. Wong, J. T. C. Yeh, and D. N. Langenberg, Phys. Rev. Lett. **37**, 150 (1976).
21. A. A. Abrikosov and L. P. Gor'kov, Sov. Phys. JETP **12**, 1243 (1961).
22. M. Johnson and R. H. Silsbee, Phys. Rev. Lett. **55**, 1790 (1985).
23. M. J. DeWeert and G. B. Arnold, Phys. Rev. B**39**, 11307 (1989) have shown this to be the case for an insulating ferromagnet.
24. A. F. Andreev, Sov. Phys. JETP **19**, 1228 (1964).
25. G. E. Blonder, M. Tinkham, and T. M. Klapwijk, Phys. Rev. B **25**, 4515 (1982).
26. M. J. M. de Jong and C. W. J. Beenakker, Phys. Rev. Lett. **74**, 1657 (1995).
27. J. Byers, private communication.
28. Jian-Xin Zhu, B. Friedman, and C. S. Ting, to be published.
29. J. Hauser, H. C. Theuerer, and N. R. Werthamer, Phys. Rev. **142**, 118 (1966).
30. S. Amelinckx, G. Van Tendeloo, and J. Van Landuyt, in *Oxygen Disorder Effects in High-T_c Superconductors*, edited by J. L. Moran-Lopez and I. K. Schuller (Plenum Press, New York, 1990), p. 9.
31. D. de Fontaine, in *Oxygen Disorder Effects in High-T_c Superconductors*, edited by J. L. Moran-Lopez and I. K. Schuller (Plenum Press, New York, 1990), p. 75.

PARTICIPANTS

Gabriel Aeppli
NEC Research Institute, Inc.
4 Independence Way
Princeton, NJ 08540
USA
gabe@research.nj.nec.com

Keun Hyuk Ahn
The Johns Hopkins University
Department of Physics and Astronomy
3400 N. Charles Street
Baltimore, MD 21218
USA
kahn@eta.pha.jhu.edu

Daniel Arovas
University of California, San Diego
Physics Department 0319
La Jolla, CA 92093
USA
dpa@borges.ucsd.edu

Yaroslaw Bazaliy
Stanford University
Department of Physics
Stanford, CA 94305-4060
USA
yar@leland.stanford.edu

Simon Billinge
Michigan State University
Physics and Astronomy Department
East Lansing, MI 48824
USA
billinge@pa.msu.edu

Corwin H. Booth
Los Alamos National Lab.
MST-10, MS K764
Los Alamos, NM 87545
USA
cbooth@mst.lanl.gov

Emil Bozin
Michigan State University
Physics and Astronomy Department
East Lansing, MI 48824
USA
bozin@pa.msu.edu

Viktor Cerovski
Michigan State University
Physics and Astronomy Department
East Lansing, MI 48824
USA
cerovski@pa.msu.edu

Sang-W. Cheong
Lucent Technologies, Bell Labs.
Innovations
Room 1C-211
600 Mountain Avenue
Murray Hill, NJ 07974
USA
scheong@lucent.com

Jerry Cowen
Michigan State University
Physics and Astronomy Department
East Lansing, MI 48824
USA
cowen@pa.msu.edu

Elbio Dagotto
Florida State University
Department of Physics
Magnet Lab., Innovation Park
Tallahassee, FL 32306
USA
dagotto@Fangio.Magnet.FSU.EDU

Pengcheng Dai
Oak Ridge National Lab.
9809 Martha Knight Circle #45
Knoxville, TN 37932
USA
piq@ornl.gov

Timir Datta
University of South Carolina
Physics/Astronomy Superconduct
Institute
Columbia, SC 29208
USA
datta@psc.psc.sc.edu

Sushanta Dattagupta
Jawaharlal University
New Delhi
INDIA
dgupta@jnuriv.ernet.in

Remo DiFrancesco
Michigan State University
Physics and Astronomy Department
East Lansing, MI 48824
USA
difrance@pa.msu.edu

Phillip M. Duxbury
Michigan State University
Physics and Astronomy Department
East Lansing, MI 48824
USA
duxbury@pa.msu.edu

Nobuo Furukawa
Department of Physics
Aoyama Gakuin University
6-16-1 Chitose-dai
Setagaya, Tokyo 157-8572
JAPAN
furukawa@phys.aoyama.ac.jp

Ted H. Geballe
Stanford University
Department of Applied Physics
Stanford, CA 94305
USA
geballe@loki.stanford.edu

Brage Golding
Michigan State University
Physics and Astronomy Department
East Lansing, MI 48824
USA
golding@pa.msu.edu

Allen M. Goldman
University of Minnesota
School of Physics and Astronomy
116 Church St., SE
Minneapolis, MN 55455
USA
goldman@physics.spa.umn.edu

Denis Golosov
Univ. of Chicago & Argonne Nat. Lab.
James Franck Institute
5640 S. Ellis Avenue
Chicago, IL 60637
USA
golosov@control.uchicago.edu

John B. Goodenough
University of Texas, Austin
Center for Materials Science and
Engineering, ETC 9.104
Austin, TX 78712-1063
USA
Jgoodenough@mail.utexas.edu

James Harrison
Michigan State University
Chemistry Department
East Lansing, MI 48824
USA
harris13@pilot.msu.edu

Peter Horsch
F. Festkoerperforschung Max-Planck-
Institut Stuttgart
Heisenbergstr. 1
D-70569 Stuttgart
GERMANY
horsch@audrey.mpi-stuttgart.mpg.de

Sumio Ishihara
Institute for Materials Research
Tohoku University
Katahira 2-1-1
Sendai 980-77
JAPAN
ishiha@imr.tohoku.ac.jp

Marcelo Jaime
Los Alamos National Laboratory
MST-10, MS K764
Los Alamos, NM 87545
USA
mjaime@lanl.gov

Viktoria Jane
Michigan State University
Physics and Astronomy Department
East Lansing, MI 48824
USA
jane@pa.msu.edu

Jong Hoon Jung
Seoul National University
Dept. of Physics, c/o Prof. T.W. Noh
Room 25-324
Seoul 151-742,
KOREA
hoon@phya.snu.ac.kr

Thomas Kaplan
Michigan State University
Physics and Astronomy Department
East Lansing, MI 48824
USA
kaplan@pa.msu.edu

Takuro Katsufuji
Lucent Technologies, Bell Labs.
Innovations
1C-211
600 Mountain Avenue
Murray Hill, NJ 07974
USA
tkatsuf@clockwise.lucent.com

Kenn Kubo
University of Tsukuba
Institute of Physics
Ibaraki 305
JAPAN
kkubo@cm.ph.tsukuba.ac.jp

Henry P. Kunkel
University of Manitoba
Department of Physics and Astronomy
Winnipeg, MB R3T 2N2
CANADA
hkunkel@cc.umanitoba.ca

Jeffrey Lynn
University of Maryland
Department of Physics
Center for Superconductivity Research
College Park, MD 20742
USA
jeff.lynn@nist.gov

Sadamichi Maekawa
Tohoku University
Institute for Materials Research
Sendai 980-8577
JAPAN
maekawa@imr.tohoku.ac.jp

S. D. Mahanti
Michigan State University
Physics and Astronomy Department
East Lansing, MI 48824
USA
mahanti@pa.msu.edu

Hakim Meskine
University of Missouri, Columbia
Columbia, MO 65211
USA
hakim@agni.physics.missouri.edu

Andrew Millis
The Johns Hopkins University
3400 North Charles Street
Baltimore, MD 21218
USA
millis@landau.pha.jhu.edu

Charles Moreau
Michigan State University
Physics and Astronomy Department
East Lansing, MI 48824
USA
moreau@pa.msu.edu

Tae W. Noh
Seoul National University
Department of Physics
Building 27-119B
Seoul, 151-742
KOREA
twnoh@phya.snu.ac.kr

Amra Peles
University of Manitoba
Department of Physics and Astronomy
Winnipeg, MB R3T 2N2
CANADA
umpelesa@cc.umanitoba.ca

Valeri Petkov
Michigan State University
Physics and Astronomy Department
East Lansing, MI 48824
USA
petkov@pa.msu.edu

Warren E. Pickett
University of California, Davis
Department of Physics 1050
One Shields Avenue
Davis, CA 95616
USA
pickett@physics.ucdavis.edu

William Pratt
Michigan State University
Physics and Astronomy Department
East Lansing, MI 48824
USA
pratt@pa.msu.edu

Arthur Ramirez
Lucent Technologies, Bell Labs.
Innovations
Room 1D-216
600 Mountain Avenue, P.O. Box 636
Murray Hill, NJ 07974
USA
apr@bell-labs.com

Anne Reilly
Michigan State University
Physics and Astronomy Department
East Lansing, MI 48824
USA
schaefer@pa.msu.edu

James J. Rhyne
University of Missouri, Columbia
Research Reactor Center
Department of Physics and Astronomy
Columbia, MO 65211
USA
jrhyne@showme.missouri.edu

Sashi S. Satpathy
University of Missouri
Department of Physics
Columbia, MO 65211
USA
satpathy@agni.physics.missouri.edu

Peter Schiffer
University of Notre Dame
Department of Physics
Notre Dame, IN 46556
USA
schiffer.1@nd.edu

Pedro Schlottmann
Florida State University
Department of Physics
Tallahassee, FL 32306
USA
schlottm@phy.fsu.edu

Yen-Sheng Su
Michigan State University
Physics and Astronomy Department
East Lansing, MI 48824
USA
su_y@pa.msu.edu

Ramanathan Suryanarayanan
Laboratoire de Chimie des Solides,
CNRS
Bat. 414, Univ. de Paris-sud
91405 Orsay
FRANCE
suryan@isma.isma.u-psud.fr

Stuart Tessmer
Michigan State University
Physics and Astronomy Department
East Lansing, MI 48824
USA
tessmer@pa.msu.edu

Yasuhide Tomioka
Joint Research Center for Atom
Technology (JRCAT)
c/o National Institute for Advance
Interdisciplinary Research (NAIR)
1-1-4 Higashi, Tsukuba 305
JAPAN
tomioka@jrcat.or.jp

Wayne Tonjes
Michigan State University
Physics and Astronomy Department
East Lansing, MI 48824
USA
tonjes_w@pa.msu.edu

Stuart A. Trugman
Los Alamos National Lab. - MS B262
Theor. Div./Ctr. for Nonlinear Studies
Los Alamos, NM 87545
USA
sat@freja.lanl.gov

Jeroen van den Brink
Max Planck Institut FKF
Heisenbergstr. 1
70569 Stuttgart
GERMANY
brink@audrey.mpi-stuttgart.mpg.de

Steven Watts
Florida State University
P.O. Box 20779
Department of Physics
Tallahassee, FL 32316
USA
watts@martech.fsu.edu

Gwyn Williams
University of Manitoba
Department of Physics and Astronomy
Winnipeg, MB R3T 2N2
CANADA
gwill@cc.umanitoba.ca

Seiji Yunoki
Florida State University
National High Magnetic Field Lab.
Tallahassee, FL 32310
USA
yunoki@fangio.magnet.fsu.edu

Guo-meng Zhao
University of Maryland
Center for Superconductivity Research
College park, MD 20742
USA
gzhao@squid.umd.edu

Jun-Hui Zhao
University of Manitoba
Department of Physics and Astronomy
Winnipeg, MB R3T 2N2
CANADA
zhaoj@physics.umanitoba.ca

Xue-Zhi Zhou
University of Manitoba
Department of Physics and Astronomy
Winnipeg, MB R3T 2N2
CANADA
zhou@cc.umanitoba.ca

INDEX

Absorption peak
   small polaron, 184
Aharonov-Bohm loops, 253
Andreev reflection, 280, 282-283
   suppression of, 284
Andreev scattering, 279
Anomalous X-ray scattering, 58, 62, 66
Antiferromagnetic, 15, 20, 39, 60, 62, 72, 77, 87
Atomic pair distribution function method, 203

Band calculations
   density-functional methods (LDA), 245
Bilayer manganite, 152
Bipolaron, 190, 223
Bipolaronic charge carriers, 221
Bipolaronic superconductivity, 221
Bipolarons
   intersite, 223
Bottleneck effect, 230
Breathing mode distortion, 206

Cation disorder, 98
Channeling, 272
Charge and/or orbital ordering, 177
Charge gap, 167
Charge ordering, 24, 39, 246
   bipolaronic, 238
Charge stiffness, 107, 114, 124
Charge, 87
Charge/orbital ordering, 155-157, 161-162, 170
   neutron diffraction study, 156
   transition, 156, 158
Charge-density wave, 128, 132
Charge-ordering transition
   oxygen isotope shift, 235
Charge-stripe ordering, 202
Classical rotator, 2
Classical spin limit, 2
Classical spins, 40
Coherent potential approximation, 79, 98
Colossal magnetoresistance (CMR), 1, 58, 87, 91, 103, 127, 222, 269
Compressibility, 128
Conductivity, 92, 105, 108-109, 112-113
Critical current, 276
Critical field, 161
Cu-oxides, 45

Curie temperature, 1, 9, 15, 17, 25-26, 31, 103, 127
Current injection, 157

Density functional, 87-88
Density of states, 3, 12, 95, 99
Density-Matrix Renormalization Group, 41
Differential conductance, 280
DOS, 78, 80
Double exchange, 1, 13, 39, 46, 60, 68, 73-74, 77-79, 84, 87, 103, 119, 131, 135, 145, 155-157, 159, 164, 177, 196, 222, 243, 261, 269
Drude peak, 51, 110-112, 115-117, 124, 183, 188
Drude response, 50
Drude weight, 30
Dynamical mean-field approximation, 2-3, 5, 41, 46

e_g-orbital
   ordering, 158, 161
Elastic strain field, 245
Electron energy loss, 223
Electron paramagnetic resonance, 246
Electron-electron scattering, 256, 258
Electronic properties, 87
Electronic states, 71, 78, 84
Electronic structure, 57-58, 88
Electronic transport, 243
Electron-lattice interactions, 127, 132, 136, 146
Electron-magnon resistivity, 258
Electron-phonon coupling, 100, 180, 188, 196, 243, 246
   strong, 195
EPR measurements, 228
EXAF, 187

Falicov-Kimball model, 6
Ferromagnet, 87
Ferromagnetic metal-superconductor interfaces, 280
Ferromagnetic states, 57
Ferromagnetic transition
   second-order, 232
Ferromagnetism, 1, 10, 20, 24, 39, 41, 60, 62, 68, 71, 73-74, 76-77, 79, 84, 103-4, 106, 112, 117, 119, 124, 127, 135-136, 146

Ferromagnet-superconductor interface, 282
Ferromagnet-superconductor proximity, 279
Field induced transition, 164-165
First order phase transition, 57-58

Geometric tolerance factor, 127
Grain boundary scattering, 255

Half-metallic, 14, 25, 90, 255
Hall coefficient, 252
  high temperature, 252, 254
Hall effect, 253, 264
  low temperature, 255
  unusual, 202
Hall mobility, 251
Heisenberg interaction, 106
Heisenberg spin model, 9, 15, 135, 139, 142
Heterostructures, 270, 275
  CMR ferromagnets and high-T_c superconductors, 284
High-T_c superconductor, 221, 269, 270
Hund's-rule coupling, 1, 6, 39, 47, 58, 71-74, 77-78, 84, 87, 95, 129, 136, 143, 156

IM transition, 211
  percolative, 213
Incipient metallic, 216
Infrared reflectivity, 168
Insulating phase
  ferromagnetic, 213
Insulator-metal transition, 87, 164, 222
  field-induced, 157
  magnetic field induced, 159, 163, 167, 168
Interband transitions, 184
Intergrain barriers, 248
Internal phonon modes
  frequency shifts, 195
Intraatomic transition, 185, 188
Ion channeling measurements, 202
Isotope effect, 246
  in the EPR, 202
Isotope effects
  oxygen, 221, 224
I-V characteristics, 270, 276

Jahn-Teller (JT) interaction
  dynamic, 178
Jahn-Teller distortion, 201
  cooperative, 155
  local, 205, 218
Jahn-Teller effect, 25, 222, 259
Jahn-Teller lattice distortions, 246
Jahn-Teller polarons, 223
Jahn-Teller, 31, 39, 46, 68, 71, 89, 92, 100, 103, 106, 128, 130-131, 146, 156, 234

Kohn anomalies, 94, 100
Kondo lattice model, 39, 40, 44, 71, 103-105, 109, 124, 143, 178
Kondo, 40
Korringa-type relaxation, 228

(La,Ca)MnO$_3$, 178

(La,Pr)$_{0.7}$Ca$_{0.3}$MnO$_3$, 178
(La,Sr)MnO$_3$, 1
(La$_{1-y}$Nd$_y$)$_{2-x}$Sr$_x$CuO$_4$, 155
La$_{.7}$Pb$_{.3}$MnO$_3$, 135
La$_{0.67}$Ca$_{0.33}$MnO$_3$
  quasielastic scattering, 151
La$_{0.6}$Sr$_{0.4}$MnO$_3$, 181
La$_{0.79}$Ca$_{0.21}$MnO$_3$, 263
La$_{0.7-x}$Pr$_x$Ca$_{0.3}$MnO$_3$, 128
La$_{0.7-y}$Pr$_y$Ca$_{0.3}$MnO$_3$, 191
La$_{0.825}$Sr$_{0.175}$MnO$_3$, 181
La$_{1/2}$Ca$_{1/2}$MnO$_3$, 158
La$_{1-x}$Ca$_x$MnO$_3$, 23, 135, 201, 211
  local structure, 205
La$_{1-x}$Pb$_x$MnO$_3$, 23
La$_{1-x}$Sr$_{1+x}$MnO$_4$, 158
La$_{1-x}$Sr$_x$FeO$_3$, 155
La$_{1-x}$Sr$_x$MnO$_3$, 23, 211, 213
La$_{1-x}$Sr$_x$NiO$_4$, 155
La$_{2/3}$Ca$_{1/3}$MnO$_3$, 2
La$_{2/3}$Sr$_{1/3}$MnO$_3$, 2
LaMnO$_3$, 149, 244
  doped, 149
Large isotope effect
  on T_c, 203
Large polaron absorption, 189
Large polaron bands, 202
Large polaron, 191
Lattice polaron, 223, 234
Lattice-polaronic effects, 222
Linewidths
  intrinsic, 150
Local atomic structure, 246
Local density approximation, 88, 100
Local disorder, 247
Local structural measurements, 218
Local structure
  doping dependence, 211
Localized and delocalized carriers
  coexistence, 211
Longitudinal magnetoresistance, 252

Magnetic ordering, 177
Magnetic polarons, 202
Magnetic tunnel junctions, 270
Magnetization, 7, 26, 103
Magnetoconductance, 281
Magnetoelastic polarons, 251
Magnetoresistance, 3, 26-27
Magnetoresistive oxides, 149
Magnetostriction effects, 246
Magnon lifetimes, 28, 146-147, 150
Magnon-electron interaction, 150
Manganite
  optimally doped, 263
Manganite-cuprate heterostructures, 269
Manganites, 201
Metal-insulator phase transition, 248
Metal-insulator phenomena, 155
  magnetic field induced, 172
Metal-insulator transition, 1, 32, 48, 57-58, 149, 177, 230, 235, 245, 274
  induced by oxygen isotope substitution, 233

Metal-insulator phase boundary, 103
Metamagnetic transitions, 159
Midgap states, 188
Midgap transitions, 180
Molecular beam epitaxy, 270
Monte Carlo, 2, 10, 40
Muon spin relaxation, 246
Muon spin resonance, 202, 257

Nanodomain, 32-34
$Nd_{0.7}Sr_{0.3}MnO_3$, 151, 183
$Nd_{1/2}Sr_{1/2}MnO_3$, 158-159
$Nd_{1-x}Ca_xMnO_3$, 166
$Nd_{1-x}Sr_xMnO_3$, 23, 157
Neutron diffraction, 89, 246
Neutron scattering
  small angle, 151, 235
Neutron scattering, 52, 65, 147, 149, 257
Ni-oxides, 45
Nordheim-Gorter rule, 259

Optical conductivity, 3, 17, 48, 50, 103-104, 107, 109, 111, 116-118, 124, 167, 178
  doping dependent, 186
Optical properties, 177-178, 246
Optical spectroscopy, 167
Optical sum rule, 108, 124
Orbital correlations, 105, 118
Orbital degeneracy, 103, 117, 135, 143, 146
Orbital degree of freedom, 40, 57, 59, 68, 103, 106-107, 74-75
Orbital liquid, 117, 124, 179
Orbital ordering, 46, 49, 57, 62-63, 71, 84, 87, 89, 108, 114, 119, 128, 201
Orbital t-J model, 106
Orbitals, 46, 57, 72
Organic superconductors, 74
Oxygen isotope shift
  giant, 222

Pair distribution function (PDF), 202
Pair-breaking, 275, 278
  spin-polarized carriers, 284
PDF measurements, 187
PDF peak
  sharpening, 206
Phase diagram, 167
  charge/orbital ordering, 166
Phase separation, 19-20, 39, 41-42, 60, 62, 77, 127-128
Phonon dispersion, 100
Phonon frequency
  shift, 194
Phonon spectrum, 94
Photo excitation, 157
Photoemission spectroscopies, 223
Plasma energy, 91, 93, 100
Polaron absorption, 188
Polaron ordering, 177
Polaron, 31-32, 151
Polaronic carriers, 221
Polaronic, 30
Polarons, 127-128, 131, 136, 201, 218, 223, 259

in manganites, 202
local structural evidence, 203
spin, 152
Polycrystal, 24, 26
Polycrystal, 26
$Pr_{.63}Sr_{.37}MnO_3$, 135
$Pr_{0.65}(Ca_{1-y}Sr_y)_{0.35}MnO_3$, 170
$Pr_{0.7}Sr_{0.3}MnO_3$, 151
$Pr_{1-x}Ca_x MnO_3$ (x=0.3), 163
$Pr_{1-x}Ca_x MnO_3$, 23, 162, 165, 167
$Pr_{1-x}Sr_xMnO_3$, 23
Pressure effects, 246
Pressure, 105, 119
Pressure-effect coefficient, 227
Pseudo-spin operator, 59, 73
Pseudo-spin operators, 73
Pyrochlore, 152

Quantum t_{2g}-spins, 44
Quasielastic central component, 151

Raman spectroscopy, 196
$RE_{1/2}AE_{1/2}MnO_3$, 161
Resistivity, 2-3, 16, 23, 26, 31, 71, 82, 88, 99, 100, 103, 109, 112, 127
RHEED diffraction patterns, 272

Scaling behavior, 193-194, 196
Seebeck coefficient, 254, 259-260
Seebeck effects, 264
Single-magnon scattering, 256
$Sm_{1/2}Ca_{1/2}MnO_3$, 158
  charge/orbital- and spin-ordering, 159
$Sm_{1-x}Ca_xMnO_3$, 157
Small lattice polarons, 249
Small polaron, 243, 254
  activation energy, 187
Small polaron-like, 250
Small to large polarons
  crossover, 188
Small-polaronic transport, 251
Sound velocity, 196
Specific heat, 117
Spectral function, 12
Spectral weight, 7, 50, 107, 113
Spin correlations
  paramagnetic, 151
Spin diffusion length, 278
Spin diffusion, 149-150
Spin dynamics, 149-150
Spin injection, 269, 275
Spin polarized tunneling, 248
Spin polaron, 139
Spin split band model, 181
Spin stiffness coefficient, 150
Spin wave, 21, 139
  dispersion, 150
  energy gap, 150
  linewidths, 28, 146-147, 150
Spin waves, 135, 145, 149
Spin, 87
Spin-flip scattering, 279
Spin-polarized carriers, 269

Spin-polarized photoemission, 256
Spins classical, 47
Stiffness constant, 257
Strong correlations, 71, 107, 117
Strong coupling, 104
Strong electron correlation, 58, 103
Structural information
   local, 201
Structural phase transition
   magnetic field dependent, 177
Structural transitions, 87
Superconductivity
   suppression of, 275
Superconductor-ferromagnet interfaces, 274
Superexchange, 60, 68, 178
   antiferromagnetic, 156

Termination ripple, 210
Thermal expansion, 128
Thermal transport, 246
Thermal-expansion coefficient, 231
Thermoelectric power, 27, 131
Thermopower, 202
Three-site hopping terms, 104, 107-108, 114, 144

t-J model, 107-108, 111, 114-115, 122, 124
   generalized, 179
$Tl_2Mn_2O_7$, 152
   pyrochlore, 149
Tolerance factor, 156, 247, 264
Transmission electron microscopy, 282
Transport
   high temperature, 248
   low temperature, 255
Transport and tunneling
   spin-polarized, 269
Transport properties, 243, 263
Tricritical point, 262
Tunneling, 282
   spin-polarized, 269
Tunneling magnetoresistance, 24
Two-fluid approach, 261
Two-fluid behavior, 214

Variable Range Hopping, 250
Virial theorem, 127-128

X-ray absorption fine structure, 202, 246